THE hERG CARDIAC POTASSIUM CHANNEL: STRUCTURE, FUNCTION AND LONG QT SYNDROME

The Novartis Foundation is an international scientific and educational charity (UK Registered Charity No. 313574). Known until September 1997 as the Ciba Foundation, it was established in 1947 by the CIBA company of Basle, which merged with Sandoz in 1996, to form Novartis. The Foundation operates independently in London under English trust law. It was formally opened on 22 June 1949.

The Foundation promotes the study and general knowledge of science and in particular encourages international co-operation in scientific research. To this end, it organizes internationally acclaimed meetings (typically eight symposia and allied open meetings and 15–20 discussion meetings each year) and publishes eight books per year featuring the presented papers and discussions from the symposia. Although primarily an operational rather than a grant-making foundation, it awards bursaries to young scientists to attend the symposia and afterwards work with one of the other participants.

The Foundation's headquarters at 41 Portland Place, London W1B 1BN, provide library facilities, open to graduates in science and allied disciplines. Media relations are fostered by regular press conferences and by articles prepared by the Foundation's Science Writer in Residence. The Foundation offers accommodation and meeting facilities to visiting scientists and their societies.

Information on all Foundation activities can be found at
http://www.novartisfound.org.uk

Novartis Foundation Symposium 266

THE hERG CARDIAC POTASSIUM CHANNEL: STRUCTURE, FUNCTION AND LONG QT SYNDROME

2005

John Wiley & Sons, Ltd

Copyright © Novartis Foundation 2005
Published in 2005 by John Wiley & Sons Ltd,
The Atrium, Southern Gate,
Chichester PO19 8SQ, UK

National   01243 779777
International (+44) 1243 779777
e-mail (for orders and customer service enquiries): cs-books@wiley.co.uk
Visit our Home Page on http://www.wileyeurope.com
or http://www.wiley.com

All Rights Reserved. No part of this book may be reproduced, stored in a retrieval system or transmitted in any form or by any means, electronic, mechanical, photocopying, recording, scanning or otherwise, except under the terms of the Copyright, Designs and Patents Act 1988 or under the terms of a licence issued by the Copyright Licensing Agency Ltd, 90 Tottenham Court Road, London W1T 4LP, UK, without the permission in writing of the Publisher. Requests to the Publisher should be addressed to the Permissions Department, John Wiley & Sons Ltd, The Atrium, Southern Gate, Chichester, West Sussex PO19 8SQ, England, or emailed to permreq@wiley.co.uk, or faxed to (+44) 1243 770620.

This publication is designed to provide accurate and authoritative information in regard to the subject matter covered. It is sold on the understanding that the Publisher is not engaged in rendering professional services. If professional advice or other expert assistance is required, the services of a competent professional should be sought.

Other Wiley Editorial Offices

John Wiley & Sons Inc., 111 River Street, Hoboken, NJ 07030, USA

Jossey-Bass, 989 Market Street, San Francisco, CA 94103-1741, USA

Wiley-VCH Verlag GmbH, Boschstr. 12, D-69469 Weinheim, Germany

John Wiley & Sons Australia Ltd, 33 Park Road, Milton, Queensland 4064, Australia

John Wiley & Sons (Asia) Pte Ltd, 2 Clementi Loop #02-01, Jin Xing Distripark, Singapore 129809

John Wiley & Sons Canada Ltd, 22 Worcester Road, Etobicoke, Ontario, Canada M9W 1L1

Wiley also publishes its books in a variety of electronic formats. Some content that appears in print may not be available in electronic books.

Novartis Foundation Symposium 266
ix+297 pages, 52 figures, 7 tables

Library of Congress Cataloging-in-Publication Data

The hERG cardiac potassium channel : structure, function, and long QT
  syndrome / [editors, Derek J. Chadwick, James Goode].
      p. cm. – (Novartis Foundation symposium ; 266)
  Based on a symposium.
  Includes bibliographical references and index.
  ISBN 0-470-02140-3 (alk. paper)
  1. Potassium channels–Congresses. 2. Long QT syndrome–Congresses. 3. Heart–Physiology–Congresses. 4. Heart–Pathophysiology–Congresses. I. Chadwick, Derek. II. Goode, Jamie. III. Novartis Foundation. IV. Series.
  QP114.M46H44 2005
  612.1'7–dc22   2004061227

British Library Cataloguing in Publication Data

A catalogue record for this book is available from the British Library
ISBN 0 470 02140 3

Typeset in $10\frac{1}{2}$ on $12\frac{1}{2}$ pt Garamond by Dobbie Typesetting Limited, Tavistock, Devon.
Printed and bound in Great Britain by T. J. International Ltd, Padstow, Cornwall.
This book is printed on acid-free paper responsibly manufactured from sustainable forestry, in which at least two trees are planted for each one used for paper production.

Contents

Symposium on The hERG cardiac potassium channel: structure, function and long QT syndrome, held at the Novartis Foundation, London, 4–6 May 2004

Editors: Derek J. Chadwick and Jamie Goode

This symposium is based on a proposal made by John Mitcheson, Martin Gosling and Paul Nicklin

Michael C. Sanguinetti Chair's introduction 1

Gail A. Robertson, Eugenia M. C. Jones and **Jinling Wang** Gating and assembly of heteromeric hERG1a/1b channels underlying I_{Kr} in the heart 4
Discussion 15

Gea-Ny Tseng and **H. Robert Guy** Structure–function studies of the outer mouth and voltage sensor domain of hERG 19
Discussion 35

General discussion I 44

David R. Piper, Michael C. Sanguinetti and **Martin Tristani-Firouzi** Voltage sensor movement in the hERG K^+ channel 46
Discussion 52

Eckhard Ficker, Adrienne Dennis, Yuri Kuryshev, Barbara A. Wible and **Arthur M. Brown** hERG channel trafficking 57
Discussion 70

Anna Kagan and **Thomas V. McDonald** Dynamic control of hERG/I_{Kr} by PKA-mediated interactions with 14-3-3 75
Discussion 89

General discussion II 95

Arun Anantharam and **Geoffrey W. Abbott** Does hERG coassemble with a β subunit? Evidence for roles of MinK and MiRP1 100
Discussion 112

Arthur M. Brown   hERG block, QT liability and sudden cardiac death   118
Discussion   131

John Mitcheson, Matthew Perry, Phillip Stansfeld, Michael Sanguinetti, Harry Witchel and **Jules Hancox**   Structural determinants for high-affinity block of hERG potassium channels   136
Discussion   150

General discussion III   155

Michael C. Sanguinetti, Jun Chen, David Fernandez, Kaichiro Kamiya, John Mitcheson and **José A. Sanchez-Chapula**   Physicochemical basis for binding and voltage-dependent block of hERG channels by structurally diverse drugs   159
Discussion   166

Maurizio Recanatini, Andrea Cavalli and **Matteo Masetti**   *In silico* modelling — pharmacophores and hERG channel models   171
Discussion   181

Peter J. Schwartz   The long QT syndrome: a clinical counterpart of *HERG* mutations   186
Discussion   198

Steven Poelzing and **David S. Rosenbaum**   Cellular mechanisms of Torsade de Pointes   204
Discussion   217

Annarosa Arcangeli   Expression and role of hERG channels in cancer cells   225
Discussion   232

Luc M. Hondeghem   TRIad: foundation for proarrhythmia (triangulation, reverse use dependence and instability)   235
Discussion   244

Rashmi R. Shah   Drug-induced QT interval prolongation: regulatory guidance and perspectives on hERG channel studies   251
Discussion   280

Michael C. Sanguinetti   Closing remarks   286

Index of contributors   288

Subject index   290

Participants

Geoffrey W. Abbott   Weill Medical College of Cornell University, Division of Cardiology, Starr 463, 520 East 68th Street, New York, NY 10021, USA

Annarosa Arcangeli   Department of Experimental Pathology and Oncology, University of Viale G B Morgagni 50, 50134 Firenze, Italy

Arthur M. Brown   ChanTest Inc., 14656 Neo Parkway, Cleveland, OH 44128, USA

Eckhard Ficker   Rammelkamp Center for Education and Research, Metrohealth Campus, Case Western Reserve University School of Medicine, Cleveland, OH 44109-1998, USA

Martin Gosling   Novartis Research Centre, Wimblehurst Road, Horsham, West Sussex RH12 5AB, UK

Jules C. Hancox   Department of Physiology and Cardiovascular Research Laboratories, School of Medical Sciences, University of Bristol, University Walk, Bristol BS8 1TD, UK

Terry E. Herbert   Research Center, Montreal Heart Institute, 5000 Rue Bélanger Estate, Montreal, Quebec, H1T 1C8, Canada

Peter Hoffmann   F. Hoffman-La Roche Limited, Department PRBN-S, 4070 Basel, Switzerland

Luc Hondeghem   Department of Pharmacology, Katholieke Universiteit Leuven, Leuven, B-3000, Belgium

Craig T. January   Department of Medicine (Cardiology), University of Wisconsin Hospital and Clinics, Room H6/352, 600 Highland Avenue, Madison, WI 53792, USA

PARTICIPANTS

Yuliya Korolkova (*Novartis Foundation Bursar*)   Shemyakin and Ovchinnikov Institute of Bioorganic Chemistry, 1177997, GSP-7 ul. Miklukho-Maklaya, 16/10, Moscow V-437, Russia

Tom V. McDonald   Departments of Medicine (Cardiology) and Molecular Pharmacology, Albert Einstein College of Medicine, Forcheimer G35, 1300 Morris Park Avenue, Bronx, New York, NY 10461, USA

John Mitcheson   Department of Cell Physiology and Pharmacology, University of Leicester, Maurice Shock Medical Sciences Building, University Road, Leicester LE1 9HN, UK

Rainer Netzer   Senior Vice President, Business Development, Evotec OAI AG, Schnackenburgallee 114, 22525 Hamburg, Germany

Paul Nicklin   Novartis Research Centre, Wimblehurst Road, Horsham West Sussex RH12 5AB, UK

Denis Noble   University Laboratory of Physiology, University of Oxford, Parks Road, Oxford OX1 3PT, UK

Maurizio Recanatini   Department of Pharmaceutical Sciences, University of Bologna, Via Belmeloro 6, I-40126, Bologna, Italy

Gail A. Robertson   Department of Physiology, University of Wisconsin–Madison Medical School, Madison, WI 53706, USA

David S. Rosenbaum   The Heart and Vascular Research Center, MetroHealth Campus, Case Western Reserve University, 2500 Metrohealth Drive, Hamann 3, Cleveland, OH 44109-1998, USA

José A. Sanchez-Chapula   Unidad de Investigación Carlos Mendez, Centro Universitario de Investigaciones Biomedicas, Universidad de Colima, Colima, APDO Postal 199, CP 28000, Mexico

Michael C. Sanguinetti (*Chair*)   Department of Physiology, Nora Eccles Harrison Cardiovascular Research and Training Institute, University of Utah, 95 South 2000 East, Salt Lake City, UT 84112, USA

Peter J. Schwartz   Cattedra di Cardiologia, Universita degli Studi, Dipartimento di Cardiologia, IRCCS Policlinico San Matteo, Viale Golgi, 19, 27100 Pavia, Italy

PARTICIPANTS

Rashmi Shah   Senior Medical Officer, Medicines and Healthcare products Regulatory Agency (MHRA), Room 11-212, Market Towers, 1 Nine Elms Lane, London SW8 5NQ, UK

Derek A. Terrar   University Department of Pharmacology, Oxford University, Mansfield Road, Oxford OX1 3QT, UK

Dierk Thomas   Department of Cardiology, Medical University Hospital Heidelberg, Im Nemenheimer Feld 410, Heidelberg, D-69120, Germany

Martin Traebert   In Vitro Safety Pharmacology, Novartis Pharma AG, Preclinical Safety-EU, MUT-2881.205, Auhafenstrasse, CH-4132 Muttenz, Switzerland

Martin Tristani-Firouzi   Department of Pediatrics, Pediatric Cardiology, Suite 1500 PCMC, University of Utah School of Medicine, 100 N. Medical Drive, Salt Lake City, UT 84113, USA

Gea-Ny Tseng   Departments of Physiology/Cardiology (Internal Medicine), Virginia Commonwealth University, 1101 E Marshall Street, PO Box 980551, Richmond, VA 23298-0551, USA

Enzo Wanke   Dipartimento di Biotecnologie e Bioscienze, Universita di Milano-Bicocca, Pizza della Scienza 2, I-20126 Milano, Italy

Harry Witchel   University of Bristol, Cardiovascular Research Laboratories, Department of Physiology, Medical School, Bristol BS8 1TD, UK

Chair's introduction

Michael C. Sanguinetti

Department of Physiology, Nora Eccles Harrison Cardiovascular Research & Training Institute, University of Utah, Salt Lake City, UT 84112, USA

The three topics chosen for discussion by the organizers of this meeting are (1) the structural basis of function of hERG channels, (2) the physiological roles of hERG channels and (3) mechanisms of drug-induced long QT syndrome (LQTS). Due to time constraints, we will not have formal presentations on many other important topics related to hERG biology, but we plan to address some of these issues during the discussion periods.

Although an X-ray crystallographic structure is not available for hERG, we do have a model that is based on crystal structures of the bacterial channels KcsA, KvAP and MthK (Doyle et al 1998, Jiang et al 2002, 2003). We know from biophysical studies of Shaker and the X-ray structures of these bacterial channels that the S5, pore helix and S6 transmembrane domains compose the pore domain of hERG, as well as the likely structural basis of the selectivity filter. The S1–S4 transmembrane domains comprise the voltage-sensing component of the channel. The fourth transmembrane domain is by far the most important part of the voltage sensor, with several positively charged amino acids spaced three residues apart. Some of the basic residues in S4 also interact with the few acidic residues located in the S2 and S3 transmembrane domains. What is unusual about hERG is that it activates very slowly and inactivates rapidly compared to most other voltage-gated K^+ channels. The regions thought to be important for slow activation and deactivation include the N-terminal PAS domain, the S4–S5 linker and the end of the S6 domain. The structural basis for rapid inactivation is unknown, but it is presumed to be somewhat similar to C-type inactivation, involving residue interactions between S5 and the pore loop.

hERG is expressed in tissues throughout the body, including the heart, neurons and smooth muscle. Although the physiological roles of hERG expressed in tissues other than the heart will be briefly discussed, the main focus of this meeting will be the role of hERG in the repolarization of cardiac myocytes and drug-induced LQTS caused by block of hERG channels. Multiple K^+ currents mediate repolarization of the cardiac action potential. One of these currents, I_{Kr}, is conducted by hERG channels. About 15 years ago, while studying delayed rectifier K^+ currents in guinea pig myocytes, my lab identified two components based on sensitivity to block by an antiarrhythmic drug, E4031 (Sanguinetti & Jurkiewicz 1990). One component (I_{Kr}) was drug-sensitive, one (I_{Ks}) was not. This was not the first time two components of delayed rectifier current had been proposed. Twenty years earlier, Denis Noble and Richard Tsien had described two components of delayed rectification in sheep Purkinje fibres which they named I_{x1} and I_{x2}, the latter being the slower component (Noble & Tsien 1969). There were a number of problems (e.g. extracellular K^+ accumulation) with this earlier study inherent to using Purkinje fibres as a preparation for voltage clamp experiments, and subsequent studies concluded that delayed rectifier current was actually conducted by a single type of channel with complex kinetics. However, 20 years later it was proven that the original interpretation by Noble and Tsien was correct.

Normal hERG function is obviously important for action potential repolarization, but it is also important for pacemaking in the sinoatrial and atrioventricular nodes of the heart, and setting the resting potential in tumour cells, neurons and smooth muscle. In addition, hERG may serve as an oxygen sensor in glomus cells of the carotid body (Overholt et al 2000). The precise role of hERG channels expressed in many tissues is unknown.

The origins of hERG channel molecular biology can be traced back to *Drosophila* geneticists studying funny fly behaviour. When mutant flies were anaesthetized with ether they exhibited a twitching behaviour that resembled the action of a go-go dancer, hence the mutant phenotype was called ether-a-go-go (EAG). Barry Ganetsky and Jeff Warmke at the University of Wisconsin later decided to look for an EAG-like channel expressed in human tissue. They screened a human hippocampal cDNA library and found a channel gene they named hERG, human EAG-related gene (Warmke & Ganetzky 1994).

Drug-induced LQTS is a disorder of cardiac repolarization most commonly caused by block of hERG channels. Block of any of the cardiac repolarizing K^+ channels can prolong action potential duration and induce Torsade de Pointes. However, the only clinically significant way that this seems to occur is by direct block of hERG or indirect block of channels subsequent to a drug–drug interaction (e.g. inhibition of drug metabolism). The role of hERG in drug-induced LQTS explains the current intense interest in this channel.

This meeting is divided into four parts. The first part concerns hERG channel gating and we plan to address a number of questions. What is the physiological relevance of alternatively spliced forms of hERG in different organs? What is the mechanism of the different deactivation rates for hERG1a and hERG1b? How do the interactions of the acidic residues in S2 and S3 with basic residues in S4 compare with Shaker channels? How do these interactions control movement of the voltage sensor in response to transmembrane voltage changes? What is the structural basis of inactivation? We know that the S5–P loop is important for inactivation gating, but how? What is the function of, and the protein partner for the PAS domain?

The second part of the meeting concerns hERG channel trafficking, modulation of hERG and the role of accessory subunits. Again, several questions will be addressed. What is the therapeutic potential of chemical chaperones for treatment of LQTS? How is hERG regulated by cAMP, protein kinase A and reactive oxygen species? How does heterologous expression of hERG compare with I_{Kr} in myocytes? What are the tissue-dependent physiological roles for the many proposed regulatory subunits (e.g., MIRP1, MinK, 14-3-3, KvRP1)?

The third part of the symposium will address hERG channel pharmacology. One of the most important questions at the forefront of the field concerns the usefulness of pharmacophore models for *in silico* prediction of hERG blocker potency. The drug industry expends tremendous resources testing for hERG channel block by new chemical entities. If *in silico* predictions of drug block proved useful it would save a lot of time and money. What are the consequences of hERG1 (or hERG2, hERG3) block in organs other than the heart? Is there a common drug binding site located in the central cavity of hERG? If there are multiple sites, where might they be located? How far advanced are we in developing screens for hERG block with high-capacity voltage clamp assays?

The final part of the symposium will address mechanisms of drug-induced LQTS. What are the mechanisms of repolarization inhomogeneity that sustain Torsade de Pointes? What are the most useful and predictive animal and *in vitro* models for drug-induced LQTS? Is action potential instability and triangulation a better predictor for drug liability compared with measurement of hERG channel block? Has concern about drug-induced Torsade de Pointes been blown out of proportion by government regulators, or is it really a problem that has only recently been recognized?

References

Doyle DA, Morais Cabral J, Pfuetzner RA et al 1998 The structure of the potassium channel: molecular basis of K^+ conduction and selectivity. Science 280:69–77

Jiang Y, Lee A, Chen J, Cadene M, Chait BT, MacKinnon R 2002 Crystal structure and mechanism of a calcium-gated potassium channel. Nature 417:515–522

Jiang Y, Lee A, Chen J, Ruta V, Cadene M, Chait BT, MacKinnon R 2003 X-ray structure of a voltage-dependent K^+ channel. Nature 423:33–41

Noble D, Tsien RW 1969 Outward membrane currents activated in the plateau range of potentials in cardiac Purkinje fibres. J Physiol 200:205–231

Overholt JL, Ficker E, Yang T, Shams H, Bright GR, Prabhakar NR 2000 HERG-Like potassium current regulates the resting membrane potential in glomus cells of the rabbit carotid body. J Neurophysiol 83:1150–1157

Sanguinetti MC, Jurkiewicz NK 1990 Two components of cardiac delayed rectifier K^+ current: Differential sensitivity to block by class III antiarrhythmic agents. J Gen Physiol 96:195–215

Warmke JW, Ganetzky B 1994 A family of potassium channel genes related to *eag* in *Drosophila* and mammals. Proc Natl Acad Sci USA 91:3438–3442

Gating and assembly of heteromeric hERG1a/1b channels underlying I_{Kr} in the heart

Gail A. Robertson, Eugenia M.C. Jones and Jinling Wang

Department of Physiology, University of Wisconsin–Madison Medical School, 1300 University Avenue, Madison, WI 53706, USA

Abstract. Until recently, ion channels generating cardiac I_{Kr} were thought to comprise four identical α subunits encoded by the ERG1a transcript. Despite studies identifying another transcript, ERG1b, failure to identify the corresponding protein in native tissue led to the conclusion that the ERG1b subunit is not a constituent of cardiac I_{Kr} channels. Interestingly, hERG1b subunits coexpressed in heterologous systems preferentially form heteromultimers with hERG1a and modify the deactivation gating properties previously attributed to the hERG1a N-terminus. The two subunits are identical except for their divergent N-termini. Moreover, I_{Kr} kinetic properties are more closely mimicked by currents from heteromeric, compared to homomeric, channels. Studies with a new generation of antibodies now show that ERG1b subunits also contribute to I_{Kr} channels *in vivo*, likely in heteromeric assemblies with ERG1a. Bidirectional co-immunoprecipitation of ERG1a and 1b subunits from canine and human ventricle indicates that the subunits associate in native tissue, where they are also found by immunocytochemistry to localize to the same subcellular compartment. These new findings raise questions as to the role of the respective N-termini in deactivation gating and assembly *in vivo*, as well as the disease mechanisms of mutations causing hERG-linked long QT syndrome, approximately 20% of which reside in the hERG1a N-terminus and have previously been evaluated only in the context of the hERG1a homomers.

2005 The hERG cardiac potassium channel: structure, function, and long QT syndrome. Wiley, Chichester (Novartis Foundation Symposium 266) p 4–18

During the past 10 years, hERG1 channels have received much attention as the targets of acquired and type 2 long QT syndrome (LQTS). Significant advances have been made in our understanding of mechanisms underlying gating, modulation, drug block and disease due to inherited mutations. Studies have largely been carried out in cell culture, where the channels can be examined in relative isolation in a highly controlled environment.

hERG1, or more precisely, hERG1a homomeric channels, have served as a model for cardiac I_{Kr} since the *hERG1* gene was shown to encode channels producing I_{Kr}-like biophysical and pharmacological properties in heterologous expression systems (Fig. 1; Sanguinetti et al 1995, Trudeau et al 1995). The features that define hERG1a currents and I_{Kr} are:

- a fast inactivation process that suppresses current during the depolarizing phase of the action potential;
- fast recovery from inactivation followed by slow deactivation, which together trigger a resurgent current as channels revisit the open state during repolarization; and
- block by methanesulfonanilides, originally used to identify I_{Kr} in guinea pig heart (Sanguinetti & Jurkiewicz 1990).

These properties vary somewhat when the currents in native and heterologous expression systems are compared, among different expression systems, at different temperatures, and when hERG1a is coexpressed with other proteins.

hERG1b as a potential *α* subunit of I_{Kr}

The possibility that other α subunits contribute to I_{Kr} channels was raised with the discovery of an *ERG1*[1] alternative transcript abundantly expressed in mouse and human heart (London et al 1997, 1998, Lees-Miller et al 1997). This transcript, *ERG1b*, encodes a protein identical to the ERG1a subunit except for a shorter, unique N-terminus. Variations in the N-terminal region are intriguing because this region in ERG1a plays an important role in deactivation gating, modulation and assembly. Among mutations causing LQTS-2, 20% occur in the hERG1a N-terminus[2].

In contrast to ERG1a, ERG1b subunits do not efficiently produce surface membrane homomultimers. Expression of the *ERG1b* transcript in *Xenopus* oocytes produced functional channels only when an artificial Kozak consensus sequence was attached to its 5′ end and large quantities of RNA were injected (London et al 1997). The resulting currents had a fast deactivation phenotype compared to hERG1a homomers. When 1a was coexpressed with the native 1bK$^-$ (*sans* Kozak sequence) the currents deactivated with kinetics intermediate to those of 1a and 1b, suggesting the 1a subunits combined with 1b to produce channels with altered properties (Fig. 2). Curve-fitting analysis demonstrated the deactivation kinetics could not be described by a weighted algebraic summation of the 1a and 1b currents expressed independently as would be expected if the

---

[1]The term hERG1 is used to describe human ERG1 subunits and the corresponding gene; ERG1 is used more generally to encompass all homologues, including non-human species. Italics are used when describing the gene or transcript, normal text for the subunits or proteins.
[2]http://pc4.fsm.it:81/cardmoc/hergmut.htm

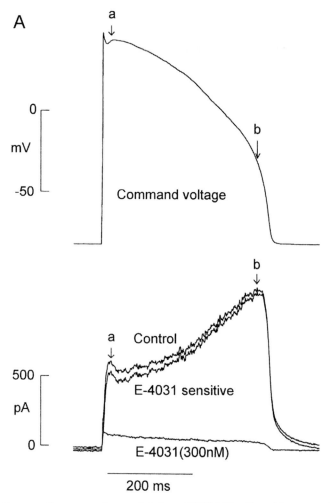

FIG. 1. Currents (lower trace) recorded from HEK-293 cells stably expressing hERG1a in response to action potential voltage clamp command (upper trace). hERG1a current is the E-4031-sensitive current trace. Current is initially suppressed during depolarization as many channels quickly enter the inactivated state (point a). A resurgent current (point b) results as channels recover from inactivation and slowly deactivate during repolarization. From Zhou et al (1998) with permission.

subunits produced two populations of homomeric channels. The most compelling evidence for heteromeric assembly is that 1a deactivation kinetics were altered by coexpression with native 1bK$^-$ injected at concentrations too low to produce currents independently. The intermediate deactivation kinetics of the 1a/1b heteromer more closely mimic those of the native currents (Sanguinetti

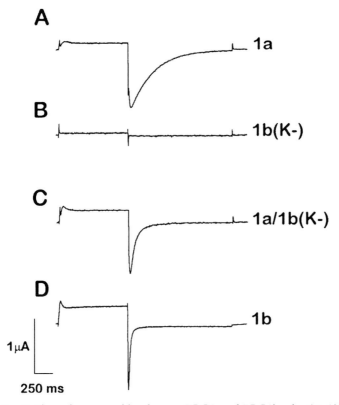

FIG. 2. Evidence for coassembly of mouse ERG1a and ERG1b subunits. (A) Expression of ERG1a is shown. (B) Expression of ERG1bK⁻ (lacking an upstream Kozak consensus sequence) did not give rise to measurable currents ($n = 12$). (C) Coexpression of ERG1a and ERG1bK⁻ resulted in currents with faster deactivation kinetics than those observed for ERG1a. (D) Expression of ERG1b (same as ERG1bK⁻ but containing a Kozak consensus sequence) produced currents with faster deactivation kinetics than observed for either ERG1a or ERG1a/1b expressed together. The voltage command was a depolarizing step to 20 mV for 1 second followed by a repolarizing step to -100 mV for 1.5 seconds, from a holding potential of -80 mV. From London et al (1997) with permission.

& Jurkiewicz 1991, Hancox et al 1998), implicating hERG1b as a potential α subunit of I_{Kr} channels.

The N-terminus of hERG1 determines deactivation kinetics

The hERG1a N-terminus encompasses at least two distinct regions involved in slow deactivation, a gating mechanism critical to the resurgent qualities characteristic of I_{Kr} and essential to its role in repolarization (Spector et al 1996,

Schonherr & Heinemann 1996, Wang et al 1998). Much attention has focused on the PAS domain, a structure emerging from analysis of N-terminal hERG1a crystals and implicated in various protein–protein interactions (Morais Cabral et al 1998). Mutations in the PAS domain accelerated channel deactivation, reflecting a disruption of N-terminal function (Morais Cabral et al 1998, Chen et al 1999).

However, deletion of only the first 16 amino acids, upstream from the PAS domain and not ordered in the crystal structure, was sufficient to phenocopy the deactivation defects of a complete N-terminal truncation mutant (Wang et al 1998). Indeed, intracellular application of a peptide corresponding to the first 16 amino acids, or 'deactivation domain', largely restored slow deactivation to the N-terminal truncation mutant in excised macropatches. The effect of the peptide was dose-dependent with a Hill coefficient of 2.2, indicating at least three peptides act to slow deactivation (Wang et al 2000). Dialysing the larger N-terminal fragment encompassing the PAS domain (aa 1–136) into oocytes resulted in a similar restoration of deactivation to the N-truncated channel, but did so slowly over a period of 3–24 h (Morais Cabral et al 1998), as if proteolytic cleavage or another time-dependent modification were required to unleash the active domain.

At the single channel level, mean open times were greater when the N-terminus was present, reflecting a stabilizing effect on the channel open state (Fig. 3) (Wang et al 2000). The action of the N-terminus was disrupted by cysteine modification of the S4-S5 linker, resulting in rapid deactivation characteristic of the N-truncated channel. These data support a model in which the N-terminus stabilizes the open state of the channel by interacting with a site near the internal mouth of the pore (Wang et al 1998). Insights into how this interaction might stabilize the open state were provided using a second-site suppressor strategy that revealed an interaction between the S4-S5 linker and residues in S6 close to or comprising the activation gate (Sanguinetti & Xu 1999, Tristani-Firouzi et al 2002).

Taken together, the findings from these studies suggest a model in which the PAS domain docks the N-terminus in the region of the internal mouth of the pore via an interaction with or near the S4-S5 linker. The most N-terminal 16 amino acids are thus positioned to interact with the voltage sensor or gate residues, or to couple these two regions, stabilizing the open state. The unique hERG1b N-terminus lacks the deactivation domain of hERG1a, yet acts in a dominant manner to alter the deactivation kinetics of the hERG1a/1b heteromeric channel.

N-terminal phosphorylation regulates hERG1 activity

The hERG1a N-terminus is also the site of phosphorylation by kinases. Two sites are phosphorylated, one by the serine-threonine protein kinase A (PKA) (Thomas

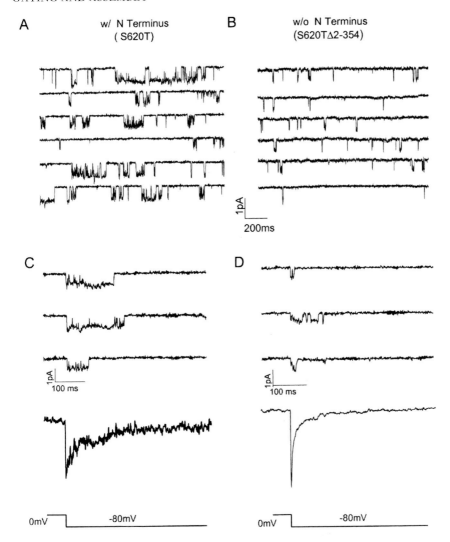

FIG. 3. Single-channel activities at negative (repolarizing) potentials display longer open times when the N-terminus is present. During a continuous voltage command of −80 mV, lasting for 3–5 min, single-channel activities were recorded in cell-attached patch with 100 mM potassium in the pipette. Representative segments of the records from inactivation-removed S620T and N-truncated S620TΔ2-354 channels are shown in A and B, with longer openings apparent in S620T channels ($n=3$ for each construct). (C and D) Ensemble tail currents and representative single trials of activities from which the ensemble averages were constructed for S620T and S620TΔ2-354 channels, respectively, during a 2 s repolarizing command to −80 mV after depolarization to 60 mV. Deactivation of the ensemble currents displays typical slow phenotype for S620T channel and fast phenotype for S620TΔ2-354 channels ($n=3$ for each construct). From Wang et al (2000) with permission.

et al 1999, Cui et al 2000) and another by Src tyrosine kinase (Cayabyab & Schlichter 2002). PKA was found to accelerate deactivation at least in part by phosphorylating the site S283 on the N-terminus. Interestingly, S283 resides downstream from both the N-terminal deactivation domain and the PAS domain, suggesting the possibility of yet another N-terminal region contributing to deactivation gating. The amplitude of hERG1 current is correspondingly reduced by PKA, which is important in the context of LQTS1 disease mutations that eliminate KvLQT1/I_{Ks} currents. Under normal circumstances, β-adrenergic stimulation of the heart increases PKA, which enhances KvLQT1/I_{Ks} current and helps regulate the QT interval (Marx et al 2002). In patients with LQTS1 who lack the stabilizing force of I_{Ks} currents, PKA is expected to further prolong the QT interval by suppressing hERG1/I_{Kr} currents (Cui et al 2000). Phosphorylation at S283 serves another purpose: it promotes the binding of 14-3-3ε, which mediates an interaction between the hERG1 N- and C-termini (Kagan et al 2002). A function of 14-3-3ε is to protect C-terminal phosphorylated sites from phosphatase, thus prolonging the effects of phosphorylation. Since 14-3-3 has been shown to promote forward trafficking of other potassium channels (O'Kelly et al 2002), phosphorylation-dependent binding between 14-3-3 and hERG1 may be a mechanism for regulating surface channel density.

In contrast to PKA, Src tyrosine kinase had the opposite effect of increasing hERG1 current by slowing deactivation, possibly by increasing the stability of the N-terminal interaction with the S4-S5 linker (Cayabyab & Schlichter 2002). The action of Src tyrosine kinase is opposed by Shp-1, a tyrosine phosphatase (Cayabyab et al 2002). Both Src and Shp-1 were found to associate with hERG1 in cultured cells. Although there is no evidence that Src regulates hERG1 or I_{Kr} in the heart, Src has been implicated in preconditioning (Krieg et al 2002); perhaps Src phosphorylation enhances hERG1 activity, bolstering repolarization and thus contributing to this cardioprotective process. Mechanisms involving phosphorylation of the hERG1a N-terminus may differ in heteromeric assemblies with hERG1b, which lacks both phosphorylation sites.

N-terminal interactions in assembly

The role of the N-terminus in assembly was first implicated in biochemical studies identifying a tetramerization domain within the hERG1a N-terminus (Li et al 1997). Subsequently, N-terminal mutations in the PAS domain were shown to disrupt trafficking, possibly by affecting assembly, in heterologously expressed hERG1a channels (Paulussen et al 2002). In the Kv family of potassium channels, N-termini interact during biogenesis even as the more C-terminal portions of the protein materialize on the ribosome (Lu et al 2001). This interaction facilitates the oligomerization of subunits, and may underlie the specificity of heteromeric interactions among closely related family members (Li et al 1992, Xu et al 1995). Because the only difference between hERG1a and 1b subunits lies within the N-terminus, sequences within the 1b N-terminus must be responsible for the preferential appearance of surface hERG1a/1b heterooligomers over homooligomers in oocytes. Whether these sequences specify heteromeric assembly, selective export from the ER or longer lifetime in the plasma membrane remains to be determined.

Association with auxiliary subunits

In addition to the alpha subunit constituency of I_{Kr} channels, other proteins have been implicated in hERG1 channel function and trafficking. Two of these are the single transmembrane-domain proteins MinK and MiRP1. When coexpressed in CHO cells, MinK stably associated with hERG1 and increased hERG1 current amplitudes (McDonald et al 1997). In atrially derived AT-1 tumour cells, antisense *MinK* reduced I_{Kr} (Yang et al 1995). MiRP1 was found to alter the functional properties of hERG1 channels in oocytes, accelerating deactivation, reducing single channel conductance and increasing sensitivity to drug block (Abbott et al 1999). It is interesting to note that the acceleration of deactivation might be explained by an interaction of MiRP1 with the hERG1a N-terminus; experiments testing whether MiRP1 similarly affects the properties of the N-truncated channel have not been reported. The sensitivity to PKA-mediated phosphorylation diminished markedly when hERG1a was coexpressed with MiRP1 (Cui et al 2000), also consistent with the hypothesis that MiRP1 binds the N-terminus, masking the PKA phosphorylation or binding site. KCR1, a membrane-associated protein, counteracted some of the functional effects of MiRP1 on hERG1 channels, as if MiRP1 could be functionally decoupled from the hERG1 channel (Kupershmidt et al 2003). The effects of MiRP1 on hERG1 currents found in oocytes (Abbott et al 1999) were not replicated in a subsequent study using CHO cells (Weerapura et al 2002). It should not be surprising that different expression systems may possess different complements of proteins modulating hERG1 function. As a result, there is increasing impetus to move toward cardiac tissue preparations to uncover the complexities of the channel in its native environment.

hERG1a/1b heteromeric channels underlie cardiac I_{Kr}

The first such advance in the study of native hERG1 or, more generally, ERG1 *in vivo* came with the generation of ERG-specific antibodies and analysis of ERG1 in rat, mouse and human heart and brain (Pond et al 2000). These landmark studies described native ERG1 isoforms and showed they were glycosylated. Contrary to

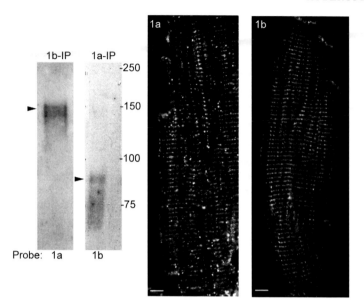

FIG. 4. ERG1a and 1b are associated *in vivo*. Western blots of myocyte membrane proteins: left, immunoprecipitated with 1b and probed with 1a antibodies; right, immunoprecipitated with 1a and probed with 1b antibodies. Immunocytochemistry of fixed canine ventricular myocytes probed with ERG1a-specific (left) and ERG1b-specific (right) antibodies shows punctate, Z-line staining characteristic of T-tubular distribution (E. M. C. Jones, E. C. Roti Roti, S. A. Delfosse and G. A. Robertson, unpublished data; Jones et al 2004).

expectation, however, these first-generation ERG antibodies failed to identify a protein corresponding to ERG1b in Western blots of native tissue. A role for the ERG1b subunit in cardiac I_{Kr} seemed unlikely.

A fuller picture has recently emerged with the ERG1-KA antibody, generated against a distinct C-terminal region for a study of hERG1-interacting proteins (Roti Roti et al 2002). In initial Western blots of rat heart, a robust band at the size predicted for ERG1b serendipitously appeared. An ERG1b-specific antibody was subsequently generated and shown on Western blots of lysate from hERG1a, hERG1b and hERG1a/1b stable cell lines to specifically interact with the 1b isoform. As expected, this antibody also identified corresponding bands in rat, canine and human ventricles. Moreover, bidirectional co-immunoprecipitation from native tissue and localization to the same subcellular compartment in isolated myocytes strongly suggests the subunits associate to form heteromeric channels *in vivo* (Fig. 4).

Concluding remarks

The finding that hERG1b contributes as another α subunit to I_{Kr} channels raises many questions. How does the hERG1b N-terminus regulate deactivation? Is the faster deactivation reported in myocytes and hERG1a/1b heteromeric channels typical of I_{Kr} under all circumstances, or might modulation via N-terminal phosphorylation play a role in adjusting the deactivation rate and therefore the magnitude of I_{Kr}? Does coexpression with hERG1b change our assessment of the disease phenotype of N-terminal LQTS-2 mutations? Will we find LQTS mutations in the hERG1b N-terminus? How do MinK, MiRP1 or KCR affect the properties of hERG1a/1b heteromers? Are all I_{Kr} channels heteromers, or do regions of homomeric hERG1a or 1b channels exist? If both populations exist, are their numbers subject to modulation in normal or disease states? Does isoform expression vary transmurally, or in different regions of the heart? The answers to these and related questions should significantly advance our understanding of the biology of the hERG1 channel and its role in disease.

References

Abbott GW, Sesti F, Splawski I et al 1999 MiRP1 forms I_{Kr} potassium channels with HERG and is associated with cardiac arrhythmia. Cell 97:175–187
Cayabyab FS, Schlichter LC 2002 Regulation of an ERG K$^+$ current by Src tyrosine kinase. J Biol Chem 277:13673–13681
Cayabyab FS, Tsui FW, Schlichter LC 2002 Modulation of the ERG K$^+$ current by the tyrosine phosphatase, SHP-1. J Biol Chem 277:48130–48138
Chen J, Zou A, Splawski I, Keating MT, Sanguinetti MC 1999 Long QT syndrome-associated mutations in the Per-Arnt-Sim (PAS) domain of HERG potassium channels accelerate channel deactivation. J Biol Chem 274:10113–10118
Cui J, Melman Y, Palma E, Fishman GI, McDonald TV 2000 Cyclic AMP regulates the HERG K$^+$ channel by dual pathways. Curr Biol 10:671–674
Hancox JC, Levi AJ, Witchel HJ 1998 Time course and voltage dependence of expressed HERG current compared with native 'rapid' delayed rectifier K current during the cardiac ventricular action potential. Pflügers Arch 436:843–853
Jones EMC, Roti Roti EC, Wang J, Delfosse SA, Robertson GA 2004 Cardiac I_K channels minimally comprise hERG1a and 1b subunits. J Biol Chem 279:44690–44694
Kagan A, Melman YF, Krumerman A, McDonald TV 2002 14-3-3 amplifies and prolongs adrenergic stimulation of HERG K$^+$ channel activity. EMBO J 21:1889–1898
Krieg T, Qin Q, McIntosh EC, Cohen MV, Downey JM 2002 ACh and adenosine activate PI3-kinase in rabbit hearts through transactivation of receptor tyrosine kinases. Am J Physiol Heart Circ Physiol 283:H2322–2330
Kupershmidt S, Yang IC, Hayashi K et al 2003 The I_{Kr} drug response is modulated by KCR1 in transfected cardiac and noncardiac cell lines. FASEB J 17:2263–2265
Lees-Miller JP, Kondo C, Wang L, Duff HJ 1997 Electrophysiological characterization of an alternatively processed ERG K$^+$ channel in mouse and human hearts. Circ Res 81:719–726
Li M, Jan YN, Jan LY 1992 Specification of subunit assembly by the hydrophilic amino-terminal domain of the Shaker potassium channel. Science 257:1225–1230

Li X, Xu J, Li M 1997 The human delta1261 mutation of the HERG potassium channel results in a truncated protein that contains a subunit interaction domain and decreases the channel expression. J Biol Chem 272:705–708

London B, Trudeau MC, Newton KP et al 1997 Two isoforms of the mouse ether-a-go-go-related gene coassemble to form channels with properties similar to the rapidly activating component of the cardiac delayed rectifier K^+ current. Circ Res 81:870–878

London B, Aydar E, Lewarchik C et al 1998 N and C-terminal isoforms of HERG in the human heart. Biophys J 74:A26

Lu J, Robinson JM, Edwards D, Deutsch C 2001 T1-T1 interactions occur in ER membranes while nascent Kv peptides are still attached to ribosomes. Biochemistry 40:10934–10946

Marx SO, Kurokawa J, Reiken S et al 2002 Requirement of a macromolecular signaling complex for beta adrenergic receptor modulation of the KCNQ1-KCNE1 potassium channel. Science 295:496–499

McDonald TV, Yu Z, Ming Z et al 1997 A minK–HERG complex regulates the cardiac potassium current I_{Kr}. Nature 388:289–292

Morais Cabral JH, Lee A, Cohen SL et al 1998 Crystal structure and functional analysis of the HERG potassium channel N terminus: a eukaryotic PAS domain. Cell 95:649–655

O'Kelly I, Butler MH, Zilberberg N, Goldstein SA 2002 Forward transport. 14-3-3 binding overcomes retention in endoplasmic reticulum by dibasic signals. Cell 111:577–588

Paulussen A, Raes A, Matthijs G et al 2002 A novel mutation (T65P) in the PAS domain of the human potassium channel HERG results in the long QT syndrome by trafficking deficiency. J Biol Chem 277:48610–48616

Pond AL, Scheve BK, Benedict AT et al 2000 Expression of distinct ERG proteins in rat, mouse, and human heart. Relation to functional I_{Kr} channels. J Biol Chem 275:5997–6006

Roti Roti EC, Myers CD, Ayers RA et al 2002 Interaction with GM130 during HERG ion channel trafficking. Disruption by type 2 congenital long QT syndrome mutations. Human Ether-a-go-go-Related Gene. J Biol Chem 277:47779–47785

Sanguinetti MC, Jurkiewicz NK 1990 Two components of cardiac delayed rectifier K^+ current. Differential sensitivity to block by class III antiarrhythmic agents. J Gen Physiol 96:195–215

Sanguinetti MC, Jurkiewicz NK 1991 Delayed rectifier outward K^+ current is composed of two currents in guinea pig atrial cells. Am J Physiol 260:H393–H399

Sanguinetti MC, Xu QP 1999 Mutations of the S4-S5 linker alter activation properties of HERG potassium channels expressed in Xenopus oocytes. J Physiol (Lond) 514:667–675

Sanguinetti MC, Jiang C, Curran ME, Keating MT 1995 A mechanistic link between an inherited and an acquired cardiac arrhythmia: HERG encodes the IKr potassium channel. Cell 81:299–307

Schonherr R, Heinemann SH 1996 Molecular determinants for activation and inactivation of HERG, a human inward rectifier potassium channel. J Physiol (Lond) 493:635–642

Spector PS, Curran ME, Zou A, Keating MT, Sanguinetti MC 1996 Fast inactivation causes rectification of the IKr channel. J Gen Physiol 107:611–619

Thomas D, Zhang W, Karle CA et al 1999 Deletion of protein kinase A phosphorylation sites in the HERG potassium channel inhibits activation shift by protein kinase A. J Biol Chem 274:27457–27462

Tristani-Firouzi M, Chen J, Sanguinetti MC 2002 Interactions between S4-S5 linker and S6 transmembrane domain modulate gating of HERG K^+ channels. J Biol Chem 277:18994–19000

Trudeau MC, Warmke JW, Ganetzky B, Robertson GA 1995 HERG, a human inward rectifier in the voltage-gated potassium channel family. Science 269:92–95

Wang J, Myers CD, Robertson GA 2000 Dynamic control of deactivation gating by a soluble amino-terminal domain in HERG K^+ channels. J Gen Physiol 115:749–758

Wang J, Trudeau MC, Zappia AM, Robertson GA 1998 Regulation of deactivation by an amino terminal domain in human ether-a-go-go-related gene potassium channels J Gen Physiol 112:637–647 (Erratum in J Gen Physiol 113:359)

Weerapura M, Nattel S, Chartier D, Caballero R, Hebert TE 2002 A comparison of currents carried by HERG, with and without coexpression of MiRP1, and the native rapid delayed rectifier current. Is MiRP1 the missing link? J Physiol (Lond) 540:15–27

Xu J, Yu W, Jan YN, Jan LY, Li M 1995 Assembly of voltage-gated potassium channels. Conserved hydrophilic motifs determine subfamily-specific interactions between the alpha-subunits. J Biol Chem 270:24761–24768

Yang T, Kupershmidt S, Roden DM 1995 Anti-minK antisense decreases the amplitude of the rapidly activating cardiac delayed rectifier K^+ current. Circ Res 77:1246–1253

Zhou Z, Gong Q, Ye B et al 1998 Properties of HERG channels stably expressed in HEK 293 cells studied at physiological temperature. Biophys J 74:230–241

DISCUSSION

Gosling: Have you tried mutating any of the 16 amino acids at the N-terminus to find out which ones are important in the deactivation time constants?

Robertson: We haven't done this yet. We are embarking on a new study to dissect the mechanism at a finer level, and mutagenesis will be part of that effort.

Gosling: With Shaker gating there were some studies performed where the chain holding the inactivation 'ball' was lengthened—this profoundly affected inactivation kinetics. Are you going to try to insert some residues in between the PAS domain and the N-terminus to see whether this affects channel kinetics?

Robertson: That is a good idea.

Traebert: Do we know how conserved the amino acid sequences are between different species with regards to gating kinetics? Is there a high degree of homology between guinea-pig and rabbit, for example?

Robertson: We don't have the full sequence for guinea-pig: the databases just have a partial sequence. Are you implying that there are differences between rabbit and guinea-pig and that this could help us?

Traebert: No, it was just that from a pharmacological point of view it could be very interesting to have different sequences. For a pharmaceutical company this is of interest.

McDonald: With regards to a potential species difference, what species did you use for your immunofluorescence studies?

Robertson: Dog and rat.

McDonald: In rabbit, which we have been studying using a range of commercial and home-made antibodies, we primarily see a band at around 100–110 kDa. We are not sure whether this is representing ERG1a or 1b. However, the sequence in the database would suggest that it is almost identical to the human. In fact, it may be two amino acids longer.

Robertson: I would guess that it is ERG1b. I will send you my antibody and you can find out. It can't really be ERG1a unless it is migrating in an anomalous way or is a degradation product, because it is much smaller than the predicted size of 125 kDa.

McDonald: When we run it next to dog or pig, or just transfected cells, those migrate at the usual higher molecular weights.

Robertson: Do you see anything with your antibody that looks like ERG1b in the dog and the pig?

McDonald: A little bit. It is not very prominent.

Robertson: Perhaps the rabbit has a surfeit of 1b.

Sanguinetti: Although I think deactivation is very slow in the rabbit.

Robertson: This fast and slow deactivation could be physiologically dependent. It could be altered during different physiological states, and there could be some regional differences.

Hancox: It is slow in ventricle. If you look at the nodes in rabbit, the deactivation is much faster than in ventricle (e.g. Howarth et al 1996) and this may also be the case for the atrium. Are there any data on this in the dog or rat?

Robertson: Not yet.

Hancox: Another key point that Mike Sanguinetti raised is that in the nodes, when you are making recordings they don't have T tubules. The current seen may be different from that seen in cells with a T tubular system.

Sanguinetti: I think the single-channel data from guinea-pig ventricular myocytes look very similar to the single-channel behaviour in nodal cells (Horie et al 1990, Shibasaki 1987). I would think that T tubule location might make the currents appear different. Related to this, why do all channels seem to be located in the T tubule? You mentioned a potential artefact; could you expand on this?

Robertson: We were initially concerned that we might have wiped out the surface membrane with our fixation process, but this doesn't seem to be the case. We do see the plasma membrane there, although it is not 100% intact. There is a recent study by Rasmussen et al (2004) showing immunogold localization of ERG. This is also primarily showing localization to the T tubules.

Gosling: There are functional ways of addressing this also. I think scanning ion conductance microscopy has been used to measure channel location around T tubules in intact tissue.

Robertson: What is that?

Gosling: It involves using a very fine patch pipette to simultaneously measure cell surface architecture and localize channels (see Korchev et al 2000).

Robertson: I don't think it is surprising that T tubules should have a lot of ion channels. This is the site of E–C coupling, and the duration of the action potential will be important in regulating calcium influx.

Gosling: You mentioned the promoter regions for the splice variants. Is anything known about what binding sites are on those promoter regions?

Robertson: Not at all. This is a wide open area for a promoter basher to come in and figure it out. For example, I_{Kr} levels differ in males versus females. This would be one place to find out whether there is different hormonal regulation of those transcripts. As it is, all we know is that the transcript has its own 5′ end which is unique from that of the 1a transcript.

Brown: In rat and dog, is the transcript structure the same as in mouse? Also, are there alternative splice sites as well as different transcripts? Finally, when you do the co-IP experiments, if you do densitometry can you get any idea of the relative amounts of protein?

Robertson: Those are very good questions. We don't know the gene structure in the dog and rat, but we expect it to be conserved as it is between mouse and human. We are interested in looking at the relative abundance of the two proteins, and possibly to get some stoichiometric insights by using native gels. They seem fairly similar in abundance at this point.

Ficker: I have problems looking at these Western blots. If you take the C-terminal antibody, which is a very good antibody, and then you blot, you don't see a lot with the regular antibodies. What I am getting at is that the epitope is there in 1a and 1b. You work under denaturing conditions and see in many preparations primarily the bigger form. Does this not mean that the abundance of 1b is rather small?

Robertson: We don't think the abundance is small. It really depends on the antibody and the accessibility of the epitope. Yes, the proteins are denatured and in an SDS gel, but they are not without some structure. On the Western blots themselves the only explanation is that the epitope is not very available. If we go back and look at the published records of the antibodies people have been using, and we look at their Western blots, often we see a band of the right size for 1b. The antibodies identify it but not very well. There is something about that epitope that is not readily accessible in those Western blots. This sort of phenomenon is not unprecedented. You saw our 1b antibody characterization and that it is specific for the 1b protein expressed in a heterologous tissue. Now we go to a native tissue and see very little background and a robust band at the same position we saw with our new C-terminal antibody. This doesn't leave much room for another explanation.

Ficker: I don't doubt that it is there, but it is more a quantitative argument. Perhaps there is little present, and this is why you detect little. Looking at the KA antibody, doesn't that detect two bands in the core or non-glycosylated region of 1b?

Robertson: Yes.

Ficker: It seems to me, then, that there is a lot of unglycosylated protein there. What does this mean for the stability of this protein?

Robertson: In the native tissue most or all the signal we see is mature protein. The large fraction of immature observed in heterologous expression systems may be an artefact of overexpression.

Hebert: Is there evidence that hERG1a and 1b interact with different populations of auxillary subunits?

Robertson: That is something we are interested in testing.

Hebert: Is there any evidence that the interaction between 1a and 1b is direct, and doesn't go through an intermediate of one of these other subunits?

Robertson: I don't have really good evidence for this yet. We have some GST fusion protein interactions. We can see interactions between the 1a and 1b in the N-terminal regions. We can readily pull out of lysate the other subunit with a 1a or 1b N-terminus, but this doesn't rule out the involvement of an interacting protein.

Arcangeli: I am happy to hear that 1b is expressed in the heart. When we published that it was expressed in cancer cells it was a big surprise that it wasn't thought to be expressed in the heart. It is the same result as in tumour cells: the 1a and 1b form heterotetramers. Do you have data suggesting that there could be a modulation in the expression of the two isoforms in the heart, as happens in tumours, for example, in embryonic hearts?

Robertson: Those studies haven't been done.

Arcangeli: You really need a monoclonal antibody against the N-terminus of 1a or 1b, just to do the co-expression experiments, and the double immunofluorescence.

Robertson: We have tried doing double immunofluorescence with the antibodies we have, but we haven't succeeded. The 1a antibodies, including ours, are not that great.

References

Howarth FC, Levi AJ, Hancox JC 1996 Characteristics of the delayed rectifier K current compared in myocytes isolated from the atrioventricular node and ventricle of the rabbit heart. Pflügers Arch 431:713–722

Horie M, Hayashi S, Kawai C 1990 Two types of delayed rectifying K^+ channels in atrial cells of guinea pig heart. Jap J Physiol 40:479–490

Korchev YE, Negulyaev YA, Edwards CR, Vodyanoy I, Lab MJ 2000 Functional localization of single active ion channels on the surface of a living cell. Nat Cell Biol 2:616–619

Rasmussen HB, Moller M, Knaus HG, Jensen BS, Olesen SP, Jorgensen NK 2004 Subcellular localization of the delayed rectifier K^+ channels KCNQ1 and ERG1 in the rat heart. Am J Physiol Heart Circ Physiol 286:H1300–1309

Shibasaki T 1987 Conductance and kinetics of delayed rectifier potassium channels in nodal cells of the rabbit heart. J Physiol 387:227–250

Structure–function studies of the outer mouth and voltage sensor domain of hERG

Gea-Ny Tseng and H. Robert Guy*

*Departments of Physiology/Internal Medicine (Cardiology), Virginia Commonwealth University, 1101 E. Marshall Street, Richmond, VA 23298 and *Laboratory of Experimental and Computational Biology, National Cancer Institute, National Institutes of Health, 12 South Drive, Bethesda, MD 20892-5567, USA*

> *Abstract.* hERG has uniquely fast inactivation but slow activation processes. We study the structural basis for these unique gating properties using the following approaches: site-specific mutagenesis, the MTS accessibility test, disulfide bond formation, thermodynamic mutant cycle analysis, peptide toxin 'foot-printing', NMR spectroscopy, and molecular modelling. We propose the following: (1) two structural features in hERG's outer mouth contribute to its fast inactivation rate: a lack of 'open mouth'-stabilizing hydrogen bonds and an unusually long extracellular 'S5-P' linker that contains an α-helix. During membrane depolarization, four such 'S5-P helices' from the tetramer channel come near each other to occlude the outer mouth. This occurs rapidly due to the dynamic nature of the S5-P helices. (2) Two structural features in hERG's voltage sensor domain contribute to its slow activation rate: hERG's major voltage-sensor, S4, has three (instead of four as in Shaker) positive charges involved in gating charge transfer, and hERG has six (instead of three as in Shaker) negative charges in the other transmembrane segments (S1–S3) of the voltage sensor domain. Thus a less voltage-sensitive S4, in conjunction with more surrounding negative charges (some of which can form salt-bridges with S4's positive charges in the pre-open state), retards channel activation.
>
> *2005 The hERG cardiac potassium channel: structure, function, and long QT syndrome. Wiley, Chichester (Novartis Foundation Symposium 266) p 19–43*

The hERG channel shares the basic design of structure-function relationship with other voltage-gated K (Kv) channels (Fig. 1A). Activation of the hERG channel is mediated by depolarization-triggered S4 movements that lead to conformational changes around the inner mouth and opening of the pore (Tristani-Firouzi et al 2002). Inactivation of the hERG channel is mediated by conformational changes around the outer mouth that occlude ion flux through the pore, similar to the C-type inactivation process first described for the Shaker channel (Smith et al 1996, Hoshi et al 1991). However, the kinetics of the activation and inactivation

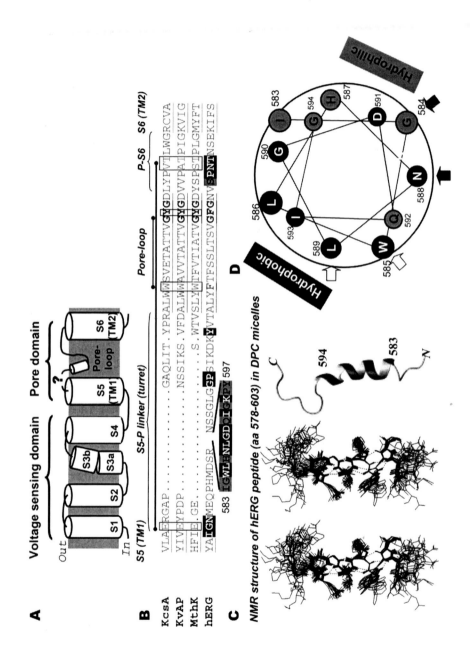

FIG. 1. (A) Two-dimensional transmembrane topology of a voltage-gated K (Kv) channel subunit, with voltage-sensing and pore domains, as well as transmembrane helices (S1–S6), labelled. '?' points to S5-P linker that lines the outer mouth of K channels. S5 and S6 correspond to TM1 and TM2 of two-transmembrane domain K channels. (B) Partial amino acid sequence alignment of three bacterial K^+ channels whose crystal structures are solved (Doyle et al 1998, Jiang et al 2002, 2003a) and hERG. Sequences are from the end of S5 (TM1) to the beginning of S6 (TM2). S5 (TM1), pore-loop, and S6 (TM2) are underlined. The '. . .' denote gaps introduced to improve alignment. Residues shown in bold denote K channel signature sequence (G-Y/F-G). Boxed residues connected by lines with dots denote H-bonding donor-acceptor pairs (glutamate 'E' carboxylate at the outer end of S5/TM1 with peptide backbone amido nitrogen in P-S6 linker, tryptophan 'W' nitrogen with tyrosine 'Y' hydroxyl at two ends of pore-loop). The hERG S5-P linker is much longer than those of the other three, and the 583–597 segment is shown as an insert. Highlighted hERG residues denote high-impact positions (white letters on black background) and intermediate-impact positions (black letters on grey background) (see text for explanation). (C) NMR structure of a hERG peptide (corresponding to residues 578–603) in DPC micelles. *Left and middle:* Stereoview of the 20 best structures of the hERG peptide superimposed on the backbone atoms of the 583–593 region. Backbone and side chain heavy atoms of the 580–596 region are shown. Hydrogen bonds, detected in more than 10 of the 20 best structures are denoted by dotted lines. *Right*: Ribbon representation of hERG peptide structure (numbers denote the borders of helical region). (D) Helical wheel representation of residues 583–594 (S5-P helix). High- and intermediate-impact positions are highlighted as in (B). Hydrophobic and hydrophilic surfaces of the helix are marked. Black arrows point to G584 and N588 residues hypothesized to be in the centre of a tightly packed bundle of four S5-P helices in the inactivated conformation. White arrows point to W585 and L589 residues that are slightly farther apart in adjacent subunits in this model (see Fig. 2C).

processes in the hERG channel is quite unique among Kv channels. For example, hERG's inactivation process operates in the millisecond time scale, vs. the second time scale of C-type inactivation in Shaker or Shaker-like channels (Smith et al 1996, Hoshi et al 1991). In the voltage range of 0 to +60 mV, the time constant of hERG activation ranges from 250 to 60 ms, while the time constant of activation is less than 2 ms for the Shaker channel (Zagotta et al 1994) and less than 10 ms for the EAG channel (in the fast-gating mode, Silverman et al 2000), a close relative of hERG (Fig. 3B). The combination of a fast inactivation process with a much slower activation process results in an inward rectification in the current-voltage relationship of the hERG channel, which is important for its physiological function (Spector et al 1996, Zhou et al 1998). Our experiments were designed to address 2 questions: (1) what is the structural basis for the uniquely fast inactivation process of the hERG channel? (2) what are the factors contributing to the slow activation process of the hERG channel?

Experimental procedures

The procedures of site-specific mutations of hERG and oocyte expression/voltage clamping have been described in detail before (Liu et al 2002, 2003). Molecular modelling procedures and criteria are described in Durell et al (1998).

Results and discussion

hERG's outer mouth and the fast inactivation process

Figure 1B aligns the amino acid sequences of KcsA, KvAP, MthK (three prokaryotic K^+ channels whose crystal structures have been solved (Doyle et al 1998, Jiang et al 2002, 2003a) and hERG. These sequences line the outer mouths of these channels (Fig. 1A). This alignment reveals two features separating hERG from the other K^+ channels. First, two H-bonds can be formed between residues within each subunit around the outer mouth and within the pore-loop of these prokaryotic (and also eukaryotic) K^+ channels (H-bond donor and acceptor residues are boxed and connected by horizontal lines in Fig. 1B) (Larsson & Elinder 2000, Doyle et al 1998). In hERG, the equivalent positions are occupied by residues that cannot form such H-bonds. Thus, the outer mouth of the hERG channel may be easier to collapse than in Shaker or other Kv channels, where H-bonds can help maintain the outer mouth in the open conformation (Larsson & Elinder 2000). Second, the S5-P linker in hERG is much longer than in other K^+ channels (Liu et al 2002). In the crystal structures of K^+ channel pore domains, the corresponding 'turret' domains have a coil type secondary structure (Doyle et al 1998, Jiang et al 2002, 2003a). Can portions of the much longer S5-P linker in the hERG channel assume a helical secondary structure and what is its role in channel function?

Cysteine (Cys) scanning mutagenesis identifies high- and intermediate-impact positions lining hERG's outer mouth. We replace each of the residues along hERG's S5-P linker, as well as those of P-S6 linker, with Cys, one at a time, and study the impact of Cys substitution as well as the effects of thiol-modifying reagents (MTSET and MTSES, increasing Cys side chain volume and adding a positive or negative charge) on hERG channel function (Liu et al 2002). The positions can be divided into three groups:

- high-impact positions, where Cys substitution severely disrupts channel function
- intermediate-impact positions, where Cys substitution alone is well tolerated but subsequent thiol side chain modification by MTS reagents suppresses or disrupts channel function, and
- low-impact positions, where neither Cys substitution nor MTSET treatment alters channel function.

The high- and intermediate-impact positions are highlighted by black and grey shades, respectively, in the hERG sequence in Fig. 1B. Importantly, Cys mutants at high-impact positions share the same 'mutant' phenotype: a disruption of inactivation and K^+ selectivity, and a negative shift in the voltage-dependence of activation.

Importantly, there are 15 consecutive positions in the middle of hERG's S5-P linker, from 583 to 597, that are either high- or intermediate-impact positions. In one-dimensional sequence this segment is far from the pore entrance. However, their mutations markedly affect the inactivation process and the K^+ selectivity, suggesting that this segment may be close to the pore and selectivity filter in three-dimensional space. We therefore focus further on this 583–597 segment: what is its secondary structure, and what is its relation to the pore? Perturbation analysis using a Fourier transform method suggests that the impact of Cys substitution in this segment on the activation gating process manifests an α-helical pattern (Liu et al 2002). More direct evidence comes from an NMR spectroscopy study. It shows that a synthetic peptide corresponding to hERG residues 578–603 can assume a helical structure from positions 583–594 when placed in a membrane-mimetic environment (detergent micelles) (Fig. 1C). A similar conclusion was reached by Torres et al (2003a). Figure 1D depicts a helical wheel representation of this segment. The helix has a hydrophobic surface occupied by high-impact positions. The opposite side is a hydrophilic surface occupied mainly by intermediate-impact positions. How is this S5-P helix oriented with respect to the pore of the channel?

The N-terminus of the S5-P helix is close to the central axis of the hERG pore. Among all the Cys substituted mutants along the S5-P and P-S6 linkers that are sensitive to

MTSET or MTSES (high- and intermediate-impact positions), MTSET or MTSES modification affects channel conductance but not gating processes or K^+ selectivity. However, three positions respond in a unique fashion: T613C, S631C and G584C. Both MTSET and MTSES modifications make these Cys substituted channels switch from 'WT-like' to the mutant phenotype: a disruption of inactivation and K^+ selectivity. T613 and S631 are two residues flanking the pore-loop, situated right at the edge of the external pore entrance (Fig. 1A,B). Thus, in three-dimensional space their side chains are close to the pore's central axis and thus to counterparts from other subunits. MTSET or MTSES modification of these Cys side chains would add several positive or negative charges around the pore entrance. Like charges would repel each other, thus limiting the flexibility of the outer mouth structure and disrupting the pore function (fast inactivation and K^+ selectivity). G584 is at the N-terminal end of the S5-P helix. The similarities between 584C and 613C or 631C in the response to MTSET and MTSES suggest that 584 may also be close to the central axis of the pore, placing the N-terminus of the S5-P helix close to the pore entrance. This orientation is further supported by data from disulfide bond formation experiments discussed below.

Peptide toxins distinguish a unique hERG outer mouth. There are several scorpion peptide toxins that show a high specificity for the hERG channel (Korolkova et al 2002, Pardo-Lopez et al 2002, Lecchi et al 2002, Nastainczyk et al 2002). One of them, BeKm-1, belongs to the α-KTx family and shares 3D scaffold and Cys disulfide bridge patterns with those of ChTx or AgTx2 (Tytgat et al 1999). However, unlike ChTx or AgTx2 whose interaction surface is formed by a β-strand, BeKm-1 uses another structure, a short α-helix, as its interaction surface for binding to hERG (Korolkova et al 2002). Furthermore, the interaction surface of BeKm-1 is hydrophobic in nature (lined by Y11 and F14), followed by positive charges at the carboxyl end (K18 and R20). Another scorpion toxin, ErgTx1, belongs to a different (γ-KTx) family and has a different disulfide bridge pattern than BeKm-1 or other α-KTx members (Tytgat et al 1999). ErgTx1 shares the high specificity for hERG with BeKm-1 (Pardo-Lopez et al 2002). It is possible that ErgTx1 also uses its α-helix as the interaction surface for binding to hERG (Torres et al 2003b). Mutant cycle analysis of hERG and BeKm-1 mutants indicates that hydrophobic residues along the 583–597 segment (I583 and Y597), as well as S631 at the pore entrance, are critical for toxin binding. These data help us develop models of how BeKm-1 binds in the outer entrance of the hERG channel.

Cys side chains introduced into the 583–597 segment formed intersubunit disulfide bonds and at least four positions close to the N-terminus of this segment form disulfide bonds with counterparts from other subunits. A unique feature of the 583–597 segment is that Cys side chains introduced anywhere within this segment spontaneously form disulfide bonds under non-reducing conditions. This indicates proximity between these introduced Cys side chains and counterparts from neighbouring subunits or native Cys. Identifying these disulfide-bonding partners is a powerful way to deduce relationships between these residues in 3D space. We ask whether Cys side chains introduced into the 583–597 segment form inter- or intra-subunit disulfide bonds. Western blot experiments show that all these introduced Cys side chains can form intersubunit disulfides. To determine whether these intersubunit disulfides are formed with counterparts or with native Cys, we remove all five native Cys and reintroduce Cys into positions 584, 585, 588 and 589 in the N-terminal half of the 583–597 segment. Removing native Cys does not prevent intersubunit disulfide bond formation at these positions. These data are consistent with the notion that the N-termini of the S5-P helices are close to each other, and to the pore entrance.

Cd^{2+} bridge formation between Cys side chains is another sensitive measure of inter-Cys distance. It also allows us to compare inter-Cys distances by comparing the apparent affinities of Cd^{2+} binding sites formed between different pairs of Cys side chains. Among the four positions where Cys side chains can form intersubunit disulfide bonds with counterparts, 584C and 588C can form high-affinity Cd^{2+} binding sites, while 585C and 589C form low-affinity Cd^{2+} binding sites. Our data further suggest that the hydrophilic surfaces (where G584 and N588 reside, Fig. 1D) of 2 S5-P helices from adjacent subunits face each other.

A model of hERG's outer mouth structure and proposed mechanism for hERG's fast inactivation process. Figure 2 summarizes our working hypothesis for hERG's inactivation mechanism mediated by conformational changes involving four S5-P helices around the outer mouth. This inactivation process differs from that described for the Shaker channel in important ways. In the Shaker channel, it has been proposed that C-type inactivation is triggered by S4 movements during membrane depolarization, which allow an interaction of the extracellular end of S4 with that of S5 (Gandhi et al 2003, Laine et al 2003, Broomand et al 2003). This then disrupts H-bonds between outer mouth residues (Fig. 1B), resulting in a collapse of the outer pore (Larsson & Elinder 2000). The lack of such H-bonding capability around hERG's outer mouth, in conjunction with the dynamic nature and proximity between the N-termini of the S5-P helices will support an extremely fast inactivation process. We propose that the four S5-P helices together form an 'extended' pore extracellular to and in series with the 'primary' pore lined by the pore-loops (Fig. 1A and 1B). Intimate interactions between the 'primary' and the 'extended' pores explain why mutations in the pore-loop at position 631 (Fan et al 1999) and S620T (Herzberg et al 1998) also disrupt the inactivation process.

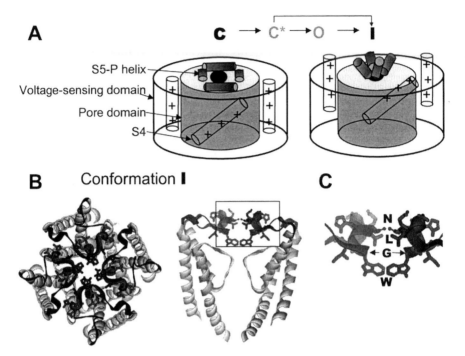

FIG. 2. (A) Cartoon illustrating the working model of conformational changes around hERG's outer mouth during membrane depolarization. Gating states are diagrammed on top: C, C*, O and I are closed, pre-open, open and inactivated states. Marked on left are structural elements. (B) Molecular structure of hERG in the inactivated state. The model was built based on KvAP pore domain. *Left*: top view from the extracellular perspective. *Right*: side view with 2 subunits removed for clarity. The boxed area is enlarged in (C). (C) Close-up view of S5-P helices from two subunits across the central axis of the pore. Four residues are labelled: G(584) and N(588) of adjacent subunits are near each other in the centre of the bundle; the carbons of W(585) and L(589) of adjacent subunits are slightly farther apart.

hERG's voltage-sensor domain and the slow activation process

The slow activation rate in hERG could be due to slow S4 motions during membrane depolarization, slow conformational changes of S6's cytoplasmic halves (the 'activation gate'), an inefficient coupling between S4 motions and the activation gate, or a combination of the above. The landmark work by Piper, Tristani-Firouzi, and Sanguinetti (Piper et al 2003) showed that hERG's gating currents have two distinct kinetic components: one fast and one slow. The slow component correlates with channel opening in voltage-dependence and kinetics. This was corroborated by work from Yellen's group, who conjugate fluorophores to Cys side chains engineered to the outer end of hERG's S4 and use alterations in fluorescence (ΔF) during voltage clamp protocols as a measure of local conformational or environmental changes (Smith & Yellen 2002). They detect a slow component of ΔF that correlates with channel opening in its voltage-dependence and kinetics (although there is an additional fast component of ΔF at two of the three positions examined). Therefore, both direct gating current measurements and indirect fluorescence measurements implicate slow S4 motions as the major cause for the slow activation kinetics of the hERG channel.

Several factors can slow S4 motions upon membrane depolarization. First, S4 with fewer positive charges involved in gating charge transfer will be less sensitive to membrane depolarization, and thus may move more slowly than S4 with more positive charges (Logothetis et al 1992, Islas & Sigworth 2000, Ahern & Horn 2004). Second, negative charges in the other transmembrane helices (S1, S2 and S3) of the voltage-sensing domain may slow S4 motions, by forming salt bridges with S4's positive charges preferentially in closed or pre-open states, or by forming divalent cation binding sites in the external crevices around S4 that can resist outward movements of S4 during membrane depolarization (Silverman et al 2000, Schonherr et al 2002). We designed experiments to investigate these possibilities. The recently proposed 'paddle' voltage-sensing mechanism based on the crystal structure of KvAP (Jiang et al 2003b) has significant inconsistencies with many experimental data from eukaryotic Kv channels (Horn 2004, Cohen et al 2003, Gandhi et al 2003, Starace & Bezanilla 2004). Therefore, our discussion will be based on the 'conventional' voltage-sensing mechanism based on studies of eukaryotic Kv channels.

Figures 3A–C compare the pattern of charge distribution in the voltage-sensing domains of hERG, Shaker and EAG. The KvAP sequence is also included for reference. Note that there is no consensus as to how the S4 positive charges in hERG (or EAG) should be aligned with those in the Shaker (Smith & Yellen 2002, Silverman et al 2003, Schonherr et al 2002), because a direct comparison of functional roles of these S4 positive charges among these channels has not been available.

Which positive charges in hERG's S4 domain are responsible for transferring gating charges during channel activation? To test the functional role of each of the positive charges in hERG's S4 domain, we neutralize them to Cys one at a time. We ask two questions:

- how does such charge neutralization impact on the number of gating charges transferred during hERG's activation?
- what is the accessibility of these Cys side chains to external and internal MTSET in different gating states? The rationale is that if a positive charge in the S4 is important for voltage-sensing during hERG's activation, then neutralizing it should reduce the number of gating charges transferred during gating, and the

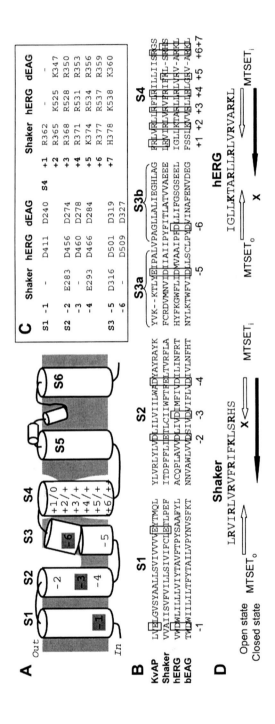

FIG. 3. (A) A Kv channel subunit diagram, highlighting negative charges in S1, S2 and S3 (grey shades: negative charges unique to EAG and ERG channels), and positive charges in S4 (+1 to +6 in Shaker/0 or + for ERG and EAG). (B) Partial amino acid sequence alignment of KvAP, Shaker, hERG and b(ovine)EAG. Channel regions (S1, S2, S3a, S3b, and S4) are denoted on top. Charged residues are boxed, and their generic numbers are marked below (−1 to −6, +1 to +7). (C) Corresponding charged residues in S1–S4 and their position numbers of Shaker, hERG and d(rosophila)EAG. (D) Sidedness and state-dependence of MTSET accessibility to cysteine side chains introduced into Shaker's S4 positions (based on Larsson et al 1996, Baker et al 1998) and hERG's S4. White arrows from left and right denote MTSET accessibility from outside and inside, respectively, in open state. Black arrows denote the same in closed state. 'X' indicates no accessibility.

Cys side chain introduced into its position should change accessibility to MTSET in a state-dependent fashion (e.g. from internally accessible in the closed state to inaccessible or even externally accessible in the open state, corresponding to an outward S4 movement at depolarized voltages).

We choose to use the 'limiting slope' method to deduce the number of gating charges per channel (Zagotta & Aldrich 1990). The numbers of equivalent gating charges transferred per channel during activation gating for wild-type (WT) hERG, K525C, R528C, R531C, R534C, R537C and K538C are (number in parenthesis is that of measurements): 6.4±0.7 (5), 4.8±0.4 (5), 3.8±0.4 (6), 2.6±0.1 (5), 6.4±0.3 (6), 6.5±0.2 (4) and 6.4±0.2 (4).

Results from the MTSET accessibility test are summarized in Fig. 3D (right panel). In the closed state, 525C is not accessible to either $MTSET_o$ or $MTSET_i$, while 528C and 531C are accessible to $MTSET_i$. In the open state, 525C becomes accessible to $MTSET_o$, while 528C and 531C have reduced accessibility to $MTSET_i$ but are still not accessible to $MTSET_o$. On the other hand, Cys side chains at 534, 537 and 538 are always accessible to internal MTSET but not accessible to external MTSET, indicating that they do not change their position relative to the membrane barrier or the electrical field during hERG activation. These data indicate that during transitions from the closed to open states, S4 movements transfer K525 from a 'buried position' to a position accessible to the external aqueous phase, and at the same time transfer R528 and R531 from internally accessible positions to more 'buried positions'.

Therefore, results from the limiting slope analysis and from the MTSET accessibility tests are consistent. Both indicate that the three positive charges at the N-terminal half of hERG's S4 (K525, R528, and R531) are involved in voltage-sensing and gating charge transfer. This supports the S4 aa alignment between hERG and Shaker shown in Fig. 3B and 3C. Fig. 3D further shows that the pattern of MTSET accessibility of S4 side chains is quite similar between Shaker (Larsson et al 1996, Baker et al 1998) and hERG.

Interactions between negative charges in S1, S2 and S3 and positive charges in S4 of the hERG channel. We also examine the functional role of negative charges in the transmembrane helices of hERG's voltage-sensing domain by replacing them with neutral Cys or positively charged lysine or arginine, one at a time, and studying the impact of these mutations on hERG channel function (Liu et al 2003). The results suggest that all six negative charges contribute to stabilizing the hERG channel in the open conformation. This is likely due to salt-bridge formation between these negative charges and S4's positive charges in gating states very close to, or in, the open conformation.

Neutralizing (Cys substitution) or reversing (aspartate substitution) positive charges in hERG's S4 domain also affects the voltage-dependence and kinetics of activation. Reversing the positive charge of R531 or R537 causes a positive shift in the voltage-dependence of activation. Reversing R531 also greatly slowed channel activation. These data suggest that these positive charges preferentially engage in salt-bridge formation in the open state. In contrast, reversing the positive charges at the two ends of hERG's S4, K525 or K538, shifts the voltage-dependence of activation in the negative direction and accelerates channel activation, suggesting that these positive charges preferentially engage in salt-bridge formation in closed or pre-open states, thus retarding channel activation. Reversing the positive charge at 528 (R528D) caused a 10-fold slowing of channel activation and more than a 100 fold slowing of deactivation. These marked kinetic effects, in conjunction with the knowledge that R528 is the middle one of the three positive charges in hERG involved in voltage-sensing and S4 movements, suggest that R528 has the important role of anchoring the S4 with nearby negative charges during its helical screw motions in the outward direction during activation as well as in the inward direction during deactivation. We use the approach of 'mutant cycle analysis' to further explore which negative charges form salt bridges with K525 and K538 in the closed or pre-open state of the hERG channel (Yifrach & MacKinnon 2002). Results indicate that K525 interacts with D456 at the outer end of S2 and that K538 interacts with D411 at the inner end of S1.

Helix packing in hERG's voltage-sensing domain and interactions between charges during gating. We build structural models of hERG's voltage-sensing domain using the crystal structure of KvAP's isolated voltage-sensing domain as the template. The open state of hERG corresponds to the KvAP crystal structure. S4 is postulated to move via the helical screw mechanism (Guy & Seetharamulu 1986), in which the backbone is repositioned by three residues for each helical screw step. S4 moves inward by one helical screw step for the C* conformation, and by two more helical screw steps to reach the closed (C) state. The predicted interactions between negative charges in S1, S2 and S3 and the positive charges in S4 in different conformations are listed in the legend to Fig. 4. Comparable charge–charge interactions have been described for the Shaker channel (Papazian et al 2002): -2 with $+3$ in the pre-open state, -2 with $+4$ in the open state; and for the EAG channel (Silverman et al 2003): -2 with $+4$ in the open state. These similarities in charge–charge interactions between different Kv channels support the notion that the basic design in Kv channels' voltage-sensing mechanism is well conserved.

Is there a divalent cation (M^{2+}) binding site in the hERG channel as in the EAG channel? An M^{2+} binding site has been described for the EAG channel, that is formed between

STRUCTURE–FUNCTION STUDIES

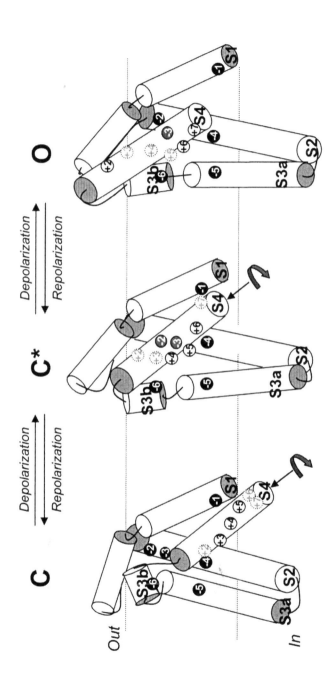

FIG. 4. Hypothesized helix packing and charge–charge interactions between α-helices in the voltage-sensing domain of hERG. *Top*: Gating state diagram (C, C* and O: closed, pre-open and open states). *Bottom*: hERG S1–S4 helices in one subunit of each of the gating states, viewed from the pore domain. The helix numbers (S1, S2, S3a, S3b and S4) are marked. Positive and negative charges are denoted by their generic numbers (positive charges shown in grey dotted circles on S4 are those facing away from the viewer; negative charges shown in white on grey background are those behind S4). S4 motions triggered by membrane depolarization (translating toward the extracellular side and rotating clockwise around its long axis) are also denoted. Charge–charge interactions occurring in different gating states include: (1) in 'C' state, +2 to −4 or −5, +3 with −4, (2) in 'C*' state, +3 with −2 or −6, +4 with −2, −3 or −6, +5 with −3, +6 with −5, +7 with −4, (3) in 'O' state, +4 with −2 or −6, +5 with −2, −3 or −6, +6 with −4 or −5, and +7 with −4.

'−3' and '−6' in S2 and S3, respectively (Fig. 3A). Elevating $[Mg^{2+}]_o$ from 2 to 10 mM, by binding to this M^{2+} binding site and restricting outward S4 movements during depolarization, can markedly slow both the early transitions and the late opening step during EAG activation (Silverman et al 2000, Schonherr et al 2002). hERG shares these negative charges (Fig. 3B). It is possible that there is an extremely high-affinity M^{2+} binding site in the hERG channel. Elevating $[Ca^{2+}]_o$ in 3–100 μM range specifically slows the rate of hERG activation (Johnson et al 2001). Only at much higher $[Ca^{2+}]_o$ (above 300 μM) does the voltage-dependence of channel activation shift in the positive direction ('surface charge' effect) (Johnson et al 2001). The apparent difference between hERG and EAG in their modulation by extracellular divalent cations may be due to different set points in the divalent cation concentration at which the rate of channel activation is affected.

Summary

The fast rate of hERG inactivation has two contributing factors: a weakened H-bonding capability around the outer mouth and a dynamic structure of the four S5-P helices that are close to each other and interact across the central axis of the pore. The alignment in Fig. 3B indicates that hERG's S4 is missing 1 positive charge relative to that of Shaker. This would contribute to a smaller number of gating charges, a lower voltage-sensitivity, and a slower activation rate. The additional negative charges in the other transmembrane segments of hERG's voltage-sensing domain should further impact on the rate of channel activation (Schonherr et al 2002, Silverman et al 2000).

Two questions about the structure–function relationship of the hERG channel remain unanswered:

- what is the origin of a component of voltage-sensitivity in the inactivation process that is independent of S4 movements? and
- what is the relationship between this voltage-sensitivity to the fast component of gating current or ΔF signals reported by fluorophores attached to the outer end of S4?

Acknowledgements

The authors would like to thank the following people for their contributions to the projects: Drs E. V. Grishin, Y. V. Korolkova, A. Arseniev, I. V. Meslannikov, Min Jiang, Mei Zhang, Jie Liu and Kailas Sonawane. The projects were supported by HL-46451 from National Institutes of Health and a Grant-in-Aid from American Heart Association/Mid-Atlantic Affiliate (to GNT).

References

Ahern CA, Horn R 2004 Specificity of charge-carrying residues in the voltage sensor of potassium channels. J Gen Physiol 123:205–216

Baker OS, Larsson HP, Mannuzzu LM, Isacoff EY 1998 Three transmembrane conformations and sequence-dependent displacement of the S4 domain in *Shaker* K^+ channel gating. Neuron 20:1283–1294

Broomand A, Mannikko R, Larsson HP, Elinder F 2003 Molecular movement of the voltage sensor in a K channel. J Gen Physiol 122:741–748

Cohen BE, Grabe M, Jan LY 2003 Answers and questions from the KvAP structures. Neuron 39:395–400

Doyle DA, Cabral JM, Pfuetzner RA et al 1998 The structure of the potassium channel: molecular basis of K^+ conduction and selectivity. Science 280:69–77

Durell SR, Hao Y, Guy HR 1998 Structural models of the transmembrane region of voltage-gated and other K^+ channels in open, closed, and inactivated conformations. J Struct Biol 122:263–284

Fan J-S, Jiang M, Dun W, McDonald TV, Tseng G-N 1999 Effects of outer mouth mutations on *hERG* channel function: a comparison with similar mutations in *Shaker*. Biophys J 76:3128–3140

Gandhi CS, Clark E, Loots E, Pralle A, Isacoff EY 2003 The orientation and molecular movement of a K^+ channel voltage-sensing domain. Neuron 40:515–525

Guy HR, Seetharamulu P 1986 Molecular model of the action potential sodium channel. Proc Natl Acad Sci USA 83:508–512

Herzberg IM, Trudeau MC, Robertson GA 1998 Transfer of rapid inactivation and sensitivity to the class III antiarrhythmic drug E-4031 from HERG to M-eag channels. J Physiol 511:3–14

Horn R 2004 How S4 segments move charge. Let me count the ways. J Gen Physiol 123:1–4

Hoshi T, Zagotta WN, Aldrich RW 1991 Two types of inactivation on Shaker K^+ channels: effects of alterations in the carboxy-terminal region. Neuron 7:547–556

Islas LD, Sigworth FJ 2000 Voltage sensitivity and gating charge in *Shaker* and *Shab* family potassium channels. J Gen Physiol 114:723–741

Jiang Y, Lee A, Chen J, Cadene M, Chait BT, MacKinnon R 2002 Crystal structure and mechanism of a calcium-gated potassium channel. Nature 417:515–522

Jiang Y, Lee A, Chen J et al 2003a X-ray structure of a voltage-dependent K^+ channel. Nature 423:33–41

Jiang Y, Ruta V, Chen J, Lee A, MacKinnon R 2003b The principle of gating charge movement in a voltage-dependent K^+ channel. Nature 423:42–48

Johnson JP, Balser JR, Bennett PB 2001 A novel extracellular calcium sensing mechanism in voltage-gated potassium ion channels. J Neurosci 21:4143–4153

Korolkova YV, Bocharov EV, Angelo K et al 2002 New binding site on the old molecular scaffold provides selectivity of HERG-specific scorpion toxin BeKm-1. J Biol Chem 277:43104–43109

Laine M, Lin M-CA, Bannister JPA et al 2003 Atomic proximity between S4 segment and pore domain in Shaker potassium channels. Neuron 39:467–481

Larsson HP, Elinder F 2000 A conserved glutamate is important for slow inactivation in K^+ channels. Neuron 27:573–583

Larsson HP, Baker OS, Dhillon DS, Isacoff EY 1996 Transmembrane movement of the shaker K^+ channel S4. Neuron 16:387–397

Lecchi M, Redaelli E, Rosati B et al 2002 Isolation of a long-lasting *eag*-related gene-type K^+ current in MMQ lactotrophs and its accommodating role during slow firing and prolactin release. J Neurosci 22:3414–3425

Liu J, Zhang M, Jiang M, Tseng G-N 2002 Structural and functional role of the extracellular S5-P linker in the HERG potassium channel. J Gen Physiol 120:723–737

Liu J, Zhang M, Jiang M, Tseng G-N 2003 Negative charges in the transmembrane domains of the HERG K channel are involved in the activation- and deactivation gating processes. J Gen Physiol 121:599–614

Logothetis DE, Movahedi S, Satler C, Lindpaintner K, Nadal-Ginard B 1992 Incremental reductions of positive charge within the S4 Region of a voltage-gated K^+ channel result in corresponding decreases in gating charge. Neuron 8:531–540

Nastainczyk W, Meves H, Watt DD 2002 A short-chain peptide toxin isolated from *Centruroides sculpturatus* scorpion venom inhibits *ether-a-go-go*-related gene K^+ channels. Toxicon 40:1053–1058

Papazian DM, Silverman WR, Lin M-CA, Tiwari-Woodruff SK, Tang C-Y 2002 Structural organization of the voltage sensor in voltage-dependent potassium channels. In: Ion channels: from atomic resolution physiology to functional genomics (Novartis Found Symp 245), p 178–192

Pardo-Lopez L, Zhang M, Liu J, Jiang M, Possani LD, Tseng G-N 2002 Mapping the binding site of a HERG-specific peptide toxin (ErgTx) to the channel's outer vestibule. J Biol Chem 277:16403–16411

Piper DR, Varghese A, Sanguinetti MC, Tristani-Firouzi M 2003 Gating currents associated with intramembrane charge displacement in HERG potassium channels. Proc Natl Acad Sci USA 100:10534–10539

Schonherr R, Mannuzzu LM, Isacoff EY, Heinemann SH 2002 Conformational switch between slow and fast gating modes: allosteric regulation of voltage sensor mobility in the EAG K^+ channel. Neuron 35:935–949

Silverman WR, Tang C-Y, Mock AF, Huh K-B, Papazian DM 2000 Mg^{2+} modulates voltage-dependent activation in Ether-a-go-go potassium channels by binding between transmembrane segments S2 and S3. J Gen Physiol 116:663–677

Silverman WR, Roux B, Papazian DM 2003 Structural basis of two-stage voltage-dependent activation in K^+ channels. Proc Natl Acad Sci USA 100:2935–2940

Smith PL, Yellen G 2002 Fast and slow voltage sensor movements in HERG potassium channels. J Gen Physiol 119:275–293

Smith PL, Baukrowitz T, Yellen G 1996 The inward rectification mechanism of the HERG cardiac potassium channel. Nature 379:833–836

Spector PS, Curran ME, Zou A, Keating MT, Sanguinetti MC 1996 Fast inactivation causes rectification of the I_{Kr} channel. J Gen Physiol 107:611–619

Starace DM, Bezanilla F 2004 A proton pore in a potassium channel voltage sensor reveals a focused electric field. Nature 427:548–553

Torres AM, Bansal P, Sunde M et al 2003a Structure of the HERG K^+ channel S5P extracellular linker: role of an amphipathic α-helix in C-type inactivation. J Biol Chem 278:42136–42148

Torres AM, Bansal P, Alewood PF, Bursill JA, Kuchel PW, Vandenberg JI 2003b Solution structure of CnErg1 (Ergtoxin), a HERG specific scorpion toxin. FEBS Letter 539:138–142

Tristani-Firouzi M, Chen J, Sanguinetti MC 2002 Interactions between S4-S5 linker and S6 transmembrane domain modulate gating of HERG K^+ channels. J Biol Chem 277:18994–19000

Tytgat J, Chandy KG, Garcia ML et al 1999 A unified nomenclature for short-chain peptides isolated from scorpion venoms: α-KTx molecular subfamilies. Trends Pharmacol Sci 20:444–447

Yifrach O, MacKinnon R 2002 Energetics of pore opening in a voltage-gated K^+ channel. Cell 1111:231–239

Zagotta WN, Aldrich RW 1990 Voltage-dependent gating of shaker A-type potassium channels in drosophila muscle. J Gen Physiol 95:29–60

Zagotta WN, Hoshi T, Dittman J, Aldrich RW 1994 Shaker potassium channel gating II: Transitions in the activation pathway. J Gen Physiol 103:279–319

Zhou J, Gong Q, Ye B et al 1998 Properties of HERG channels stably expressed in HEK 293 cells studied at physiological temperature. Biophys J 74:230–241

DISCUSSION

McDonald: If you think that part of the slowness of activation is due to the lack of one of the positive charges in S4, do you think if you added in the Shaker equivalent of that first charge that you would speed up activation?

Tseng: That's a good idea. I think the negative charges should also be taken into account. Perhaps a good experiment would be to remove the extra negative charges from S1 to S3, and add the positive charge in to see whether we can create a fast-activating channel.

Robertson: You might not be able to remove some of the negative charges. Diane Papazian showed in Shaker that biogenesis requires charge pairing to take place early on.

Tseng: Interestingly, the hERG channel is different. We can mutate each of the charged residues without impairing protein formation and trafficking.

Tristani-Firouzi: I'd like to make a comment about how many charges there are and how this would be influencing voltage sensing. If those charged residues are not a part of the electrical field, then they cannot participate in voltage sensing. One example is HCN (cyclic nucleotide-gated, hyperpolarization activated channel), which has five or six additional charged residues compared to Shaker, but these residues are not within the electrical field and likely do not participate in voltage sensing. Within the membrane, where the charges are actually moving from one side of the electrical field to the other, it is where the charge is carried that gives the voltage sensitivity. If it is not in that field it doesn't matter.

Tseng: We showed that the first three charged residue in S4 can change their side chain accessibility. Therefore these three positive charges in S4 are within the membrane electrical field.

Tristani-Firouzi: Correct. That means that it is not accessible at one time or another, but it doesn't mean that it is within the focused electrical field. We don't know where that electrical field is focused within the membrane. It can be within the membrane but not in the electrical field.

Tseng: My vision of the S4 movement is the conventional model. I think we have a thin rim of protein or membrane structure around the S4, where most of the membrane field drops. Then the first charged residue is moving from within the membrane or protein structure towards the actual crevice. I would say that this is probably experiencing a fraction of the membrane electrical field.

Tristani-Firouzi: If that is true, then if the charged residue is neutralized the amount of gating charge is reduced or the voltage dependence of activation is shifted. If that first charge is mutated to cysteine, for example, is the voltage dependence shifted?

Tseng: It is shifted in the hyperpolarizing direction. I don't know whether by itself this says anything about its role in voltage sensing. I would say that this reflects the interaction between this positive charge and negative charges in the S2 domain. When we remove this positive charge we destabilize the pre-open state and therefore the channel can activate more readily. I don't think it is very useful to use the shift of the activation curve to try to deduce what the residue may be doing in the gating charge transfer.

Recanatini: I have been impressed by the study you have done with the cysteine mutation on the S5-P linker. It might be quite informative about the structure of this linker. You mentioned that you did some molecular dynamics simulation in order to find the most stable conformation. How did you take into consideration the solvation of the structure?

Tseng: I don't do the molecular modelling myself: it is done by our collaborators, Bob Guy and his colleagues. I think they put water molecules in the system when they do the molecular dynamics stimulation. There are water molecules in the environment.

Recanatini: This might be quite important because the helix is amphiphilic, so water molecules could be critical for the physicochemical arrangement. I have what might be a naïve idea: you have shown that disulfide bonds can be introduced to link helices of S5-P linkers belonging to different monomers. Would it be possible to isolate that piece of molecule and crystallize it? It would be useful to isolate and study just this linker.

Tseng: My feeling is that it is too small a structure to do this.

Mitcheson: My understanding is that you need a lot of protein to do this.

Sanchez-Chapula: There is a paper where the investigators isolated the S5-P linker or a peptide encompassing that linker (Torres et al 2003). Then perhaps you could work on that and mutate the residues to create an environment to lose the disulphide bonds.

Tseng: My view about the S5-P linker is that it is a very dynamic structure. I am not sure whether crystallization, giving us a snapshot of the structure, would tell us very much about the kind of structure the channel would have in the different gating modes. Carefully designed mutagenesis experiments coupled with molecular modelling work is likely to be the most promising direction to take.

Robertson: Is the channel permeable when it is cross-linked?

Tseng: I can't say whether this is the case for all of our cysteine mutants, but some cysteine mutants can form disulphide bonds and pass current at the same time.

They can form disulphide bonds with their neighbour but not across the central axis.

Robertson: Does this affect voltage dependence for activation?

Tseng: It disrupts inactivation, but I have no data on activation.

Robertson: If C-type inactivation of hERG is linked with S4 movement, and it gets some of its voltage sensitivity from that process, then you might expect disulfide bonds to be constraining.

Tseng: This is difficult to study, because these structures are dynamic. We can't keep them in a steady state to study the activation gating process. There are very dynamic conformational changes in the outer mouth region, so it is difficult to keep one pair of cysteines in a disulphide-bonded state and the other in free thiol state, and measure the current to study the activation process.

Sanguinetti: Do you think that slow activation is in part due to less charge movement; that the outer lysine is not playing an important role? If you do an alignment of hERG with Shaker the outer lysine is missing. There is also one very troubling residue we have studied in our lab by mutation. D540, located in the S4-5 linker is thought to extend into the cytosol. When this residue is mutated to alanine, activation is extremely fast. This is supposedly outside the S1-S4 voltage sensor complex. Perhaps this residue interacts with another residue outside S4.

Tseng: So that single mutation can accelerate the activation.

Sanguinetti: Yes.

Tseng: Bob Guy has included the D540 residue in the S4-5 linker of his structural model. He suggests that there may be some salt bridge formation between this aspartate and a positive charge somewhere else.

Sanguinetti: Marty Tristani-Firouzi has experimental evidence for charge pairing of Asp540 with Arg665 in the C-terminal end of S6. But that is outside the voltage sensor field. What I am wondering is whether we can attribute this characteristic property of very slow activation simply to interactions between S4 and other parts of the voltage sensor.

Tseng: Maybe not. But I do feel that this high density of negative charges contributes to the slow activation. Perhaps changing the aspartate to alanine would change the conformation, making it difficult for these negative charges to come close to the positive charges.

Sanguinetti: Certainly the shift in voltage depends on activation, but the rate of activation doesn't change very much. This doesn't answer the question of how fast the voltage sensor is moving. I am talking about aspartates in S2/S3. Within the voltage sensor itself, if you neutralize some of the acidic residues, although this does shift the voltage dependence of activation up to 40 mV, the kinetics are not that dramatically altered. This finding suggests that you are stabilizing states of the voltage sensor by electrostatic interactions, but not allowing those to happen by neutralization of the aspartates doesn't change the rate at which the channel opens very much. This suggests that there is something else that is impeding the movement of S4.

Tseng: I don't know whether you can say that. The negative charges interact with the positive charges in different stages of S4 movement. Perhaps S2 is also moving in an inward direction. It is frustrating sometimes to look at the data and try to understand at a molecular level what is happening. Many things can happen and what you are seeing is the net result of all these changes in interactions.

Brown: You showed that you reduced the cysteine interaction, the disulphide bridge, between units 3 and 4. You got a big change in current. Is that a gating effect or a conductance effect?

Tseng: I would say this is a gating effect because we lose the negative slope in the I–V relationship.

Brown: If this is a gating shift, is that because of coupling to activation? These things are not in the membrane.

Tseng: By gating, I mean inactivation gating. It is not activation gating.

Brown: So the inactivation gating in this model is not independent of the closed–open transition.

Tseng: This is a question that I would pose to the audience. For the inactivation, my proposal is that there is a dynamic interaction between the extracellular S5-P helices from different subunits that can close the outer mouth, causing channel inactivation. These conformational changes in the extracellular S5-P helices may or may not be coupled to the S4 movement required during activation.

Brown: I am trying to figure out whether or not unit conductance has been changed.

Tseng: We don't know.

Gosling: Michael Sanguinetti, when you look at S6 31A, is the activation kinetics altered when you make the inactivation-deficient mutant?

Sanguinetti: Not really. S6 31A is a few residues outside the selectivity filter. When the serine is mutated to an alanine, inactivation is not removed but its voltage dependence becomes about 100 mV more positive. The activation kinetics remain the same. This is based on the triple-pulse protocol which is a confusing way to measure inactivation, but it is about the only way we can do it. It is actually voltage dependence of recovery from activation. This reminds me again of a question I wanted to ask earlier. In your model you are proposing that the S5-P α helices are basically tilting inwards by hydrophobic interactions which are enhanced. That's your model of inactivation, as opposed to C-type inactivation where the selectivity filters are coming together and collapsing a bit: is that correct? What difference is there between 'typical' C-type inactivation and hERG inactivation, other than the voltage dependence?

Tseng: For the Shaker C-type inactivation, the Na^+ permeability is actually higher in the C-type inactivated state than in the normal open state. But for the hERG channel, the opposite seems to be true: you can have an increase in Na^+ permeability when you disrupt the inactivation process by mutations.

Sanguinetti: The onset of C-type inactivation for Kv2.1 is pretty slow. You go from a K^+-selective pore, to a Na^+-selective pore, to no conductance. This is the origin of the idea that the selectivity filter must be collapsing slightly, so the aperture becomes smaller. With hERG inactivation being so fast, I don't know whether we could measure such a transition. I can't think of another difference other than the voltage dependence and perhaps the rate although there are mutations in residue 449 of Shaker where C-type inactivation becomes very fast through mutation of just a single residue.

Tseng: Let me add one point. I am not saying that the pore loop or the selectivity filter is not playing any role in the inactivation of the hERG channel, because a very conservative mutation, S620T, in the middle of the pore loop can disrupt the inactivation of the hERG channel completely. This suggests to me that the pore-loop must also be playing a role in hERG inactivation.

Ficker: This is a substantial change in model from a constriction of the selectivity filter to a discrete extracellularly located C-type inactivation gate. If I recall the TEA (tetraethylammonium) experiments of MacKinnon correctly (Heginbotham & MacKinnon 1992), then TEA block was voltage-dependent. TEA senses about 10–20% of the transmembrane field. However, the novel extracellular inactivation gate you postulate is far above the selectivity filter of the channel protein.

Tseng: In the Shaker channel, T449 forms the TEA binding site that you are referring to. In the hERG channel the equivalent position is Serine 631. Interestingly, S631 is not involved in TEA binding to the hERG channel. If we mutate S631 to lysine, tyrosine or glutamate, the TEA sensitivity was not changed at all. I think the TEA binding site in the hERG channel is actually somewhere in the S5-P helix.

Ficker: Then TEA should not sense the electrical field of the membrane. Consequently TEA binding should not be voltage dependent.

Tseng: Although the S5-P helix may be outside the membrane electrical field, its conformation may be regulated an adjacent intramembrane domain that can sense changes in the membrane voltage. Here I am proposing that the pore-helix may be the intramembrane domain that senses the membrane voltage and regulates the S5-P helix motion.

Abbott: Is TEA (tetraethylammonium) binding really voltage dependent, or is it just related to movement of ions through the pore?

Tseng: That is another possible explanation for the voltage sensitivity in TEA binding: the TEA binding site does not have to be in the membrane. I am not really proposing that the S5-P helix movement itself is a result of voltage depolarization. Actually, the conformational change will be in conflict with how this helix seems to be oriented. We believe the N-terminus of the S5-P helix is facing towards the membrane electrical field. So if there is a membrane depolarization, the S5-P helix should be moving in the outward direction. I think there must be something else happening around the outer mouth that pushes the S5-P helices into a conformation so that they can come close to each other and help occlude the pore.

Robertson: Did your structures give you any clues as to what is conferring the voltage sensitivity?

Tseng: Not yet. Marty Tristani-Firouzi, you might have some ideas: you proposed the S4 as the voltage sensor for hERG inactivation.

Tristani-Firouzi: This pore helix has to be important. It is so much longer than in Shaker channels and other K^+ channels. What could be happening is rather than the S5-Ps coming together to block the conduction of ions, it could just be influencing the region right below, the pore domain. You have already shown that when you modify with MTS you change the selectivity. You have altered the selectivity filter. This means that the pore helix can alter the selectivity filter. It may be that in these conformations as the channel inactivates, the S5-P helix sits above the pore domain and those interactions are stabilized or destabilized, and then the pore collapses, just as we think occurs with Shaker.

Tseng: We have data indicating that this might not be the case. Our data showed that when we introduced a cysteine into certain positions in the S5-P helix, it can form an intersubunit disulfide bond with its counterpart from an adjacent subunit spontaneously. This disulfide bond can be stronger than when we mutate serine 631 to cysteine. According to your argument the serine 631s should be very close to each other.

Tristani-Firouzi: I am not making any comment about serine 631 at all. I am just commenting on the fact that the turret that you described can influence the structure right below, which is the pore helix. Any modification in the pore helix is going to alter the ability to coordinate K^+ ions. Your modification showed that you altered the selectivity. This selectivity is not going to be altered by the turret helix above: that is, you are not proposing that K^+ is coordinated in any way there. So you know that your effects are at the selectivity filter. Those interactions could influence how the selectivity filter is arranged and then collapsing in the process of inactivation. The voltage dependence of that process could be driven by S4 movement initially, which then is transduced through a series of steps that no one really understands, up into the pore helix or up into the turret. Maybe the turret is secondary.

Tseng: Can I turn that argument around? You have the channel movement that changes the pore loop configuration, which in turn somehow moves the S5-P helices outside. This makes them come close to each other and the hydrophobic interaction helps occlude the pore.

STRUCTURE–FUNCTION STUDIES 41

Ficker: May I suggest another argument for a constriction of the hERG pore? It is based on the hERG mutation S620T. The homologous residue in KcsA is part of a dicarboxylate pair. The corresponding amino acid residues in hERG are S620 and N629. Now, if you make a mutation in S620 you may alter the amino acid clamp which holds the selectivity filter in place and expect fundamental changes in C-type inactivation. This is exactly what has been observed in hERG S620T.

Mitcheson: There are a number of different positions that affect inactivation on the intracellular side that are some distance from the turret. From our own studies there is T623 on the intracellular side of the selectivity filter and G648 on S6. Perhaps these two residues are close enough to one another that they are causing a direct collapse of the selectivity filter by a mechanism that is remote from the turret.

Tseng: No doubt there are interactions. Even the N-terminal domain can bind to somewhere in the mouth and change the conformation of the outer mouth. I don't want to create the impression that the S5-P helix is the only domain that can cause inactivation: this is not what I am trying to say. I am proposing a model for the inactivation, and we will have to see how well the model holds. We have to do more mutations to see whether the interactions between the S5-P helix and other channel domains are consistent with what the structural models show.

Sanguinetti: There were tryptophans at the 585 position which were the ones closest to each other in the proposed inactivated state. Have you tried mutating these residues to other hydrophobic residues, preferably non-aromatic ones? Does this affect inactivation? I presume this is the closest point of interaction between the single subunits. This should be the most important interaction that determines when conduction stops in this model. Tryptophan is quite large, so you could shorten the side-group, while keeping it aromatic.

Tseng: We have only used cysteine. Jaime Vandenburg used phenylalanine to replace that tryptophan. He said that the channel does not express (Torres et al 2003). I don't know whether he has tried other residues.

Sanguinetti: Is it permanently inactivated, or is it just not getting to the surface membrane?

Tseng: I don't remember the data very clearly. I remember he said that when the mutant is expressed with the wild-type channel it has a dominant-negative effect. This would suggest that the mutant can fold correctly and can reach the cell surface.

Mitcheson: What is the stoichiometry for inactivation? Do you think all four subunits need to be in an inactivated conformation?

Tseng: Based on the Shaker data, C-type inactivation requires concerted movements of all four subunits. In the G584C mutant of hERG, we can have two cysteine side chains at 584 forming an intersubunit disulfide bond between two adjacent subunits and the other two cysteine side chains at 584 in free thiol state, and the inactivation is disrupted. There needs to be flexibility of the structures around the outer mouth for inactivation to occur. Thus, I would say that we also need concerted movements of all four subunits of hERG to create inactivation.

Wanke: Classically, there is a general way to study ion channel biophysics and dissect whether gating originates mainly from protein conformational changes or from the sensed membrane electric field. This was used before site-directed mutagenesis was possible. Alterations of the membrane electric field can be obtained by changing the quantity of ions outside or inside. Everyone knows that if you remove the divalent cations outside you change the electric field and produce strong changes in the voltage dependence of the gating. If one is doing this type of experiment on hERG (Faravelli et al 1996), you can completely inhibit the closing of the channel, which is equivalent to saying that you can shift the activation voltage dependence leftwards by 60–80 mV. These data could fit your model or not. This is a very simple experiment that could help us to understand whether the gating is more connected with the machinery in the protein, or in the sensing of the electric field present in the membrane.

Tseng: Lowering the extracellular divalent cation concentration from millimolar down to a micromolar range can cause a shift of the activation curve and create extremely fast activation of the hERG channel.

Wanke: What about the deactivation?

Tseng: That is slower.

Wanke: When you used cadmium for changing the properties, you showed that there is a blocking effect. How can you distinguish in this experiment in which you test only one value of the membrane potential, between a simple shift and a blockade in the current? It would not be a blocking effect but a shift of the voltage dependence. You can use very different drugs, such as cadmium or lanthanum, and see that there are shifts and not blocking mainly because their action alters the membrane electric field.

Tseng: With 1 M Cd^{2+} outside, there was no shift in the activation curve. Therefore, the Cd^{2+} effects on the cysteine-substituted mutants of hERG in our experiments were explained by forming a metal ion bridge between two introduced cysteine side chains around the outer mouth, not by a shift in the voltage-dependence of channel activation.

Wanke: In our hands with lanthanum we observed only a shift (Faravelli et al 1996).

References

Faravelli L, Arcangeli A, Olivotto M, Wanke E 1996 A HERG-like K^+ channel in rat F-11 DRG cell line: pharmacological identification and biophysical characterization. J Physiol (Lond) 496:13–23 (Erratum in: J Physiol (Lond) 502:715)

Heginbotham L, MacKinnon R 1992 The aromatic binding site for tetraethylammonium ion on potassium channels. Neuron 8:483–491

Torres AM, Bansal P, Sunde M et al 2003 Structure of the HERG K^+ channel S5P extracellular linker: role of an amphipathic alpha-helix in C-type inactivation. J Biol Chem 278: 42136–42148

General discussion I

Sanguinetti: We have some time free for general discussion of issues relating to the first two papers. There is one issue that I'd like to raise. When we go to very high external K^+ we find that the rate of deactivation becomes very slow. Gail Robertson, you showed that the deactivation is the same or a little faster. Why is there this difference? We did our work in oocytes.

Tseng: I would worry about the voltage clamp conditions. Sometimes with low expression of the wild-type hERG channel, we don't see this great prolongation of tail current with high $[K^+]_o$.

Sanguinetti: It is curious because it is current magnitude-dependent. We have done the control experiments where we overexpress KCNQ1/KCNE1, with huge currents, and see only slight changes in deactivation. There is nothing like the slamming open of the channel that we see in very high K^+ for hERG. Have other people seen this, and is it expression dependent? We worried a lot about whether it is an artefact, but it doesn't look like it is. I know that it also occurs in mammalian cells.

Tseng: The degree of prolongation of hERG tail current in high $[K^+]_o$ is very variable. There isn't a persistent pattern: from one oocyte to the next you can have variations.

Sanguinetti: And it is not a simple exponential decay anymore. It is very striking.

Roberston: I have not actually observed this slow deactivation. Is there a change in the gating charge movement under these conditions?

Tristani-Firouzi: When we measure gating current we have impermeant ions.

Sanguinetti: Have you tried high K^+ with MK-499 to block the channels?

Tristani-Firouzi: No. There are other endogenous K^+ channels in the membrane which, if there is K^+ around, will conduct and contaminate measurement of gating currents.

Abbott: Which Ca^{2+} concentration were you using?

Sanguinetti: Normal: 1 mM.

Abbott: Do you observe slower deactivation when you lower external Ca^{2+}?

Sanguinetti: Yes.

Wanke: We did an experiment eight years ago on a neuroblastoma clone expressing hERG channels intrinsically. We repeated the same experiment in a stable clone of F-11 cells. If you do a classical experiment in low K^+, there are HERG currents which are very small. If we change just the K^+ concentration outside to 120 mM there is a dramatic increase in the tail currents, but the deactivation is still present and the activation curve is slightly left-shifted. If we also remove completely the divalent cations by using EGTA, we no longer see activation because it was left-shifted beyond -100 mV.

Sanguinetti: This is similar to what Earm and Noble showed in myocytes, but I don't remember whether they needed high K^+.

Wanke: Practically, high K^+ is not doing much. The big effect is due to the lack of divalent cations. The channel is changing its sensitivity because the electric field is changed due to the negative charges outside the membrane.

Sanguinetti: I think it is actually an effect on gating. Bill Gilly and Clay Armstrong reported this many years ago in *Nature* for squid K channels; when they removed external divalent cations, the channels became non-selective and always open (Gilly & Armstrong 1984). It depends on the channel type whether you have to remove intracellular and extracellular divalents, and whether K^+ has to be present or not.

Robertson: Did your deactivation get a little faster in the high K^+?

Wanke: The deactivation is somewhat slower than in control, but not much. These data might help see whether your model fits with this type of experiment. Also, does the EAG channel, which is somewhat structurally similar but biophysically different, fit with your model or not?

Tseng: I am not sure.

Tristani-Firouzi: Aren't the linkers similar in length?

Tseng: Yes, the S5-P linkers are of similar lengths in hERG and EAG.

Reference

Gilly WF, Armstrong CM 1984 Threshold channels — a novel type of sodium channel in squid giant axon. Nature 309:448–450

Voltage sensor movement in the hERG K⁺ channel

David R. Piper*†, Michael C. Sanguinetti*† and Martin Tristani-Firouzi‡[1]

*Department of Physiology, †Nora Eccles Harrison Cardiovascular Research & Training Institute, and ‡Department of Pediatrics, University of Utah, 100 N. Medical Drive, Suite 1500 PCMC, Salt Lake City, UT 84113, USA

Abstract. The critical role of hERG in the maintenance of normal cardiac electrical activity derives from its unusual gating properties: slow channel activation and fast inactivation. To characterize voltage sensor movement associated with slow activation and fast inactivation, we measured gating currents from wild-type and mutant hERG channels. Fast and slow gating components were observed that differed 100-fold in their kinetics. The slow component constituted the majority of gating charge associated with channel opening and accounted for the sluggish rate of hERG activation. Gating currents from an inactivation-deficient mutant (S631A) were indistinguishable from wild-type, despite a +100 mV shift in the voltage dependence of inactivation, suggesting that a small fraction of total gating charge is devoted to the final transitions that inactivate the channel. Ala-scanning mutagenesis in S4 identified residues that perturbed both charge movement and channel opening. Residues in the S4–S5 linker perturbed channel opening without altering charge displacement, suggesting a role for coupling S4 movement to channel opening. Finally, inactivation-sensitive residues localized to a helical face of S4 adjacent to the activation-sensitive residues. We conclude that S4 acts as the voltage sensor for hERG activation and inactivation and that S4 movement is translated to the activation gate via the S4–S5 linker.

2005 The hERG cardiac potassium channel: structure, function, and long QT syndrome. Wiley, Chichester (Novartis Foundation Symposium 266) p 46–56

hERG is a six-transmembrane voltage-gated K⁺ channel that is vital for repolarization of cardiac action potentials. All voltage-gated channels undergo conformational rearrangements that open or close the channel in response to changes in membrane potential. The highly conserved, basic residues in the S4 domain constitute the primary voltage sensor responsible for detecting transmembrane potential (Papazian et al 1991, Perozo et al 1994). Voltage sensor

---

[1]This paper was presented at the symposium by Martin Tristani-Firouzi to whom correspondence should be addressed.

VOLTAGE SENSOR MOVEMENT 47

movement within the membrane is measurable as gating current (Bezanilla 2000). Gating currents provide valuable information on conformational changes within the channel that precede opening of the pore. The properties of hERG ionic current are very different than other voltage-gated K$^+$ channels, suggesting that hERG voltage sensor movement is also different.

The critical role of hERG in the maintenance of normal cardiac electrical activity derives from its unusual gating properties: slow channel activation and fast inactivation (Tristani-Firouzi & Sanguinetti 2003). Unlike other voltage-gated K$^+$ channels, hERG channel inactivation is intrinsically voltage sensitive and appears uncoupled from activation, suggesting that inactivation and activation gating are derived from distinct voltage-sensing mechanisms (Smith & Yellen 2002). The goal of our current studies is to define the structural basis of voltage sensor movement related to hERG activation and inactivation. To accomplish this goal, we measured gating and ionic currents to determine the voltage-dependence of charge movement, activation and inactivation for wild-type and mutant channels using the cut-open oocyte (Stefani & Bezanilla 1998) and two microelectrode voltage clamp techniques. Our initial findings suggest that the S4 acts as the voltage sensor for both activation and inactivation gating of hERG channels, but that distinct regions of S4 impart the specific voltage sensitivities of these gating processes.

Properties of hERG gating current

hERG channels are unique among voltage-gated channels in that gating currents representing intramembrane charge displacement are composed of two distinct components that differ in their kinetics by nearly 100-fold. hERG channel ON gating current (I_{gON}) elicited by membrane depolarization is composed of a fast component (fast I_{gON}) followed by a slow, smaller amplitude component (slow I_{gON}) that comprises ~90% of the total gating charge (Piper et al 2003) (Fig. 1, *left*). The fast I_{gON} is visible in the ionic current traces as a transient peak that precedes activation of ionic current (Fig. 1, *right*), indicating that voltage sensor movement precedes the opening of hERG channels. The slow I_{gON} and I_{gOFF} gating components are responsible for the slow activation and deactivation kinetics of hERG ionic current.

Representative hERG gating currents recorded in response to variable test potentials are shown in Fig. 2A. The fast component of I_{gON} was first detected near $-80\,\text{mV}$, well below the apparent threshold for hERG channel opening. The slow component of I_{gON} was first detected near the threshold for channel opening, around $-40\,\text{mV}$, and became larger with stronger depolarizations. The integral of either I_{gON} (Q_{ON}) or I_{gOFF} (Q_{OFF}) provides a measure of the charge moved for a given step in voltage. The $V_{1/2}$ for the Q_{OFF}–V relationship was

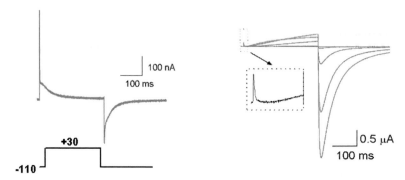

FIG. 1 (*Left*) hERG channel ON gating current (Ig_{ON}) elicited by membrane depolarization is composed of a fast component (fast Ig_{ON}) followed by a slow, smaller amplitude component (slow Ig_{ON}) that comprises $\sim 90\%$ of the total gating charge. (*Right*) The fast Ig_{ON} is visible in the ionic current traces as a transient peak that precedes activation of ionic current, indicating that voltage sensor movement precedes opening of hERG channels. (Reprinted with permission from Piper et al 2003.)

shifted -29 mV relative to the isochronal conductance-voltage (G–V) relationship for 300 ms pulses (Fig. 2B). Thus, similar to other voltage-gated channels, hERG charge movement occurs at voltages more negative than channel opening.

Is there a unique component of gating current associated with fast inactivation of hERG? Based on kinetics alone, an obvious candidate is the fast component of Ig_{ON}. Fig. 2C illustrates the gating current elicited by a 2 ms depolarizing voltage step that effectively isolates the fast Ig_{ON} component. The Q–V relationship for the fast Q_{ON} had a $V_{1/2}$ of $+28 \pm 4.4$ mV and a z of 0.7 ± 0.02 (Fig. 2D), markedly different from the voltage dependence of hERG inactivation which has a $V_{1/2}$ of -90 mV. These findings suggest that fast Ig_{ON} is unlikely to be directly linked to channel inactivation, but instead represents transitions between closed states early in the activation pathway.

Properties of gating current from the inactivation-deficient S631A hERG channel

To determine whether a component of Ig is uniquely related to fast inactivation, we turned our attention to the inactivation-deficient S631A hERG channel. The S631A mutation shifts the voltage dependence of channel inactivation by $+100$ mV. If a major component of charge displacement is associated with inactivation, then differences in the voltage dependence and kinetics for Ig of wild-type and S631A hERG channels should be measurable. In contrast to this prediction, Ig_{ON} and Ig_{OFF} for S631A channels and the voltage dependence of the Q_{OFF}–V relationship were similar to wild-type hERG.

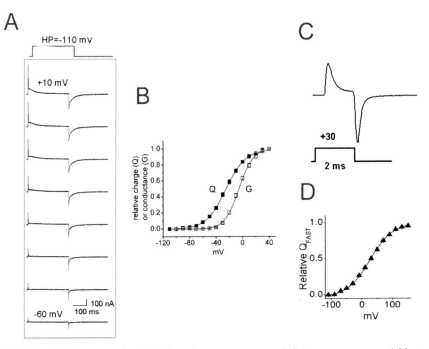

FIG. 2. (A) Representative hERG gating currents recorded in response to variable test potentials. The fast component of $I_{g_{ON}}$ was first detected near $-80\,mV$, well below the apparent threshold for hERG channel opening. The slow component of $I_{g_{ON}}$ was first detected near the threshold for channel opening, around $-40\,mV$, and became larger with stronger depolarizations. (B) $I_{g_{OFF}}$ was integrated (Q_{OFF}) and plotted versus test potential. The $V_{1/2}$ for the Q_{OFF}–V relationship was shifted $-29\,mV$ relative to the isochronal conductance-voltage (G–V) relationship for 300 ms pulses. (C) The gating current elicited by a 2 ms depolarizing voltage step that effectively isolates the fast $I_{g_{ON}}$ component. (D) The integral of fast $I_{g_{ON}}$ (fast Q_{ON}) is plotted versus test potential. The Q–V relationship for the fast Q_{ON} had a $V_{1/2}$ of $+28$, markedly different from the voltage dependence of hERG inactivation which has a $V_{1/2}$ of $-90\,mV$. (Reprinted with permision from Piper et al 2003.)

The most striking difference between wild-type and S631A hERG ionic current is revealed using a triple-pulse voltage protocol to assess the voltage dependence of channel inactivation. This pulse protocol was used to compare gating currents of wild-type and S631A channels (Fig. 3). Gating currents were elicited by an initial 0.3 s pulse to $+40\,mV$ followed by a 10 ms hyperpolarizing pulse to a variable potential to recover charge associated with the transitions from the inactivated to the open states, while minimizing charge movement associated with transitions through the closed states (i.e. channel deactivation). A third test pulse to $+40\,mV$ was applied to assess charge movement primarily derived from transitions into the

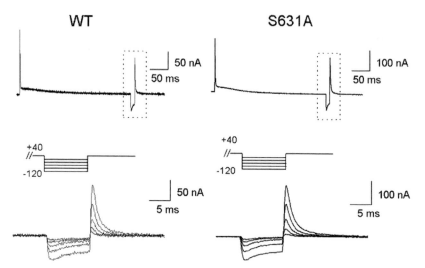

FIG. 3. Gating currents from wild-type and S631A hERG channels elicited using the same triple-pulse voltage protocol used to determine the voltage dependence of ionic current inactivation. Lower panel represents gating currents on a magnified scale. Despite a large depolarizing shift in the voltage dependence of ionic current inactivation, S631A gating currents were not appreciably different from wild-type gating currents, implying that S631A uncouples voltage sensor movement from rearrangments that inactivate the channel. (Reprinted with permission from Piper et al 2003.)

inactivated state. In contrast to the marked difference in the voltage dependence of ionic current inactivation, the Q–V relationship determined from gating currents elicited by the third pulse was not different between wild-type and S631A hERG. The inability to detect any measurable difference in gating currents between wild-type and S631A hERG using multiple voltage protocols suggests that this mutation uncouples voltage sensor movement from rearrangements in the pore domain that inactivate the channel.

Ala-scanning mutagenesis of the hERG S4 domain

To further characterize the voltage sensor movement associated with hERG activation and inactivation, we focused our attention on the primary voltage sensor, the S4 domain. We performed an Ala-scan of the S4 domain and the flanking S3–S4 and S4–S5 linkers to test the hypothesis that individual substitutions might differentially alter the equilibrium between the resting and activated states of the voltage sensor, the closed and open activation gate or the open and inactivated pore. To quantify the energetic perturbation produced by the individual mutations, we calculated the free energy difference (ΔG) between three states in wild-type channels and compared these values to the energy difference induced by Ala-substitution. The difference in ΔG between wild-type and mutant channels (ΔΔG) reflects the perturbation in a gating transition produced by the Ala-substitution. We examined the effect of Ala-substitutions on three simplified equilibrium distributions reflecting charge movement, channel opening and inactivation.

Ala-substitution at five positions clearly perturbed the ΔΔG of charge movement more than the other mutations (defined as |ΔΔG| > 1.0 kcal/mol). When mapped onto a 3D homology model of S4, these charge-perturbing residues formed a spiral thread that turns around the length of S4. This observation is consistent with previously published data implicating a helical twist or rotation of S4 induced by membrane depolarization (Bezanilla 2000). Mutation of residues in the S3–S4 and S4–S5 linkers perturbed activation (|ΔΔG| > 1.0 kcal/mol) without altering charge movement. This finding implies that residues in the linkers couple S4 movement to opening of the activation gate or stabilize the open or closed state of the channel.

Residues where Ala-substitution impacted inactivation formed a stripe along a helical face of S4 that corresponded to low-impact activation positions. The distribution of the inactivation-sensitive subset of residues across the entire length of the S4 helix suggests that the whole surface experiences a rearrangement associated with inactivation. A rigid body movement of S4, such as a small tilt, would be consistent with this longitudinal distribution. In this respect hERG inactivation may be different from Shaker, where coupling of the voltage sensor to C-type inactivation is believed to occur through discreet interactions between the N-terminal portion of S4 and the turret linking S5 to the pore helix (Larsson & Elinder 2000, Loots & Isacoff 2000).

In summary, we speculate that membrane depolarization induces a helical twist or rotation of S4 that is coupled to channel opening via the S3–S4 and S4–S5 linkers. Inactivation may involve a rigid body movement of S4 that is transduced to the pore domain by interactions with other domains within the channel that remain to be identified. Our findings support the idea that the S4 domain acts as the voltage sensor for both activation and inactivation gating of hERG channels, but that distinct regions on the S4 voltage sensor contribute differentially to these processes.

References

Bezanilla F 2000 The voltage sensor in voltage-dependent ion channels. Physiol Rev 80:555–592
Larsson HP, Elinder F 2000 A conserved glutamate is important for slow inactivation in K^+ channels. Neuron 27:573–583
Loots E, Isacoff EY 2000 Molecular coupling of S4 to a K^+ channel's slow inactivation gate. J Gen Physiol 116:623–636

Papazian DM, Timpe LC, Jan YN, Jan LY 1991 Alteration of voltage-dependence of Shaker potassium channel by mutations in the S4 sequence. Nature 349:305–310

Perozo E, Santacruz-Toloza L, Stefani E, Bezanilla F, Papazian DM 1994 S4 mutations alter gating currents of Shaker K channels. Biophys J 66:345–354

Piper DR, Varghese A, Sanguinetti MC, Tristani-Firouzi M 2003 Gating currents associated with intramembrane charge displacement in HERG potassium channels. Proc Natl Acad Sci USA 100:10534–10539

Smith PL, Yellen G 2002 Fast and slow voltage sensor movements in HERG potassium channels. J Gen Physiol 119:275–293

Stefani E, Bezanilla F 1998 Cut-open oocyte voltage-clamp technique. Methods Enzymol 293:300–318

Tristani-Firouzi M, Sanguinetti MC 2003 Structural determinants and biophysical properties of HERG and KCNQ1 channel gating. J Mol Cell Cardiol 35:27–35

DISCUSSION

Tseng: Have you looked at the rate of inactivation of these S4 mutants to see whether, for those mutants that significantly perturb the free energy involved in the inactivation gating process, the rate of inactivation is also altered?

Tristani-Firouzi: Some mutants, such as V535, really altered the kinetics of the voltage sensor movement. We have a lot of mutants that alter the rate of channel opening, the rate of activation and the rate of deactivation. But most of these mutants did not really perturb the rate of onset of inactivation. We haven't looked carefully at the recovery from inactivation, but qualitatively the rate of inactivation was similar and strikingly different from the other processes.

Robertson: Does that surprise you?

Tristani-Firouzi: I haven't really thought about this too much. It was enough just to try to understand the differences in the perturbations of the equilibrium, rather than bringing in the kinetics.

Tseng: It is somewhat surprising to me that when you have a mutation that is powerful in perturbing the equilibrium of inactivation, that it doesn't do too much to the rate of inactivation. There should be some correlation there. I still don't understand why, when the S4 domain has reached a fully activated state, say at $+20\,mV$, the rate of hERG inactivation is still voltage sensitive all the way up to $+60\,mV$.

Tristani-Firouzi: The S4 could move within the membrane, and this could be coupled to getting the channel open. Then there could be another movement which is associated with inactivation. These small movements, possibly a rigid body movement of S4, would continue to deliver that voltage sensitivity at higher voltages, at membrane potentials where the channel is already open. In this way, activation and inactivation voltage sensitivity would be derived from S4. The voltage sensor is the same primary structure, but in different conformations it is conferring the information in slightly different ways.

Sanguinetti: It is important to keep in mind that you can have voltage sensor movement without significant charge displacement; for example, tilting. Granted, you would expect to see some charge movement, but attempts to see differences in inactivation-deficient mutants have shown nothing, rather surprisingly.

Robertson: During the activation phase you have the twisting the helix, and you have one face which has the inactivation-sensitive residues. Do your results say that when the twisting movement occurs and gives rise to the activation charge movement, there is no perturbation in the environment of those residues that line one face, even though it is part of a whole helix that is twisting?

Tristani-Firouzi: The first part is going to be the twist. When we look at the residues that are perturbed in the charge movement relationship, it is just a few discrete residues. The other residues are still moving somehow, but their environment is not being altered in a way that we can pick up with our test. How does this happen? S4 moves and there is a helical twist, and then almost at the same time is there this rigid body movement? We don't know that for sure, but the idea would be that inactivation requires the voltage sensor to move into an 'activated' conformation. You are not going to get channels inactivated at a holding potential of $-80\,\text{mV}$, because the voltage sensor hasn't moved enough. When we say that the half-point of inactivation with the triple pulse protocol is $-80\,\text{mV}$, this is after we have fully activated all the channels and inactivated them with an initial depolarizing step. The measure itself is deceptive from the very beginning: it seems wrong to say that the voltage dependence of inactivation is $-80\,\text{mV}$ but the voltage dependence of activation is $-20\,\text{mV}$. How can this be? It is a fault of our measurement technique. What we have done with the triple-pulse protocol is move the voltage sensor into the activated position with the first depolarizing step and then briefly hyperpolarize to move the voltage sensor 'in' and then depolarize (in the third voltage step) to move the sensor back across the electrical field to assess the voltage dependence of inactivation.

Sanguinetti: One of the main topics we should be discussing here is what is unique about hERG compared with any other voltage-gated K^+ channel. From the papers we have heard so far, what is different about hERG compared to Shaker? Is the helical translocation and tilt really just a restatement of what has already been proposed for both Shaker and EAG? Certainly, it is a lot different than the paddle model for KvAP. It is generally agreed upon now that in the KvAP structure, as elegant as it is, the paddle probably doesn't really represent what is going on in the voltage sensor in the majority (if not all) voltage-gated K^+ channels. This may be, in part, distortion due to the crystallization process. I would ask everyone here: is hERG really any different than any other voltage-gated K^+ channel, with the exception of its kinetics?

Tseng: Have you tried to find pore mutations that will block ionic currents but leave the gating current unaltered? Is this possible?

Tristani-Firouzi: Absolutely. We have looked at the equivalent in Shaker W434.

Tseng: But there is no tryptophan in the pore loop of the hERG channel?

Tristani-Firouzi: That is right, but when you mutate that particular residue, I don't remember whether it fully blocks current. We looked at a long QT mutant that looked like it was essentially a poison pill, in the sense that all ionic current was removed. This isn't the case when you look with the cut-open technique: you can see that there is a little bit of current and the channels are still able to conduct. This can be blocked with MK-499. We haven't found the position yet that would allow us to use permeant ions on both sides.

Robertson: I am confused about a few points here. Gea-Ny asked a question about the change in activation kinetics, concerning the $\Delta\Delta G$ for the mutations that affect the later transition (at more positive levels). What about the half-maximal voltage?

Tristani-Firouzi: This is what we use to calculate the ΔG, along with the z. The $\Delta\Delta G$ gives you the advantage of saying in addition to the $V_{1/2}$ we are going to look at the slope, because the slope of this relationship could be altered as well. This is related to charge.

Robertson: Is the slope altered?

Tristani-Firouzi: For some but not all mutants.

Robertson: The fact that you had a shift with the $V_{1/2}$ that was not reflected in activation kinetics suggests that you have more than a two-stage process going on.

Tristani-Firouzi: That is definitely true. There are multiple states. Forcing it to be a single Boltzmann relationship implies that there are only two states there, and we know that there are more. Inherent in the equilibrium analysis is the assumption that gating transitions can be simplified by considering only two states (e.g. closed vs. open). Because the equilibrium distributions measured can be relatively well fitted by single Boltzmann functions. Simplified two state models have been commonly used to analyze the impact of substitutions on steady-state gating properties for other K^+ channels.

Robertson: Are there multiple inactivation states?

Tristani-Firouzi: We don't know that. There are certainly multiple conformations of the voltage sensor. This is well described.

Tseng: The fact that you have such a fast gating current component, much faster than that of the EAG channel, and also a much slower gating current component, suggests that conformational changes during membrane depolarization must be very different between hERG and EAG. Do we have to say that it is the S4 that is moving so fast, causing the fast component of gating current? Couldn't it be S2–S3 movement in an inward direction?

Tristani-Firouzi: We are recording gating current that reflects movement of charges within the membrane. Because the S4 is the dominant charged domain we assume that S4 movement is generating gating current. One of the mutations, V535A, markedly reduces the amplitude of the fast gating component without altering channel inactivation. This supports the idea that this fast gating transient is not related to inactivation. It seems more likely that this is movement in the early steps of the activation pathway; steps where the voltage sensors are moving independently of each other. Dave Piper generated a kinetic scheme to explain our gating and current data and it predicts that the early transitions are responsible for the fast gating component.

Ficker: If I understand that correctly, the coupling to inactivation starts by some sort of strain after the voltage sensor moves out and presses on S5. Now if you make S620T, which is a tiny mutation, the channel is still selective and I would assume that the conformational change is minimal. Why do you uncouple inactivation from that setting?

Tristani-Firouzi: We don't know what bonds are formed in stabilizing the open state that then allow the inactivated state to occur. In the case of the S620 mutant one has obviously disrupted a very important interaction so that the rearrangements cannot close the selectivity filter down, or prevent whatever causes inactivation. After the S4 moves and is sensed by S5, if that is true, there are multiple bonds that are going to be broken. In the S620 mutant, the breaking of those bonds results in a channel that wants to go all the way down, but is stuck, or new interactions are made which keep the channel open.

Ficker: That mutation must make the pore incredibly rigid. The mechanics are still the same but suddenly it doesn't collapse any more.

Sanguinetti: One detail that needs to be examined, and which hasn't been looked at all, is that most models assume (and single channel recording supports this idea) that there are multiple open states. How do the transitions between the final closed states and those open states differ in these mutants which seem to eliminate inactivation? It could be transitions between open states that are changing, and not this final and very minor change. For most closed–open transitions in voltage-gated channels, there is a role proposed for a glycine residue that is essentially in the middle of the S6 transmembrane domain. hERG has a glycine there as well. Does anyone have any opinions about whether hERG bends at that glycine hinge, or whether it is somewhat different? Gary Yellen has a recent paper (Webster et al 2004) about the PVP domain in channels, but hERG doesn't have a PVP. Are there any ideas about the glycine hinge versus some other gating mechanism?

Mitcheson: When we mutate that glycine to an alanine in hERG we are left with a channel that basically activates relatively normally. If anything the voltage dependence of activation is slightly left-shifted. But then the activation is also, so perhaps it is hinging somewhere else. When we mutate the other glycine residue we also produced a channel that gates relatively normally, so maybe they compensate for one another, or changing it to an alanine still allows enough movement for it to hinge. It seems unlikely.

Robertson: I had the idea that it could be involved in the slow activation, so that the step between the movement of the S4 and the movement of the S6, because glycines can be kind of floppy, gave a high energy barrier to opening. We started out using what we called glycine scanning mutagenesis to see whether we could shift that point and alter the coupling. We didn't finish it.

Sanguinetti: EAG naturally has an alanine there.

Mitcheson: In the *Drosophila* EAG it is, but in other organisms it is commonly a glycine.

Sanguinetti: We have done a little bit of mutation in KCNQ1 which naturally has an alanine there. If you make it a glycine it seems to activate faster and more channels are open. If you mutate to other residues, the channel still gates, suggesting that the glycine hinge is not really important. But this is a channel that has something similar to a PVP motif, a PAG motif. It is important to keep in mind the different models of gating, no matter what part of the protein you are talking about.

Robertson: Actually, I remember that we had a mutation to cysteine in that spot which made activation much faster. That's where I got the idea.

Mitcheson: That is what we see with the alanine mutation as well. It is faster. hERG has IFG in place of the PXP motif. The residues C-terminal to IFG also affect deactivation. It is this little spot that seems to alter deactivation but not in a way that is obviously consistent with a hinge.

Shah: I have a rather naïve question. In these amino acid substitutions that you are doing, you are substituting one with another that are all L isomers. What would happen if you substituted L-tryptophan, for example, with D-tryptophan?

Sanguinetti: I don't think this has been looked at. There are ways to produce unnatural amino acids.

Ficker: McKinnon did some experiments putting D-amino acids in the selectivity filter. It's a very difficult and expensive experiment because you have to synthesize the protein.

Reference

Webster SM, Del Camino D, Dekker JP, Yellen G 2004 Intracellular gate opening in Shaker K^+ channels defined by high-affinity metal bridges. Nature 428:864–868

hERG channel trafficking

Eckhard Ficker, Adrienne Dennis, Yuri Kuryshev, Barbara A. Wible* and Arthur M. Brown

*Rammelkamp Center for Education and Research, MetroHealth Campus, Case Western Reserve University, Cleveland OH 44109, USA, and *ChanTest, Inc., Cleveland OH 44128, USA*

Abstract. Mutations in the cardiac potassium channel hERG/I_{Kr} cause inherited long QT syndrome with increased susceptibility to ventricular arrhythmias. Several mutations in hERG produce trafficking-deficient channels that are retained in the endoplasmic reticulum (ER). Surface expression of certain mutations (i.e. hERG G601S) can be restored by specific channel blockers. Although hERG currents have been studied extensively, little is known about proteins in the processing pathway. Using biochemical and electrophysiological assays we show that the cytosolic chaperones Hsp70 and Hsp90 interact transiently with wild-type hERG. Inhibition of Hsp90 prevents maturation and reduces hERG/I_{Kr} currents. Trafficking-deficient mutants remain tightly associated with chaperones in the ER until trafficking is restored, e.g. by channel blockers. hERG/chaperone complexes represent novel targets for therapeutic compounds with cardiac liability such as arsenic, which is used in the treatment of leukaemias. Arsenic interferes with the formation of hERG/chaperone complexes and inhibits hERG maturation causing ECG abnormalities. We conclude that Hsp90 and Hsp70 are crucial for productive folding of wild-type hERG. Therapeutic compounds that inhibit chaperone function produce a novel form of acquired long QT syndrome not by direct channel block but by reduced surface expression due to an acquired trafficking defect of hERG.

2005 The hERG cardiac potassium channel: structure, function, and long QT syndrome. Wiley, Chichester (Novartis Foundation Symposium 266) p 57–74

Mutations in the cardiac potassium channel gene hERG (KCNH2) cause inherited long QT syndrome (LQT2) characterized by a prolonged QT interval on the electrocardiogram (ECG), Torsade de Pointes (TdP) arrhythmias and sudden cardiac arrest (Sanguinetti et al 1995, Keating & Sanguinetti 2001). Mutations in hERG reduce the repolarizing cardiac potassium current I_{Kr} thereby prolonging the cardiac action potential. Loss-of-function may be produced by different mechanisms including kinetic alteration of channel function or insertion of non-functioning, tetrameric channels into the plasma membrane (e.g. Sanguinetti et al 1996, Chen et al 1999). An ever-expanding group of LQT2 mutations, however, produces trafficking-deficient channels that are retained in the endoplasmic reticulum (ER) and are thought of as expressing conformational defects recognized by cellular quality control mechanisms (Zhou et al 1998, Thomas et al 2003). Consequently, many misfolded mutants cluster in hERG domains with complex protein folds such as the PAS domain (Morais-Cabral et al 1998, Paulussen et al 2002), the highly structured potassium conduction pathway (Doyle et al 1998, Zhou et al 1999a, Kagan et al 2000, Ficker et al 2000a, 2002) and the cyclic nucleotide binding domain (cNBD) (Zagotta et al 2003, Ficker et al 2000b, 2002). Trafficking-deficient LQT2 mutants often give rise to channels capable of conducting I_{Kr} currents if they can escape the endoplasmic reticulum and reach the cell surface. More recently, a variety of strategies have been explored to restore channel trafficking *in vitro* as a first step to implementing similar strategies in patients. Briefly, trafficking of several LQT2 mutants including hERG N470D and G601S could be restored by incubation at low temperature or with chemical chaperones such as DMSO or glycerol (Zhou et al 1999a). Most notably, channel blockers such as E4031 or astemizole act as pharmacological chaperones to rescue surface expression (Zhou et al 1999a, Ficker et al 2002, Paulussen et al 2002). However, hERG channel blockers are not of therapeutic value as they interfere with ion conduction. In addition, rescue by pharmacological chaperones is highly domain dependent in that they are unable to rescue LQT2 mutants in the cNBD domain including hERG R752W and F805C (Ficker et al 2002). The first example for a possible dissociation of hERG block and rescue was provided by the antihistamine fexofenadine, which restored trafficking in two LQT2 mutants in the transmembrane region of hERG, N470D and G601S at a concentration too low to block hERG currents (Rajamani et al 2002). More recently, the endoplasmic reticulum ATPase inhibitor thapsigargin has been described as another pharmacological chaperone without channel block (Delisle et al 2003). Thapsigargin was able to restore trafficking for a different subset of LQT2 mutants including a mutant in the C-terminal cyclic nucleotide-binding domain, hERG F805C, which was not rescued by any conventional hERG channel blocker.

Thus, a complicated picture arises where certain LQT2 mutants are rescued by pharmacological chaperones while others are unaffected. To resolve this patchwork and provide a solid mechanistic basis for future rescue attempts in patients, we need a better understanding of how rescue molecules interact with hERG channels to stabilize certain trafficking-deficient LQT2 mutants for export to the plasma membrane but fail to stabilize others. Here we will focus on the rescue of hERG G601S by specific channel blockers and explore how block and rescue are related on a structural level.

While drug binding to the hydrophobic core of the channel protein seems to be crucial for rescue of hERG G601S, it is not clear how trafficking deficient mutations are retained in the endoplasmic reticulum (ER) in the first place. Given the large number of misfolded LQT2 mutants the question arises as to how cells recognize and retain abnormally folded mutant channels and how they move wild-type channels along productive folding pathways. In mammalian cells, the primary level of quality control for membrane proteins such as hERG is comprised of molecular chaperones that repeatedly interact with incompletely folded proteins to facilitate native conformations. We will explore how cytosolic chaperones interact with newly synthesized wild-type and LQT2 mutant channels and determine how pharmacological chaperones modify this interaction.

Just as the channel can be directly blocked by therapeutic compounds leading to QT prolongation and TdP arrhythmias the possibility exists that proteins in the processing pathway such as chaperones may provide additional targets whose inhibition will lead to a defect in hERG processing, a reduction in cell surface expression of I_{Kr} and ultimately a novel form of acquired LQTS. Arsenic, which is used clinically for the treatment of acute promyelocytic leukaemia and is known for its cardiotoxicity, is introduced as a first member of this novel compound class. Arsenic represents a first example where drug-induced LQTS is not caused by direct channel block as described for the majority of non-cardiac compounds with cardiac liability but by an acquired trafficking defect.

Restoration of hERG G601S trafficking by the specific blocker astemizole

hERG G601S is an LQT2 mutation in the S5–S6 linker region of the channel protein with most channels retained in the ER at physiological temperatures (Furutani et al 1999). At low incubation temperatures trafficking resumes and functional channels appear at the cell surface. Astemizole is an antihistamine that blocks hERG currents with an IC_{50} of 0.9 nM. It is metabolized into norastemizole, which blocks hERG at much higher concentrations (Zhou et al 1999b). We asked whether differences observed in current block were preserved for pharmacological rescue of hERG G601S. We found that astemizole rescued hERG G601S more efficiently than norastemizole. Assuming that mutations in the drug-binding site of hERG should abolish restoration of trafficking similar to the weak channel blocker norastemizole we mutated the drug-binding site of hERG (Mitcheson et al 2000, Tseng 2001). We found that the doubly mutated channel hERG G601S/F656C was not rescued even by high astemizole concentrations. These results provide evidence that block and rescue correlate with each other and that rescue involves interactions with the inner vestibule of hERG.

Quaternary ammonium block and rescue

To pinpoint the inner vestibule of hERG as the side of pharmacological chaperone action we used N-alkyl triethylammonium (TEA) derivatives as generic hERG

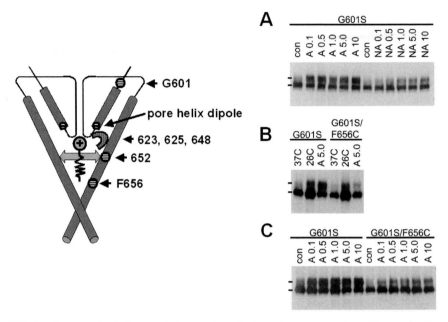

FIG. 1. Pharmacological rescue, left part of panel: schematic representation of the conduction pathway of hERG. Indicated are residues critical for the binding of high affinity hERG blockers (in the S6 transmembrane domain, F656, Y 652 and G648; in the pore helix, T623 and V625, which are depicted as a S6/pore helix binding pocket formed together with G648). Mutation of hERG F656 located in the S6 transmembrane helix abolishes or reduces binding of a wide variety of hERG blockers. Dipoles formed by pore helices contribute to stabilization of cationic blocking molecules in the inner vestibule of hERG. Location of hERG G601S mutation is depicted in the extracellular part of the S5 linker region. Inner vestibule is shown with schematic hERG blocker consisting of positively charged head group and a lipophilic tail (core) region, which is thought to interact with S6 transmembrane domains. (A) Western blot analysis of hERG G601S expressing cells treated with increasing concentrations of either astemizole (A, in μM) or norastemizole (NA, in μM). Treatment with astemizole induces maturation of hERG G601S into fully glycosylated, 155 kDa cell-surface form. NA produced much less fully glycosylated hERG G601S protein. (B) Western blot of hERG G601S incubated at physiological temperatures, at low temperatures, and with 5 μM astemizole. Both a low temperature and astemizole restore trafficking of hERG G601S. The double mutant G601S/F656C with part of the drug-binding site destroyed is only rescued by a low incubation temperature and not by astemizole. (C) Western blot analysis of hERG G601S and hERG G601S/F656C after overnight treatment with increasing concentrations of astemizole. The double mutant hERG G601S/F656C is poorly rescued by astemizole. Molecular weight markers correspond to 135 and 155 kDa.

blockers that combine a protonated nitrogen head atom with lipophilic tail regions. These experiments were developed with more recent crystallographic data in mind that demonstrated binding of a tetrabutylammonium-like compound (TBA) to the central cavity of the generic potassium channel KcsA

(Zhou et al 2001). The structural data in KcsA show that TBA not only binds to the inner cavity but also re-orients the inner (S6) helices towards the centre of the inner vestibule. On the basis of these structural data we hypothesized that N-alkyl TEA derivatives should rescue hERG G601S trafficking. We tested a whole series of N-alkyl TEA derivatives and found for all of them concentration-dependent rescue of hERG G601S cell surface expression.

Rescue of hERG trafficking with channel blockers is domain-dependent

Our results support a model where pharmacological chaperones interact with S6 transmembrane helices of hERG to initiate rescue. Here it is very important to ask whether the binding of channel blockers to the inner cavity produces conformational changes that are strictly localized to the binding site, which is the transmembrane domain of the channel protein, or whether binding may also affect more distant independent protein domains such as the cyclic nucleotide-binding region. To answer this question, we tested three trafficking deficient mutants in the cyclic nucleotide domain, hERG R752W, hERG F805C and hERG R823W. We found none of the three could be rescued by astemizole or C8-TEA, while both treatments restored the maturation of hERG N470D, an LQT2 mutant in the voltage-sensing region of the transmembrane domain. Taken together, our results highlight the significance of interactions in the central cavity of hERG K^+ channels. Blocker binding appears to stabilize the native state of the hERG G601S channel protein analogous to enzyme/ligand complexes. Since binding to the inner cavity appears to rely on assembled channel tetramers, impaired assembly might explain best why certain temperature-sensitive hERG proteins such as hERG R752, F805C or R823W with mutations in the cyclic nucleotide binding domain cannot be rescued with channel blockers.

Newly synthesized hERG wild-type channels form complexes with heat-shock proteins Hsp/c70 and Hsp90

While drug binding to the inner cavity of the channel seems to be a prerequisite for pharmacological rescue, the mechanism by which LQT2 mutants such as hERG G601S are retained and how trafficking is restored upon drug binding remains to be elucidated. In mammalian cells, quality control for membrane proteins consists of ER-resident and cytoplasmic chaperones that recognize incompletely folded proteins and promote productive folding (Ellgard et al 1999). Since the large N- and C-terminal portions of hERG are exposed to the cytosol, we explored the cytoplasmic chaperones Hsp90 and Hsp70 as candidates for interaction with hERG. Our studies have shown that:

- hERG wild-type channels associate transiently with chaperones during maturation in the ER
- most LQT2 mutant channels remain permanently associated with chaperones while retained in the ER, and
- correction of the conformational defect with pharmacological chaperones results in the dissociation of channel/chaperone complexes prior to the resumption of trafficking (Ficker et al 2003).

To test whether hERG wild-type channels interact with cytosolic chaperones, we performed co-immunoprecipitation experiments. In these experiments we found core-glycosylated, ER-resident hERG immediately after biosynthesis in association with Hsp70. In marked contrast, hERG–Hsp90 complexes could only be isolated after chemical cross-linking, which may point towards a more dynamic, less stable interaction. Formation of hERG–Hsp90 complexes was nevertheless specific since the Hsp90 inhibitor geldanamycin (GA) blocked complex formation (Roe et al 1999). Interestingly, we were not able to isolate hERG protein together with ER resident chaperones such as Grp94 or BiP, which represent ER-luminal homologues of Hsp90 and Hsp70, respectively. As expected for chaperones which are thought to facilitate the folding of ER resident precursor substrates into mature native protein conformations, both Hsp70 and Hsp90 interacted exclusively with newly synthesized core-glycosylated hERG but not with the fully glycosylated cell surface form of hERG. To study the conversion of newly synthesized ER-resident hERG into the cell surface form directly, we performed pulse–chase experiments in the absence and presence of the specific Hsp90 inhibitor GA. Under control conditions hERG was initially synthesized in the ER as a core-glycosylated protein, which was then slowly converted into the fully glycosylated cell surface form. In marked contrast, in the presence of GA only small amounts of fully-glycosylated cell surface protein were produced. Treatment with GA reduced not only the expression of fully-glycosylated hERG but also tail current amplitudes measured in an overexpression system. Similarly, in ventricular cardiomyocytes native, I_{Kr} channels were reduced after inhibition of Hsp90 with GA. In these experiments GA reduced tail current densities in guinea pig cardiomyocytes from 0.72 ± 0.09 pA/pF to 0.32 ± 0.04 and 0.19 ± 0.03 pA/pF upon exposure to 1 or 2 μg/ml (1.8 or 3.6 μM) GA, respectively.

Interactions between chaperones and LQT2 mutant hERG channels are prolonged

To determine whether wild-type channels and LQT2 mutant channels are handled differentially by cellular chaperones, we analysed the interaction of the LQT2

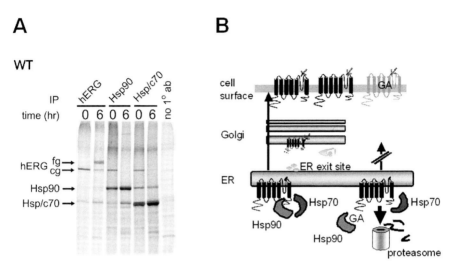

FIG. 2. Chaperones form transient complexes with wild-type hERG. (A) HEK cells stably expressing wild-type hERG were radiolabelled, lysed and immunoprecipitated with anti-hERG antibody. Only the core-glycosylated form of hERG was immunoprecipitated immediately after radiolabelling. After a 6 hour chase period, the fully-glycosylated mature form of hERG was immunoprecipitated. Both Hsp70 and Hsp90 could be isolated in complex with the core-glycosylated form of hERG immediately after radiolabelling but not with the fully glycosylated mature form of hERG. Arrows indicate the position of the respective protein bands. No 1° ab indicates control with no primary antibody added. (B) Trafficking scheme depicting different subcellular compartments involved in hERG processing and export to the cell surface. hERG is synthesized as a core-glycosylated 135 kDa protein in the ER where it associates with the cytosolic chaperones Hsp70 and Hsp90 until its native conformation is reached. After the native conformation has been adopted, hERG is transported to the Golgi apparatus where complex sugar chains are added to produce the fully glycosylated 155 kDa form. After processing in the Golgi has been completed, the fully glycosylated hERG is shipped out to the cell surface. Geldanamycin interferes with Hsp90 function and prevents adoption of the native hERG conformation. Thus geldanamycin produces immature protein that is not exported from the ER but instead degraded via proteasomal pathways. Overall, interference with Hsp90 function reduces cell-surface expression of hERG.

mutants hERG R752W and hERG G601S with Hsp70 and Hsp90. The temperature-sensitive, trafficking-deficient LQTS2 mutant hERG R752W is located in the C-terminal cNBD domain (Ficker et al 2000b). In immunoprecipitation experiments identical to the ones performed with wild-type, we found that hERG R752W formed complexes with both Hsp90 and Hsp70 that were stable for its entire lifetime in the ER. To determine the abundance of channel–chaperone complexes and time-dependent changes in their composition, we quantified the amount of hERG R752W that could be

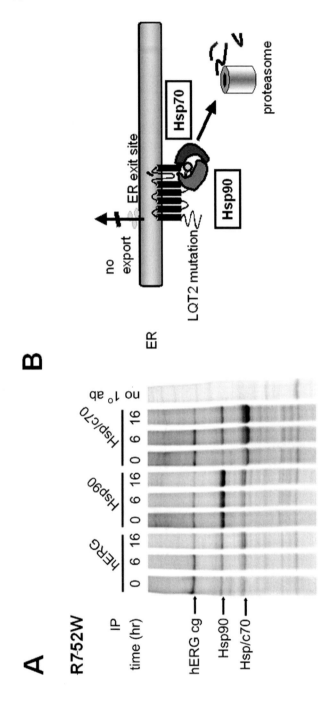

FIG. 3. The LQTS2 mutant hERG R752W forms long-lasting complexes with the cytosolic chaperones Hsp70 and Hsp90. (A) hERG R752W/ chaperone complexes were analysed using radiolabelling and immunoprecipitation. At all time points studied hERG R752W was isolated in association with Hsp70 and Hsp90. (B) Trafficking scheme depicting retention of hERG R752W in the ER. hERG R752W is synthesized as a core-glycosylated protein in the ER that forms long-lasting complexes with the cytosolic chaperones Hsp70 and Hsp90. Since hERG R752W is not able to reach an exportable conformation, the failed attempts of Hsp70 and Hsp90 to re-fold hERG R752W are terminated by degradation in the proteasome.

immunoprecipitated in complex with either Hsp70 or Hsp90. We found that chaperone association remained constant over time for R752W while both Hsp90 and Hsp70 interacted transiently with wild-type hERG.

Based on our experiments with hERG R752W, we conclude that chaperones provide an assessment of hERG conformation leading to ER retention of the mutant protein by mechanisms, which remain to be defined. As a direct consequence, we predict that LQTS2 mutants for which trafficking can be restored are tied up in chaperone complexes under control conditions but will be released from these complexes after correction of conformational defects by pharmacological chaperones. To test this hypothesis directly we studied changes in Hsp70/Hsp90 association with hERG G601S upon exposure to the pharmacological chaperone astemizole.

At 37 °C, hERG G601S was present mainly in its core-glycosylated form with only a small fraction of newly synthesized protein spontaneously converted into mature, fully glycosylated cell surface channels as expected for a hypomorphic mutation that generates only small currents at 37 °C. hERG G601S was isolated in complex with both Hsp70 and Hsp90 and remained stably associated with both chaperones just as described for hERG R752W. After incubation with astemizole, however, trafficking of hERG G601 was largely restored with channel–chaperone complexes being dissociated and mature channel reaching the cell surface.

Taken together, our data show that productive folding and maturation of the cardiac potassium channel hERG/I_{Kr} are controlled by the cytosolic chaperone Hsp90. Complex formation with chaperones is not only important for the productive folding of hERG wild-type channels but also for the handling of LQT2 mutant channels that are retained in the ER. For two different LQT2 mutations, hERG R752W and hERG G601S, we demonstrated prolonged association with Hsp90 and Hsp70 in the ER. In rescue experiments with G601S, we showed that initiation of productive folding by incubation with pharmacological chaperones led to the dissociation of channel–chaperone complexes, followed by successful channel export to the cell surface. However, it is not clear at present how dissociaton of channel/chaperone complexes is tied to successful export from the ER. The prolonged chaperone decoration of trafficking-deficient mutants is a clear indication that a conformational defect is the major culprit preventing export. Furthermore, our data identify an entirely new pharmacology for hERG channels. Ansamycin antibiotics such as GA that block Hsp90 function are used to target oncogenic kinases for degradation and are being tested in preclinical as well as clinical trials against various cancers (Smith et al 1998, Neckers 2002). This raises the possibility that pharmacological inhibition of Hsp90 may produce a novel form of acquired LQTS.

Arsenic trioxide: a first example of a therapeutic compound that causes acquired LQTS by inhibition of hERG channel maturation

Drug-induced, acquired LQTS is most often caused by direct blockade of the cardiac potassium channel hERG by a wide variety of structurally diverse therapeutic compounds (Vandenberg et al 2001, Redfern et al 2003). Arsenic trioxide (As_2O_3) produces dramatic remissions in patients with relapsed or refractory acute promyelocytic leukaemia (Zhang et al 2001). We found that As_2O_3, despite its well-known cardiotoxic side effects including QT prolongation and TdP (Unnikrishnan et al 2001, Westervelt et al 2001), did not directly block hERG. Neither 10 nor 100 μM As_2O_3 applied extracellularly reduced hERG currents. Similarly intracellular application of As_2O_3 did not alter hERG currents. Since As_2O_3 did not block hERG directly, we tested the possibility that As_2O_3 may block hERG trafficking. Stably transfected HEK/hERG cells were exposed overnight to increasing concentrations of As_2O_3 and tested for effects on hERG processing. In these experiments we found that hERG currents were reduced in a concentration-dependent manner with an IC_{50} of about 1.5 μM. Western blot analysis was used to corroborate our electrophysiological results. Incubation with As_2O_3 for 24 hours produced a time- and concentration-dependent decrease in the amount of fully-glycosylated hERG at the cell surface. The cell surface form of hERG was suppressed with an IC_{50} of 1.5 μM, which is similar to the effect on currents.

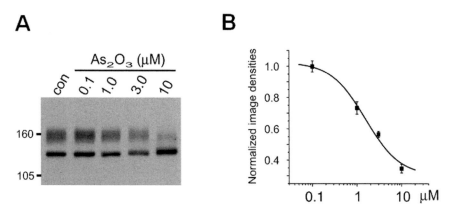

FIG. 4. Arsenic trioxide (As_2O_3) inhibits maturation of wild-type hERG. (A) Western blot analysis of HEK cells stably expressing wild-type hERG after overnight exposure to increasing concentrations of As_2O_3. As_2O_3 reduces expression of the fully glycosylated 155 kDa cell surface form of hERG. (B) Quantitative analysis of concentration-dependent reduction of the fully glycosylated 155 kDa form of hERG on exposure to As_2O_3. IC_{50} is 1.5 μM. Image densities of the 155 kDa band of hERG were analysed directly on a Storm PhosphoImager.

To gain first insight into possible mechanisms underlying the As_2O_3-induced trafficking defect of hERG we studied the kinetics of the trafficking block and the interaction of hERG channels with the cytosolic chaperones Hsp70 and Hsp90 in the absence and presence of As_2O_3. In pulse–chase experiments we found that the conversion of newly synthezised hERG into the fully glycosylated cell surface form was reduced in the presence of As_2O_3 with only small amounts of cell surface protein being generated. In addition, hERG–chaperone complexes were less stable in the presence of As_2O_3 just as described for GA. Overall, productive folding of hERG seems to be impaired by the disruption of functional hERG chaperone complexes.

Arsenic trioxide reduces native I_{Kr} currents in ventricular myocytes

To test whether As_2O_3 may indeed prolong the cardiac action potential as expected from its pro-arrhythmic effects reported in the clinic, we studied the effects of extracellularly applied As_2O_3 on action potentials in guinea-pig ventricular myocytes. As expected, action potential duration was not affected by acute As_2O_3 application. However, when ventricular myocytes were cultured overnight in the presence of $3\,\mu M$ As_2O_3, APD_{90} was significantly prolonged from 459 ± 15 to 1014 ± 127 ms. Since our experiments in an overexpression system pointed towards hERG/I_{Kr} as a major target for As_2O_3, we analysed current densities of the cardiac rapid, delayed rectifier current I_{Kr} in voltage clamp experiments. In these experiments 1.5 and $3\,\mu M$ As_2O_3 reduced I_{Kr} tail current densities in guinea pig ventricular myocytes from 0.7 to 0.3 and 0.03 pA/pF, respectively.

Since chronic exposure to As_2O_3 suppresses trafficking of hERG, we propose that this mechanism underlies the reduction in I_{Kr} and contributes to the prolongation of action potential duration in ventricular myocytes. Trafficking block of hERG was observed at clinically relevant concentrations of 0.1 to $1.5\,\mu M$ As_2O_3 (Shen et al 1997). This report is the first to show acquired trafficking block as the basis for QT prolongation and TdP and is consistent with other reports that ECG changes develop gradually and are observed most often after multiple rounds of treatment with As_2O_3 (Ficker et al 2004).

Up to now it has been thought that acquired long QT syndrome is nearly exclusively produced by non-cardiac drugs that block hERG/I_{Kr} currents. In the past, adverse cardiac events led to the withdrawal of several very successful drugs such as terfenadine (Seldane), astemizole (Hismanol), cisapride (Propulsid) and sertindole (Serlect). As a direct consequence many regulatory agencies developed safety-testing standards that recommend a hERG 'blocking' assay with patch clamp recordings being the gold standard. Our molecular analysis of hERG trafficking identified the molecular chaperone Hsp90 as a novel target for acquired LQTS in the processing pathway of hERG. With As_2O_3 we presented a first example where an acquired trafficking block of hERG contributes to QT prolongation and TdP in patients. Thus, the important question arises as to whether we would have identified the 'true' cardiac liability of As_2O_3 using common hERG safety assays. And even more importantly, how many other drugs are right now being developed or are already on the market that impair hERG trafficking? Given current safety testing standards, we just cannot know!

Acknowledgements

This work was supported by National Institutes of Health grants HL 71789 and CA 106028 (to E.F.).

References

Chen J, Zou A, Splawski I, Keating MT, Sanguinetti MC 1999 Long QT syndrome-associated mutations in the Per-Arnt-Sim (PAS) domain of HERG potassium channels accelerate channel deactivation. J Biol Chem 274:10113–10118

Delisle BP, Anderson CL, Balijepalli RC, Anson BD, Kamp TJ, January CT 2003 Thapsigargin selectively rescues the trafficking defective LQT2 channels G601S and F805C. J Biol Chem 278:35749–35754

Doyle DA, Morais Cabral J, Pfuetzner RA et al 1998 The structure of the potassium channel: molecular basis of K^+ conduction and selectivity. Science 280:69–77

Ellgard L, Molinari M, Helenius A 1999 Setting the standards: quality control in the secretory pathway. Science 286:1882–1888

Ficker E, Dennis AT, Obejero-Paz CA, Castaldo P, Taglialatela M, Brown AM 2000a Retention in the endoplasmic reticulum as a mechanism of dominant-negative current suppression in human long QT syndrome. J Mol Cell Cardiol 32:2327–2337

Ficker E, Thomas D, Viswanathan PC et al 2000b Novel characteristics of a misprocessed mutant HERG channel linked to hereditary long QT syndrome. Am J Physiol 279:H1748–H1756

Ficker E, Obejero-Paz C, Zhao S, Brown AM 2002 The binding site for channel blockers that rescue misprocessed human long QT syndrome type 2 ether-a-gogo-related gene (HERG) mutations. J Biol Chem 277:4989–4998

Ficker E, Dennis AT, Wang L, Brown AM 2003 Role of the cytosolic chaperones Hsp70 and Hsp90 in maturation of the cardiac potassium channel hERG. Circ Res 92:e87–e100

Ficker E, Kuryshev YA, Dennis AT et al 2004 Mechanisms of arsenic-induced prolongation of cardiac repolarization. Mol Pharmacol 66:33–44

Furutani M, Trudeau MC, Hagiwara N et al 1999 Novel mechanism associated with an inherited cardiac arrhythmia. Circulation 99:2290–2294

Kagan A, Yu Z, Fishman GI, McDonald TV 2000 The dominant negative LQT2 mutation A561V reduces wild-type HERG expression. J Biol Chem 275:11241–11248

Keating MT, Sanguinetti MC 2001 Molecular and cellular mechanisms of cardiac arrhythmias. Cell 104:569–580

Mitcheson JS, Chen J, Lin M, Culberson C, Sanguinetti MC 2000 A structural basis for drug-induced long QT syndrome. Proc Natl Acad Sci USA 97:12329–12333

Morais-Cabral JH, Lee A, Cohen SL, Chait BT, Li M, MacKinnon R 1998 Crystal structure and functional analysis of the HERG potassium channel N terminus: a eukaryotic PAS domain. Cell 95:649–655

Neckers L 2002 Hsp90 inhibitors as novel cancer chemotherapeutic agents. Trends Mol Med 8:S55–S61

Paulussen A, Raes A, Matthijs G, Snyders DJ, Cohen N, Aerssens J 2002 A novel mutation (T65P) in the PAS domain of the human potassium channel HERG results in the long QT syndrome by trafficking deficiency. J Biol Chem 277:48610–48616

Rajamani S, Anderson CL, Anson BD, January CT 2002 Pharmacological rescue of human K^+ channel long-QT2 mutations. HERG rescue without block. Circulation 105: 2830–2835

Redfern WS, Carlsson L, Davis AS et al 2003 Relationships between preclinical cardiac electrophysiology, clinical QT interval prolongation and torsade de pointes for a broad range of drugs: evidence for a provisional safety margin in drug development. Cardiovasc Res 58:32–45

Roe SM, Prodromou C, O'Brien R, Ladbury JE, Piper PW, Pearl LH 1999 Structural basis for the inhibition of the Hsp90 molecular chaperone by the antitumor antibiotics radicicol and geldanamycin. J Med Chem 42:260–266

Sanguinetti MC, Jiang C, Curran ME, Keating MT 1995 A mechanistic link between an inherited and an acquired cardiac arrhythmia: HERG encodes the I_{Kr} potassium channel. Cell 81:299–307

Sanguinetti MC, Curran ME, Spector PS, Keating MT 1996 Spectrum of HERG K^+-channel dysfunction in an inherited cardiac arrhythmia. Proc Natl Acad Sci USA 93:2208–2212

Shen ZX, Chen GQ, Ni JH et al 1997 Use of arsenic trioxide (As_2O_3) in the treatment of acute promyelocytic leukemia (APL): II. Clinical efficacy and pharmacokinetics in relapsed patients. Blood 89:3354–3360

Smith DF, Whitesell L, Katsanis E 1998 Molecular chaperones: biology and prospects for pharmacological intervention. Pharmacol Rev 50:493–513

Thomas D, Kiehn J, Katus HA, Karle CA 2003 Defective protein trafficking in hERG-associated hereditary long QT syndrome (LQT2): molecular mechanisms and restoration of intracellular protein processing. Cardiovasc Res 60:235–241

Tseng G 2001 I_{Kr}: the hERG channel. J Mol Cell Cardiol 33:835–849

Unnikrishnan D, Dutcher JP, Varshneya N et al 2001 Torsade de pointes in 3 patients with leukemia treated with arsenic trioxide. Blood 97:1514–1516

Vandenberg JI, Walker BD, Campbell TJ 2001 HERG K^+ channels: friend and foe. Trends Pharmacol Sci 22:240–246

Westervelt P, Brown RA, Adkins DR et al 2001 Sudden death among patients with acute promyelocytic leukemia treated with arsenic trioxide. Blood 98:266–271

Zagotta WN, Olivier NB, Black KD, Young EC, Olson R, Gouaux E 2003 Structural basis for modulation and agonist specificity of HCN pacemaker channels. Nature 425: 200–205

Zhang TD, Chen GQ, Wang ZG, Wang ZY, Chen SJ, Chen Z 2001 Arsenic trioxide, a therapeutic agent for APL. Oncogene 20:7147–7153

Zhou Z, Gong Q, Epstein ML, January CT 1998 HERG channel dysfunction in human long QT syndrome. J Biol Chem 273:21061–21066

Zhou Z, Gong Q, January CT 1999a Correction of defective protein trafficking of a mutant hERG potassium channel in human long QT syndrome. Pharmacological and temperature effects. J Biol Chem 274:31123–31126

Zhou Z, Vorperian VR, Gong Q, Zhang S, January CT 1999b Block of HERG potassium channels by the antihistamine astemizole and its metabolites desmethylastemizole and norastemizole. J Cardiovasc Electrophysiol 10:836–843

Zhou M, Morais-Cabral JH, Mann S, MacKinnon R 2001 Potassium channel receptor site for the inactivation gate and quaternary amine inhibitors. Nature 411:657–661

DISCUSSION

Traebert: I think this is an interesting approach. But you can have problems with hERG if you interfere with the transcription level or expression level and don't just use the secretory pathway. You can have problems at all these cell biological steps. Do you know a compound which interferes with the endocytotic pathway binding to the inner pore of the hERG protein leading to an internalization and maybe lysosomal degradation? If so, do you know some cell biological signals that lead to lysosomal degradation?

Ficker: We thought for a long time that we had such a drug. This was pentamidine. But we were not able to repress the decrease in surface expression using a dynamin mutant. Now we think that pentamidine uses a different mechanism which we don't understand. There might be multiple mechanisms: trafficking defects, defects in endocytic pathways but possibly also changes in gene expression that have to be analysed using genomic methods.

Traebert: Genomics? For sure, we are looking only at events that take place within minutes.

Ficker: That is not the way that patients are treated. There might be a whole pharmacology out there which is neglected. A lot of this work is done in an acute way. Many of the assays are done instantaneously. The cell biology requires a completely different frame of mind. Not many people are thinking in this dimension. I can imagine where it is going to lead if I suggest we have to look at more than one channel in terms of cost, but currently we have a narrow view of what is going on out there.

Shah: As regulators, we have been looking for a drug that is hERG negative but which is known to be torsadogenic or significantly prolong the QTc interval in human. We thought originally that arsenic drugs might be one of these, and you have confirmed this. I have a question. How do you reconcile your findings with the recent study from Dan Roden's laboratory? They found arsenic trioxide to block I_{Kr} and I_{Ks} as well as to activate $I_{K\text{-}ATP}$ (Drolet et al 2004).

Ficker: I can't. First of all, I showed you that we don't see acute block of I_{Kr}. Dan Roden's paper was exclusively in HEK cells. We looked in myocytes and didn't see acute block. We don't see action potential prolongation under acute situations. I showed also that I_{Ks} doesn't change in myocytes. I cannot reconcile our results with theirs; all I can do is state our findings.

Shah: The other important question is this. My recollection is that many patients experience QT prolongation around two hours after starting the infusion of arsenic trioxide. This, again, is not consistent with your observations that chronic exposure is required.

Ficker: Most of the data in the literature are also not consistent with that observation. There are rabbit data showing that it takes almost 4 weeks before something is seen in the chronic model. In the guinea-pig, where it is pretty fast, it takes hours before you see something. In humans, in many cases the reports clearly state that it is after days or weeks of treatment.

Shah: Because of this delay in effect on QTc interval that is observed in a number of patients, some people have suspected that one of the metabolites of arsenic trioxide might be responsible for hERG blockade. Have you looked at any of the metabolites?

Ficker: We have spoken to the people at Merck, but these compounds are not yet in our hands.

January: I had a comment on the time dependence of rescue. This is the opposite of what Eckhard Ficker was saying. Rescue is a fairly dynamic process. When you expose mutations in cell lines to drugs that affect rescue, you can measure the beginnings of rescue in a matter of hours. Arsenic is the converse. These are processes that can produce effects that are not instantaneous, but which are measurable in a short period of time — in a matter of hours. You can see changes in current density and Western blots over this timescale. They look a lot prettier if you wait a day, but they are measurable earlier.

Sanguinetti: Does this say anything about the turnover rate of hERG? What would you estimate to be the turnover rate?

Ficker: The half-life is 11 h.

Hancox: You have succeeded where we failed, because we saw no effect of arsenic and then left it. We should have done what you did. I have a question: is the action potential effect you get similar in phenotype to the kind of action potential prolongation you would get with a blocking drug? Is it, for example, reverse rate dependent?

Ficker: No, we didn't do this. There is a huge increase in Ca^{2+} currents and this is an inherent difference compared with what happens if you just block the channel. The Ca^{2+} current gets incredibly large and it is a completely different mechanism. We know that it can happen in 30 min or 1 h, and it is most likely a direct modification of the Ca^{2+} channel property.

Hancox: I don't remember this precisely, but in the *Blood* paper (Chiang et al 2002) they talked about arsenic and did action potential measurements. I think they saw some kind of rate dependence.

Ficker: This was most likely in guinea-pig.

Brown: If they saw reverse use dependence, this doesn't mean that it is a function of the defect on hERG. That could just be reverse rate dependence, period, in the intact heart or myocyte.

Gosling: This has some parallels with CFTR, specifically the trafficking mutant ΔF508, where SERCA inhibitors also traffic channels back to the membrane which are trapped in the endoplasmic reticulum. Has anyone looked to see whether arsenic will take down CFTR? Is this a specific hERG effect or is it a generic trafficking consequence for channels in general?

Ficker: We always look at Kv1.5. This is one of the channels which is glycosylated. You need a glycosylation tag to do this fast. Kv1.5 is not very sensitive to arsenic and the currents are not changed. As you saw, I_{Ks} is not sensitive. Obviously, the Ca^{2+} current changes by a different mechanism. It is not a general effect due to toxicity.

Hoffmann: You raised the question of how we could detect a compound with such a mechanism of action in drug development. This is a good point: we basically have acute *in vitro* tests and safety pharmacology is made out typically as a single dose. However, we would see it in the repeat-dosed non-rodent. This has been a weak point in the past: it hasn't been stressed enough that these repeat-dose non-rodent studies are very important. Along with the half life of the protein on the surface, the number of general proteins sitting at the surface is determined on the one hand by their production and on the other by their degradation. There is also the possibility, therefore, that compounds influence degradation. This is a new field.

Ficker: Yes, we are opening a Pandora's box here. If we want to develop drugs, it's probably best not to even think about it!

January: You commented on hERG R752W(LQT2 mutation) that rescues with temperature but not with drug. What do you think is going on there compared with other mutations? There is differential rescue.

Ficker: Let me first say what I think happens with astemizole in hERG G601S. I think the channel sits as a tetramer in the membrane and has a binding site that is not quite right. If you put astemizole in there it interacts with the S6 transmembrane domains and this mechanical strain re-folds the protein. What is needed for many rescue experiments is some kind of preconfigured binding site. With R752W we were never able to see dominant negative suppression of wild-type. The suspicion would be that this channel does not form tetramers, so there is no binding site and therefore no astemizole rescue. It would be nice to show this in detail.

Hancox: This may be a stupid question. The binding site is the cavity. Usually, with other channels they have to open for the cavity to be accessible. What is happening here?

Ficker: The channel is depolarized because there is high K^+ in the cell. At least, that is what we think. Also, in a misfolded protein there might be other access pathways. It is difficult to picture. It should be open: I would speculate this is why many channels C-type inactivate. Otherwise in the ER channels would be open and this wouldn't be good for Ca^{2+} in the ER, so the pore collapses and this process is reversed as soon as they reach the cell surface. This would be an elegant mechanism to prevent a huge K^+ leak in the ER.

hERG TRAFFICKING 73

Hondeghem: It could be like Na^+ channels, with two access pathways, a hydrophilic one (accessible when the channel is open) and a hydrophobic pathway, also accessible when the channel is closed. It may not be so fast, but this could mean that the channel doesn't need to be opened.

Gosling: What is the lowest concentration of astemizole at which you can see a rescue effect?

Ficker: The half-maximal rescue concentration is about 50 nM. It is 10 or 15 times higher than what is seen in blocking. This is why I would speculate that the configuration of the drug binding site is not quite right. There is plenty of evidence to show that minor changes make a huge difference for drug binding.

January: I would guess that there are multiple explanations. We take the view that this is a situation with a rescue drug binding domain and a drug block domain. Are they one and the same, or are they overlapping or separate? Our sense is that they are partially overlapping domains. What mediates block may not always mediate rescue. This is one of the ways to explain why the RC_{50} and IC_{50} are not the same. The other way is to say that the binding domain is not fully formed in an internalized protein that still hasn't come to the surface membrane. This is an interesting issue because most of us grew up thinking of drugs binding at cell surface membranes, and now we are having to think about how drugs are interacting in forming channels. It is daunting to try to think about what is going on with internalized proteins, but it is an interesting area.

Shah: What is the minimum exposure needed? You used chronic and 24 h exposure, but did you try 16 h, for example? If the half-life of hERG is 11 h, I would imagine that 5 h of chronic exposure might be enough.

Ficker: We haven't done this, but in the pulse chase experiments we see that just 1 h exposure impairs the protein that is made. It is an instantaneous effect.

Arcangeli: I am very interested in your data. Arsenic is used for many patients suffering from acute leukaemias. We can hypothesize that arsenic can block the tumour cell proliferation *in vivo* by blocking hERG. Arsenic blocks tumour cell proliferation *in vitro*, so it has a direct effect on leukaemias, and hERG is expressed in primary acute myeloid leukaemias. Indeed all paediatric promyelocitic leukaemias overexpress hERG. Paediatric leukaemias are different diseases as compared to adult leukaemias; nevertheless, when we treated *in vitro* primary paediatric promyelocitic leukaemias with retinoic acid, which is the main drug used for these types of leukaemia, there is a decrease in hERG expression, especially of the full length form. Could we hypothesize that some of the drugs that are used as anti-neoplastic drugs act by blocking hERG channels or altering hERG trafficking? The second question is: as you probably remember the hERG1b isoform is overexpressed in acute leukaemias, and forms heterotetramers with hERG1 in leukemia cells. The question is: can arsenic block maturation of the hERG1b, and when hERG1a and hERG1b coassemble, do they coassemble in the ER or just the plasma membrane?

Ficker: These are very good questions, but we haven't looked at this. I would speculate that if 1b uses the chaperone mechanism then it should be suppressed.

Gosling: It has to be something more than just hERG blockade otherwise we would already have the best anti-leukaemic agents, i.e. known potent hERG blockers such as astemizole. Arsenic must have additional effects other than solely down-regulating the levels of hERG channels in the cell membrane.

Ficker: It is very toxic. There is a whole range of potential targets.

Gosling: hERG might therefore just be an innocent bystander in this situation.

Arcangeli: hERG can be envisaged as a sort of 'adaptor protein' for many signalling mechanisms. We have data showing that after integrin engagement or cell adhesion to the extracellular matrix, the hERG1 protein co-precipitates with the $\beta 1$ integrin subunit. This might be suggestive of a macromolecular complex which could be somehow involved in integrin signalling.

References

Chiang CE, Luk HN, Wang TM, Ding PY 2002 Prolongation of cardiac repolarization by arsenic trioxide. Blood 100:2249–2252

Drolet B, Simard C, Roden DM 2004 Unusual effects of a QT-prolonging drug, arsenic trioxide, on cardiac potassium currents. Circulation 109:26–29

Dynamic control of hERG/I_{Kr} by PKA-mediated interactions with 14-3-3

Anna Kagan and Thomas V. McDonald[1]

Departments of Medicine (Cardiology) and Molecular Pharmacology, Albert Einstein College of Medicine, Forchheimer G35, 1300 Morris Park Avenue, Bronx, New York 10461, USA

Abstract. I_{Ks} has been considered the potassium current most responsible for adrenergic/cAMP-mediated changes in cardiac repolarization during stress. Increasing biochemical, electrophysiological and genetic evidence however, points to a role for hERG/I_{Kr} in β-adrenergic responses. Elevations of cAMP as seen in β-adrenergic stimulation can result in PKA-dependent phosphorylation of hERG and direct binding of cAMP to the channel protein. Generally, there is a suppression of current density due to the channel phosphorylation. We recently identified a novel protein–protein interaction between hERG and the adaptor protein 14-3-3ε. Interaction sites exist on both N- and C-termini of hERG and the interaction is dynamic, requiring phosphorylation of the channel by PKA. When both sites bind to 14-3-3 proteins there is an acceleration and augmentation of current activation in contrast to the depression of current with phosphorylation alone. When sufficient 14-3-3 is available the phosphorylation state of the channel is stabilized and prolonged. Thus, 14-3-3 interactions with hERG provide a unique mechanism for plasticity in the autonomic control of stress-dependent regulation of cardiac membrane excitability. Here, we summarize our findings and report on our further efforts to analyse interactions between the native channel protein and 14-3-3 in cardiac myocytes.

2005 The hERG cardiac potassium channel: structure, function, and long QT syndrome. Wiley, Chichester (Novartis Foundation Symposium 266) p 75–94

Adaptation of the cardiac rhythm in response to varying cardiovascular demands requires dynamic, beat-to-beat regulation of membrane excitability. The timing and rate of cardiac repolarization is governed by alterations of K^+ channels. A prominent mechanism for dynamically regulating channel proteins is through stimulation of G protein-coupled receptors (GPCR) such as the β-adrenergic

---

[1]This paper was presented at the symposium by Thomas V. McDonald to whom correspondence should be addressed.

receptor. Adrenergic stimulation via the sympathetic nervous system is a well-accepted contributor to the pathophysiology of hereditary and acquired arrhythmias. Increased adrenergic tone is a normal compensatory response to cardiovascular demands but also leads to maladaptive affects in acquired and hereditary heart diseases. Rhythm disturbances in the long QT syndrome (LQTS) are often stress-induced, indicating a link between increased adrenergic stimulation and cardiac ion channel activity. Since the identification of genetic subtypes of LQTS an interesting picture has evolved concerning the triggers that initiate ventricular tachyarrhythmia. LQT2 patients seem to be more susceptible to auditory or startle stimuli while LQT1 patients (KvLQT1) experience events more closely associated with physical exertion (Wilde et al 1999, Moss et al 1999, Paavonen et al 2001, Noda et al 2002, Takenaka et al 2003, Schwartz et al 2001).

The discovery of ion channels functioning within macromolecular complexes mediated by protein–protein interactions has been increasingly reported. These interactions are manifold and may involve channel accessory subunits, kinases, phosphatases, scaffolding proteins and cytoskeleton proteins. The functional effects of these interactions with ion channels may modify their expression, trafficking or channel activity (Abbott et al 1999, Marx et al 2000). With the growing evidence of a role for hERG/I_{Kr} in stress-related regulation of cardiac excitability we began searching for novel protein–protein interactions that might be involved in dynamic regulation of the channel within a macromolecular complex.

β-Adrenergic and cAMP regulation of hERG/I_{Kr}

Originally, I_{Kr} was reported as unresponsive to adrenergic stimulus while I_{Ks} was clearly augmented by manoeuvres that elevated cAMP (Sanguinetti et al 1991, 1995). Evidence from our laboratory (Cui et al 2000, 2001) and that of Kiehn's group (Kiehn et al 1998, Thomas et al 1999) supports a cAMP/PKA-mediated regulation of hERG using heterologous expression systems. Moreover, Terrar's group (Heath & Terrar 2000, Lei et al 2000) and Kiehn's group (Karle et al 2002) have also demonstrated clear changes in I_{Kr} behaviour in guinea pig cardiac myocytes in response to acute administration of isoproterenol. The effects on heterologously expressed hERG of cAMP elevation are somewhat complicated. The most prominent include a decrease in current density, an acceleration of deactivation and a variable depolarizing shift in voltage-dependence of activation (Thomas et al 1999, Cui et al 2000). The current density and deactivation effects of cAMP are mediated by PKA-dependent phosphorylation of the channel at four distinct PKA-consensus sites. One site is located in the cytoplasmic N-terminal portion of hERG (S283) and the other three are in the C-terminal tail (S890,

T895, and S1137). The voltage-dependence of activation however, is regulated by opposing effects of phosphorylation (rightward or depolarizing shift) and a direct cAMP binding to the channel's C-terminus (leftward or hyperpolarizing shift) (Cui et al 2000). The capacity for direct binding of cAMP to hERG is an absolute requirement for channel function and its effects are amplified when MinK or MiRP2 subunits are associated (Cui et al 2001).

Evidence for a macromolecular signalling complex for cAMP regulation of hERG/I_{Kr}

To investigate new protein–protein interactions with hERG we employed a yeast two-hybrid cloning screen (Kagan et al 2002). We screened a human adult heart cDNA library with portions of hERG as the 'bait'. Since we used the LexA cloning system that requires interaction to occur in the nucleus we selected portions of either the cytosolic N- or C-terminus as the 'bait' (Fig. 1A). After screening $>10^8$ independent clones with the N-terminus 'bait' we verified positive clones by back mating two strains of yeast containing individual 'bait' and 'prey' plasmids in selective growth media. The most common clone we identified encoded 14-3-3ε (Fu et al 2000). We have extensively characterized the hERG/14-3-3ε interaction.

14-3-3 proteins comprise a conserved family of scaffolding proteins that are encoded by seven genes in humans. At least one form of 14-3-3 is expressed in nearly all tissues, often in abundance. The epsilon isoform, 14-3-3ε, is highly expressed in skeletal and cardiac muscle. 14-3-3 proteins are ~ 30 kDa and exist as pairs with a hydrophobic surface that promotes dimerization and externally facing amphipathic grooves that interact with target proteins. The interaction grooves are oriented antiparallel relative to each other. The interaction groove recognizes phosphoserine proteins within the context of a recognition motif (RSxSPxP) (Muslin et al 1996). However, this motif has been expanded and numerous exceptions have been described (Wang et al 1999). Many of the recognition sites correspond to PKA consensus sites and cAMP-dependent phosphorylation is often the stimulus for 14-3-3 binding (Muslin et al 1996). A large and growing database of 14-3-3 interacting proteins (particularly signalling molecules) has been established, however the functional effects of the interaction have only been defined for a few (Fu et al 2000). The best understood interaction is with RAF-1 kinase where 14-3-3 binds the phosphorylated kinase in two positions to alter the enzymatic activity and its duration of action (Thorson et al 1998, Zhang et al 1997). Within the heart 14-3-3ε is the most abundant isoform, but 14-3-3's functional significance in the heart has not been determined.

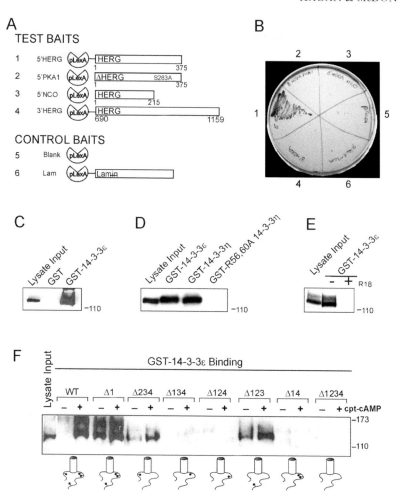

FIG. 1. Phosphorylation-dependent binding of hERG and 14-3-3 *in vitro*. (A) The schematic above shows the fusion cDNAs that we used to screen for hERG-interacting proteins in a yeast two-hybrid system. (B) Interaction of the N-terminus of hERG with 14-3-3 in yeast depends on intact S283. Numbers refer to 'bait' plasmids in A. (C) hERG is pulled down by GST-14-3-3ε but not GST. (D) hERG binds ε, η, isoforms of 14-3-3 but not mutant R56,60A 14-3-3η. (E) R18 peptide blocks 14-3-3 pull-down of hERG. (F) Immunoblot for hERG protein. The left lane shows immunoreactive hERG from whole cell lysate that was applied to GST-14-3-3 for pull down assay. The other lanes show that fraction pulled down by GST-14-3-3. The notation indicates the PKA consensus sites in hERG that have been mutated to alanine (Δ1 = S283A, Δ2 = S890A, Δ3 = T895A, Δ4 = S1137A; cartoons below show schematic of hERG PKA sites). Note the enrichment of the higher molecular weight form of hERG representing mature, glycosylated protein at the cell surface. Only those hERG mutants that have intact PKA sites 1 or 4 bind to 14-3-3. Pre-incubation with CPT-cAMP is indicated by either the − or + symbols.

hERG/14-3-3ε interactions: a cAMP/PKA-dependent physical association — a link between adrenergic stimuli and I_{Kr} channel function

Since the N-terminus of hERG that we used as bait contains one PKA consensus site (Cui et al 2000) we tested for yeast interaction with N-terminal portion of hERG with the first PKA site mutated to alanine and incapable of accepting a phosphate (hERGΔ1) or truncated to excise the PKA consensus site (5′NCO). These 'baits' failed to interact with 14-3-3ε in the yeast two-hybrid assay as did the C-terminus of hERG (Fig. 1B). We then confirmed these observations using recombinant 14-3-3ε–glutathione-S-transferase (GST) fusion proteins for *in vitro* GST pull-down experiments with hERG expressed in mammalian cultured cells. GST-14-3-3ε and GST-14-3-3η bound to hERG but the binding-deficient mutant GST-14-3-3η(R56,60A) failed to pull down hERG (Fig. 1C,D). The high affinity 14-3-3 blocking peptide, R18 (sequence: PHCVPRDLSWLDLEANMCL) (Wang et al 1999) prevented binding of hERG in the GST system (Fig. 1E). Since 14-3-3 frequently targets proteins at PKA phosphorylation sites and because yeast two-hybrid complementation showed that binding to hERG mapped to a region encompassing the PKA phosphorylation site, ^{280}RRApSSD^{285} (Thomas et al 1999, Cui et al 2000), we examined *in vitro* binding to various hERG mutants with altered PKA sites. Of the various combinations of PKA site mutants only those that lack both sites 1 and 4 (Δ1234, Δ124, Δ14) fail to bind 14-3-3 (Fig. 1F). Those mutants that contained either of the PKA sites 1 or 4, intact, (wild-type [WT], Δ1, Δ234, Δ123) were capable of binding 14-3-3. Thus, S283 and S1137 mediate 14-3-3 binding and do so independently of each other. The binding of hERG to 14-3-3 was enhanced by treatment of cells with CPT-cAMP prior to harvesting for hERG isolation indicating the phosphorylation-dependence of the interaction.

Functional effects of 14-3-3 association with hERG channel activity

GST pull-down experiment results were verified by co-immunoprecipitation (co-IP) of 14-3-3 with hERG that were co-transfected into HEK-293 cells (Fig. 2A). As in the *in vitro* pull-down, hERG with mutated PKA sites failed to co-IP with 14-3-3. When we co-transfected hERG with 14-3-3ε in CHO cells for voltage clamp analysis we observed accelerated kinetics of voltage-dependent activation and a hyperpolarizing (leftward) shift in the isochronal current voltage relationship (Fig. 2B–E). Consistent with the binding data, co-expression of 14-3-3ε had no functional effect on the hERG mutant Δ1234. The voltage dependence of activation (as measured by tail currents plotted against test potential) shifted by -11.12 ± 0.71 mV when 14-3-3 was co-expressed (Fig. 3A). Such a 14-3-3-dependent shift could potentially result in a current density increase of 20–45%

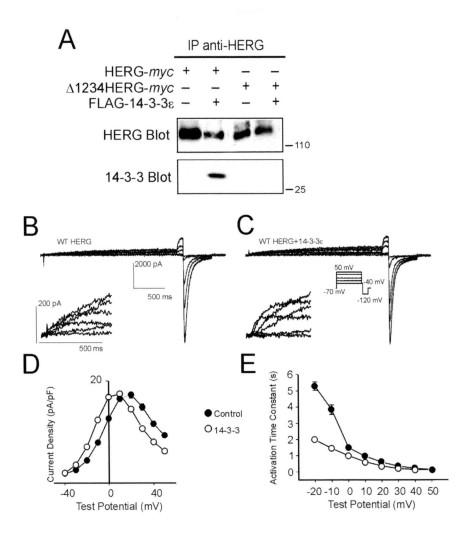

FIG. 2. Physical and functional interaction of hERG and 14-3-3 in intact cells. (A) Immunoblot of HERG from HEK293 cells transfected with either HERG-myc or Δ1234 HERG-myc, with or without FLAG-14-3-3ε. (B,C) Representative whole cell current tracings from cells expressing wild-type (WT) hERG (B) or co-expressing hERG and 14-3-3ε (WT+14-3-3ε) (C). Insets show expanded views of the early onset of outward current. (D) I-V curves for wild-type hERG with (open circles) or without (closed circles) overexpressed 14-3-3ε. (E) Time constants of onset of outward current shown for hERG with and without overexpressed 14-3-3ε.

during the plateau phase of the cardiac action potential. An analysis of the 14-3-3 dependent voltage shifts for each of the other PKA site mutants of hERG showed a pattern indicating that *both* sites 1 and 4 must be simultaneously present for the functional effect (Fig. 3B). Moreover, the 14-3-3-dependent shift in activation voltage for hERG channels with both sites 1 *and* 4 intact (wild-type and Δ23) was not the algebraic sum of effects seen in channels that had only one of either site 1 or 4 intact (Δ123, Δ234, Δ1). To examine the possibility that 14-3-3 may cross-bridge N- and C-termini, or the channel with other proteins, we employed the mutant 14-3-3η-R56,60A, a mutant that is dominant negative for cross-linking function of 14-3-3 (Thorson et al 1998). When wild-type 14-3-3ε and dominant mutant R56,60A-14-3-3η were co-expressed in CHO cells the hyperpolarizing shift in voltage dependence of hERG was abolished (Fig. 3C). Under these conditions wild-type 14-3-3ε dimerizes with mutant 14-3-3η-R56,60A to form a heterodimer with one half that can bind to hERG and another that cannot. Thus, the functional 14-3-3 mediated regulation of hERG channels requires that both binding sites within hERG are intact and that 14-3-3 be capable of forming cross-bridging dimers.

14-3-3 increases the extent of PKA-dependent phosphorylation of hERG

It has been postulated that upon binding, 14-3-3 may protect the phosphates at the binding sites of interacting proteins. Our first indication of such an activity with hERG came during attempts to phosphorylate the channel protein *in vitro* after co-expression with 14-3-3. *In vitro* phosphorylation of hERG with ^{32}P and purified PKA was greatly reduced ($\sim$ 90%) when 14-3-3ε is co-expressed in HEK 293 cells (Fig. 3D). This reduction in back-phosphorylation indicates fewer phosphate-accepting groups available on hERG due to an increase in the endogenous PKA-phosphorylated state of the protein (Bartel et al 1993, Hell et al 1995). As the balance of protein kinase and protein phosphatase activities determines the state of phosphorylation of hERG this finding suggests that 14-3-3 protects the phosphates on S283 and S1137 from the action of cellular phosphatases.

To further explore a phosphate protection function for 14-3-3 interaction with hERG we examined the effect of 14-3-3 overexpression on the protein phosphatase 1-dependent dephosphorylation of hERG. Dephosphorylation of hERG by purified protein phosphatase 1 is diminished when recombinant 14-3-3 is pre-incubated with hERG that had been phosphorylated with ^{32}P *in vitro*, by PKA (Fig. 3E). This protection is abolished in the presence of a 14-3-3 inhibitory peptide, R18. 14-3-3 overexpression with hERG also helped to shield cAMP-dependent ^{32}P-labelled hERG from subsequent application of protein phosphatase (Fig. 3F). A model where 14-3-3 protects and prolongs the

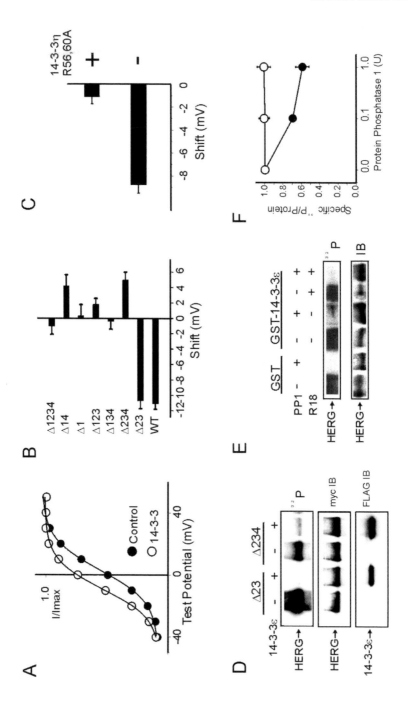

phosphorylated state of hERG is further supported by our finding that CPT-cAMP failed to significantly alter the voltage-dependence of activation or the current density from hERG when 14-3-3ε was co-expressed (basal $V_h = -8.37 \pm 0.58$ mV, CPT-cAMP stimulated $V_h = -9.52 \pm 0.12$ mV, $n = 10$).

ERG channel protein and 14-3-3 physically associate in myocardial tissue

Whenever a protein–protein interaction is detected by such methods as yeast two-hybrid screens, GST pull-downs or co-immunoprecipitations from transfected cells, one must return to the native tissue to confirm the existence of such an interaction. Knowing that both proteins co-exist in the heart and are capable of associating is not enough. The biological significance of the interaction must be determined by demonstrating physical association from native tissue by such methods as co-localization or co-immunoprecipitation from native tissue, and/or genetic evidence of functional interaction *in vivo*. The data so far (Kagan et al 2002) point to 14-3-3 as a modulator of adrenergic regulation of ERG/I_{Kr} by both altering the response to PKA-dependent phosphorylation and affecting the duration of channel phosphorylation. In our first attempts to demonstrate the possibility of an ERG/I_{Kr} interaction with 14-3-3 in the heart we probed fractionated homogenates from ventricular tissue with immobilized GST-14-3-3 and observed pull-down of ERG from membrane-rich fractions (Fig. 4A). Conversely, immobilized GST-hERG protein was capable of pulling-down 14-3-3 from canine and porcine ventricular homogenates (Fig. 4B). When detergent solubilized membrane-rich fractions of porcine ventricle were subjected to

---

FIG. 3. 14-3-3 regulation of hERG requires dimerization of 14-3-3 and binding to both N- and C-termini of the channel. (A) Normalized voltage dependent activation curves with (open circles) or without (closed circles) 14-3-3ε overexpression. (B) Summary of V_h shifts from cells expressing different PKA-mutants of hERG in the presence and absence of 14-3-3ε overexpression. Shifts are calculated as the difference in V_h between cells without 14-3-3 and with 14-3-3 overexpression. (C) V_h from cells expressing hERG and 14-3-3ε with or without R56,60A 14-3-3η (with +, without −). (D) 14-3-3 overexpression reduces capacity of PKA to *in vitro* back-phosphorylate hERG. Top panel, autoradiography signal from immunoprecipitated hERG phosphorylated with PKA and [^{32}P]ATP. Middle panel, anti-*myc* detection of total hERG protein from same gel. Bottom panel, 14-3-3 co-immunoprecipitated with hERG. (E) 14-3-3 partially shields the removal of ^{32}P-labelled of hERG by 10 U of protein phosphatase 1-α (PP1) *in vitro*. Top panel, autoradiography signal from immunoprecipitated hERG phosphorylated with PKA and [^{32}P]ATP. Bottom panel, anti-*myc* detection of total hERG protein. R18 was added at 1μM. (F) PP1-mediated dephosphorylation of intracellularly ^{32}P-labelled hERG with (open circles) or without (closed circles) overexpressed 14-3-3. Specific activity of [^{32}P]hERG was obtained as the ratio of the densitometry signal from autoradiography and anti-hERG immunoblot. Data were normalized to display relative changes with time.

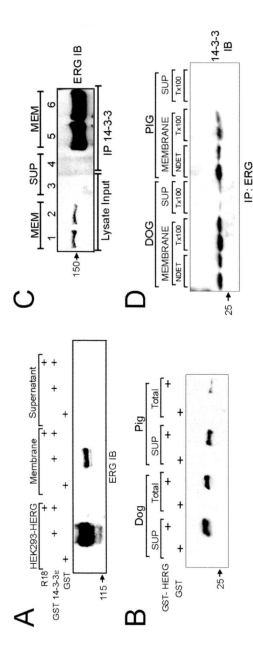

FIG. 4. Interaction of ERG and 14-3-3 in the heart. (A) Purified recombinant GST-14-3-3 on GSH agarose was incubated with detergent solubilized HERG-transfected HEK cells, porcine ventricular membrane fractions (Membrane), or ventricular cytosol/microsomal fractions (Supernatant) and analysed by SDS-PAGE-immunoblot for precipitation of ERG. ERG was precipitated by GST-14-3-3 from heart membranes and transfected cells. Negative controls: GST, and GST-14-3-3 plus the R18 peptide that blocks 14-3-3 interactions with target proteins. (B) Immobilized recombinant GST-hERG was *in vitro* phosphorylated with PKA and used to probe detergent-soluble porcine ventricular homogenate (Total) or cytosol/microsomal fraction of ventricular tissue (SUP). Precipitates were analysed by SDS-PAGE and blots were probed with anti-14-3-3 antibody. 14-3-3 was precipitated by GST-hERG but not by the control GST-GSH-Agarose. (C) Membrane-enriched fractions (MEM) and cytosol/microsomal fractions (SUP) were solubilized with either Triton X100 (lanes 1 & 5) or NP-40/DOC (lanes 2 & 6). 5% of the detergent soluble lysates were loaded into lanes 1, 2 & 3. The remaining lysate was incubated with anti-14-3-3 antibody and immunoprecipitates loaded onto lanes 4, 5 & 6. Anti-ERG immunoblot shows that ERG protein clearly co-precipitates with 14-3-3. (D) Membrane-enriched fractions (MEMBRANE) and cytosol-microsomal fractions (SUP) were solubilized with either TritonX-100 (Tx100) or NP-40/DOC (NDET). Lysates were incubated with anti-HERG antibody and immunoprecipitates were subjected to SDS-PAGE and Western blots probed with anti-14-3-3 antibody. Note that 14-3-3 co-precipitated with ERG protein from membrane-enriched fractions but very little precipitates from cytosolic fractions despite abundant 14-3-3 in the cytosol. Each condition shows two independent IPs.

immunoprecipitation of 14-3-3 we observed co-IP and enrichment of the ERG protein (Fig. 4C). Likewise, immunoprecipitation of ERG from both canine and porcine heart co-precipitated 14-3-3 (Fig. 4D). These results support an endogenous physical association of ERG channels with 14-3-3. The ability of GST fusion proteins to pull down proteins indicates that some portion of ERG is not in association under our experimental conditions, suggesting that the interaction is not constitutive but regulated.

Discussion

In summary, the data outlined above show that 14-3-3 dynamically associates with hERG in a phosphorylation-dependent manner at specific sites on both N- and C-termini of the channel. These binding sites coincide with consensus PKA phosphorylation sites. Only when both sites are capable of binding 14-3-3 and when 14-3-3 is capable of dimerizing is there a functional consequence of the interaction; a hyperpolarizing shift in activation and acceleration of channel opening kinetics. The physical association stabilizes the PKA-phosphorylated state of the channel by protection from cellular phosphatases. Lastly, the interaction between ERG/I_{Kr} and 14-3-3 is demonstrable in the heart.

The GST 14-3-3 pull-down experiments show an enrichment of the higher molecular weight form of hERG (Fig. 1F) that represents the mature glycosylated protein at the surface membrane (Zhou et al 1998). This may be interpreted as PKA preferentially targeting to the channel after it has trafficked to the plasma membrane, possibly via membrane-specific A-Kinase Anchoring Proteins (AKAPs) (Rubin 1994, Fraser & Scott 1999). AKAPs have been shown to enhance PKA regulation of another LQTS-related channel, KvLQT1 (Potet et al 2001, Marx et al 2002). The relative abundance of 14-3-3 near the plasma membrane therefore, may alter the probability of the hERG channel remaining phosphorylated. Another interpretation is that 14-3-3 alters the trafficking of the hERG protein to increase its presentation to the surface. 14-3-3 has been shown to be capable of altering surface expression of another K^+ channel, KCNK3 (O'Kelly et al 2002). However, we have not observed any obvious alteration in current density or change in total cellular hERG glycosylation with co-transfection. Subtle changes in protein processing and trafficking have not been excluded.

Given that 14-3-3 is capable of cross-bridging two different proteins or portions of a single protein and that in a single tetrameric hERG channel, there are eight potential 14-3-3 binding sites several models of the interaction are possible (Fig. 5). It may be that 14-3-3 dimers cause a different conformational change in the channel than monomers and that cross-bridging itself is not necessary. This seems unlikely since the dominant negative R56,60A-14-3-3η still dimerizes but prevents

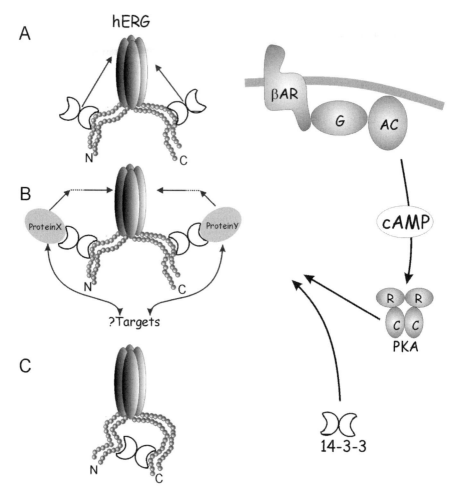

FIG. 5. Models of 14-3-3 regulation of hERG/I_{Kr}. (A) 14-3-3 acts as scaffold to approximate hERG with additional protein. (B) Binding of 14-3-3 alone has functional effects. (C) Cross-bridging of binding sites within the channel by 14-3-3 yields functional effect.

cross-linking (Fig. 3C). 14-3-3 may bring as yet, unidentified modifier proteins in contact with hERG to alter channel function (Fig. 5B). Such proteins would therefore be common to our expression system (CHO cells) and the heart and the degree of regulation would be governed by relative abundance and availability of 14-3-3 and the modifying proteins. It is possible that 14-3-3 cross-links N- and C-termini within a subunit or between subunits of the channel thereby altering channel behaviour (Fig. 5C). Each of these scenarios is possible and they are not mutually exclusive.

The most significant effect of 14-3-3 association with hERG is to alter the functional effects of cAMP-dependent phosphorylation of the channel. Phosphorylation alone decreases effective hERG/I_{Kr} by a depolarizing shift in activation and acceleration of deactivation (Thomas et al 1999, Cui et al 2000). If adequate 14-3-3 is available to interact with hERG then the PKA phosphorylation results in an increase in effective I_{Kr}. Competition by other cellular targets, therefore, may determine local concentrations of free 14-3-3 available for hERG binding (Tzivion et al 2001).

Given the different clinical profile of arrhythmia precipitants between LQT1 and LQT2 it is possible that the rate of increase in sympathetic stimulation determines which channel, I_{Ks} or I_{Kr}, is most altered. With exercise the increase in sympathetic stimulation is more slowly increased in conjunction with increased heart rate. This would lead to a greater role for I_{Ks} activation. With auditory or startle responses the sympathetic increase is much more rapid, often without concomitant increase in heart rate. It may be that hERG/I_{Kr} activation comes into play under these circumstances. Thus, a rapid rate of cAMP generation may have proarrhythmic effects on hERG/I_{Kr} unless adequate 14-3-3 is available. Moreover, the stabilization of phosphorylated hERG with 14-3-3 maintains the channel state even after the stimulus is removed. Thus 14-3-3, by dynamically associating with hERG may provide mechanism for plasticity in cardiac electrophysiological response to sudden stress.

We hypothesize that hERG/I_{Kr} exists in a macromolecular signalling complex that includes 14-3-3 and possibly AKAPs. The complex, by coordinating multiple pathways of modification of channel activity, dynamically regulates membrane excitability. Patients with either hereditary or acquired LQTS are at increased risk for fatal cardiac arrhythmias that are often triggered by acute cardiovascular demands such as stress. Therefore, mediation of increased PKA-dependent phosphorylation of hERG by 14-3-3 has important implications on the function of the channel and regulation of the cardiac action potential. Delineation of the autonomic regulation pathways of hERG/I_{Kr} is a challenge for the future but holds potential for development of new and targeted therapies.

References

Abbott GW, Sesti F, Splawski I et al 1999 MiRP1 forms I_{Kr} potassium channels with HERG and is associated with cardiac arrhythmia. Cell 97:175–187
Bartel S, Karczewski P, Krause EG 1993 Protein phosphorylation and cardiac function: cholinergic-adrenergic interaction. Cardiovasc Res 27:1948–1953
Cui J, Melman Y, Palma E, Fishman GI, McDonald TV 2000 Cyclic AMP regulates the HERG K^+ channel by dual pathways. Curr Biol 10:671–674
Cui J, Kagan A, Qin D et al 2001 Analysis of the cyclic nucleotide binding domain of the hERG potassium channel and interactions with KCNE2. J Biol Chem 276:17244–17251

Fraser ID, Scott JD 1999 Modulation of ion channels: a "current" view of AKAPs. Neuron 23:423–426
Fu H, Subramanian RR, Masters SC 2000 14-3-3 proteins: structure, function, and regulation. Annu Rev Pharmacol Toxicol 40:617–647
Heath BM, Terrar DA 2000 Protein kinase C enhances the rapidly activating delayed rectifier potassium current, IKr, through a reduction in C-type inactivation in guinea-pig ventricular myocytes. J Physiol 522:391–402
Hell JW, Yokoyama CT, Breeze LJ, Chavkin C, Catterall WA 1995 Phosphorylation of presynaptic and postsynaptic calcium channels by cAMP-dependent protein kinase in hippocampal neurons. EMBO J 14:3036–3044
Kagan A, Melman YF, Krumerman A, McDonald TV 2002 14-3-3 amplifies and prolongs adrenergic stimulation of HERG K$^+$ channel activity. EMBO J 21:1889–1898
Karle CA, Zitron E, Zhang W et al 2002 Rapid component I_{Kr} of the guinea-pig cardiac delayed rectifier K$^+$ current is inhibited by beta(1)-adrenoreceptor activation, via cAMP/protein kinase A-dependent pathways. Cardiovasc Res 53:355–362
Kiehn J, Karle C, Thomas D et al 1998 HERG potassium channel activation is shifted by phorbol esters via protein kinase A-dependent pathways. J Biol Chem 273:25285–25291
Lei M, Brown HF, Terrar DA 2000 Modulation of delayed rectifier potassium current, iK, by isoprenaline in rabbit isolated pacemaker cells. Exp Physiol 85:27–35
Marx SO, Reiken S, Hisamatsu Y et al 2000 PKA phosphorylation dissociates FKBP12.6 from the calcium release channel (ryanodine receptor): defective regulation in failing hearts. Cell 101:365–376
Marx SO, Kurokawa J, Reiken S et al 2002 Requirement of a macromolecular signaling complex for beta adrenergic receptor modulation of the KCNQ1-KCNE1 potassium channel. Science 295:496–499
Moss AJ, Robinson JL, Gessman L et al 1999 Comparison of clinical and genetic variables of cardiac events associated with loud noise versus swimming among subjects with the long QT syndrome. Am J Cardiol 84:876–879
Muslin AJ, Tanner JW, Allen PM, Shaw AS 1996 Interaction of 14-3-3 with signaling proteins is mediated by the recognition of phosphoserine. Cell 84:889–897
Noda T, Takaki H, Kurita T et al 2002 Gene-specific response of dynamic ventricular repolarization to sympathetic stimulation in LQT1, LQT2 and LQT3 forms of congenital long QT syndrome. Eur Heart J 23:975–983
O'Kelly I, Butler MH, Zilberberg N, Goldstein SA 2002 Forward transport. 14-3-3 binding overcomes retention in endoplasmic reticulum by dibasic signals. Cell 111:577–588
Paavonen KJ, Swan H, Piippo K et al 2001 Response of the QT interval to mental and physical stress in types LQT1 and LQT2 of the long QT syndrome. Heart 86:39–44
Potet F, Scott JD, Mohammad-Panah R, Escande D, Baro I 2001 AKAP proteins anchor cAMP-dependent protein kinase to KvLQT1/IsK channel complex. Am J Physiol Heart Circ Physiol 280:H2038–H2045
Rubin CS 1994 A kinase anchor proteins and the intracellular targeting of signals carried by cyclic AMP. Biochim Biophys Acta 1224:467–479
Sanguinetti MC, Jurkiewicz NK, Scott A, Siegl KS 1991 Isoproterenol antagonizes prolongation of refractory period by class III antiarrhythmic agent E-4031 in guinea pig myocytes. Circ Res 69:77–84
Sanguinetti MC, Jiang C, Curran ME, Keating MT 1995 A mechanistic link between an inherited and an acquired cardiac arrhythmia: *HERG* encodes the I_{Kr} potassium channel. Cell 81:299–307
Schwartz PJ, Priori SG, Spazzolini C et al 2001 Genotype-phenotype correlation in the long-QT syndrome: gene-specific triggers for life-threatening arrhythmias. Circulation 103:89–95

Takenaka K, Ai T, Shimizu W et al 2003 Exercise stress test amplifies genotype-phenotype correlation in the LQT1 and LQT2 forms of the long-QT syndrome. Circulation 107:838–844
Thomas D, Zhang W, Karle CA et al 1999 Deletion of protein kinase A phosphorylation sites in the HERG potassium channel inhibits activation shift by protein kinase A. J Biol Chem 274:27457–27462
Thorson JA, Yu LW, Hsu AL et al 1998 14-3-3 proteins are required for maintenance of Raf-1 phosphorylation and kinase activity. Mol Cell Biol 18:5229–5238
Tzivion G, Shen YH, Zhu J 2001 14-3-3 proteins: bringing new definitions to scaffolding. Oncogene 20:6331–6338
Wang B, Yang H, Liu YC et al 1999 Isolation of high-affinity peptide antagonists of 14-3-3 proteins by phage display. Biochemistry 38:12499–12504
Wilde AA, Jongbloed RJ, Doevendans PA et al 1999 Auditory stimuli as a trigger for arrhythmic events differentiate HERG-related (LQTS2) patients from KVLQT1-related patients (LQTS1). J Am Coll Cardiol 33:327–332
Zhang L, Wang H, Liu D, Liddington R, Fu H 1997 Raf-1 kinase and exoenzyme S interact with 14-3-3zeta through a common site involving lysine 49. J Biol Chem 272:13717–13724
Zhou Z, Gong Q, Epstein ML, January CT 1998 HERG channel dysfunction in human long QT syndrome. J Biol Chem 273:21061–21066

DISCUSSION

Schwartz: I was intrigued by your comment about triggers for activity. Whereas it is relatively easy to understand why I_{Ks} mutations lead to certain types of adrenergic-mediated arrhythmias, it has always been more puzzling for LQT2 and LQT3. I don't think that what you say in terms of the time course for parasympathetic inhibition really fits. You mentioned 30 or 40 s. In fact, the typical situation is that there is a startle and they collapse. It doesn't take 30 s. The implication is that you have a sudden release of noradrenaline because of the startle phenomenon while the heart rate is still relatively low because of the maintenance of parasympathetic activity. There is a chronic vagal activity and on top of that there is this sudden superimposition. I don't think that one needs to invoke changes in vagal activation. What really matters is this sudden startle. The situation is not totally different in the LQT3 patients. They do not respond so much to startle, but they tend to die during sleep. We know that sleep is not a fixed type of situation. There is REM sleep, which exhibits bursts of sympathetic and vagal activity. This is another situation with sudden sympathetic activation on a background of relatively low heart rate.

McDonald: I agree with you, and I probably mis-stated myself. What I meant to say is that later on we get some parasympathetic activity. The events that occur at that early point were just as you said. There is a rapid increase in sympathetic activity without the concomitant decrease in parasympathetic activity. This is much more graded and parallels with exercise. It is this sort of mismatch compared with exercise where hERG may prove important. Even as the sympathetic dies down, the effect can still last. As we have shown the 14-3-3 can maintain those phosphates even after the stimulus has gone.

Terrar: Did you see any change in the C-type inactivation?

McDonald: No. We looked very carefully for changes in deactivation. On average, there was no change.

Terrar: We (Heath & Terrar 2000) did see a change in C-type inactivation after activation of protein kinase A, but we only saw this in experiments using perforated patch electrodes or old-fashioned 'sharp' electrodes. We didn't see it with conventional whole-cell patch electrodes.

Robertson: What is the conclusion?

Terrar: We did these experiments with forskolin and isoprenaline to activate protein kinase A. Our conclusion was that indirect pathways were involved because if we did anything to stop Ca^{2+} changing in the cytosol, such as using nifedipine in the bathing solution or loading the cells with BAPTA, this prevented the action. It seemed to us that protein kinase C was involved, because PKC inhibitors did suppress the action. But there may be other downstream kinases and we don't know whether the effects are directly on the ERG channel or on accessory proteins. The observation is simply that the I_{Kr} current gets bigger and shows less C-type inactivation (after isoprenaline or forskolin).

Robertson: Is there any effect on deactivation?

Terrar: We didn't look closely at deactivation in these particular experiments (but see Connors & Terrar 1990, Lei et al 2000).

Robertson: Tom McDonald, didn't you publish that phosphorylation by PKA increases the deactivation rate?

McDonald: Yes, it does if you phosphorylate hERG in the absence of extra 14-3-3. We did not see a difference in deactivation between cells expressing hERG alone versus cells expressing hERG plus 14-3-3 at baseline. When we overexpress 14-3-3 and then try to stimulate cells with cAMP, we see very little change. We interpret this result as the channel being already phosphorylated, and the deactivation changes little with applied cAMP.

Robertson: This is what I was confused about. If the PKA phosphorylates those sites and increases the deactivation rate, this will reduce the current amplitude. Now you say 14-3-3 protects those phosphorylated sites, which suggests to me that this will stabilize that lower current amplitude.

McDonald: That is right. My point is that 14-3-3 is not only stabilizing those phosphorylation sites, it is also reinterpreting the effect of phosphorylation. Thus, there is a steady-state decrease in current amplitude due to association of hERG with 14-3-3. However, you get a left shift in the voltage dependence of activation, which is almost the opposite of what we observe with PKA-dependent phosphorylation alone. Thus, our hypothesis is that in some situations there may not be enough 14-3-3 available to associate with hERG due to the many targets for 14-3-3 in the cell. If PKA begins phosphorylating many targets there will be a decrease in available free 14-3-3. We don't know what changes the expression of 14-3-3. It might change with disease or in parallel with expression changes of the channel. But if you don't have enough free 14-3-3 we suspect that beta adrenergic stimulation would actively decrease hERG current through dynamic alterations of current density, voltage-dependence of activation and deactivation. If there is enough 14-3-3 available there won't be such acute changes in current. To the contrary, you are going to increase the amount of current above baseline because of the voltage shifts in activation.

Robertson: I noticed in your Western blots that you had differing relative levels of mature versus immature pulled down by 14-3-3 for different mutants. Is this a consequence of the phosphorylation, or is it perhaps identifying different species of hERG that normally interact with different species of 14-3-3 at different stages of trafficking.

McDonald: No matter what phosphorylation mutant form of hERG we use, if it binds to 14-3-3 it seems to enrich for the mature glycosylated form when compared to the relative abundance of the input. These were transiently transfected cells. In this case we get tremendous amounts of immature glycosylated compared with the fully glycosylated form. In comparing one mutant to another, there didn't seem to be a systematic difference in enrichment for the glycosylation among the different PKA mutants. We have looked in other studies to see whether PKA is changing trafficking, and we think that sites 1 and 4 are not involved in changes in trafficking, nor do we think that 14-3-3 is involved in changes in trafficking. We suspect that the fraction of hERG that is at the surface is more likely to be phosphorylated at sites 1 and 4. Perhaps there is an AKAP there, and perhaps there is more PKA available at that surface compared to the immediate compartments the channel traverses before it reaches the plasma membrane. Right now the working hypothesis that we are testing is whether or not there is a preferential phosphorylation of the surface pool of channels. Phosphorylation of sites 2 and 3 makes no difference to the ability of 14-3-3 to associate.

Robertson: Dierk Thomas, do you also see no changes in trafficking with phosphorylation?

Thomas: So far, we haven't had the time to investigate possible hERG trafficking changes that may be induced by different phosphorylation states of the channel protein in detail. However, if we look at our results obtained from the mutant hERG channels we generated, we find that removal of PKA-dependent phosphorylation sites does not inhibit correct hERG protein trafficking, since those mutant channels conduct hERG currents that are largely similar to wild-type hERG. Moreover, removal of PKC-dependent phosphorylation sites also does not prevent functional channel expression, with one notable exception: one out of 18 PKC-dependent phosphorylation sites at position T74 may be required for hERG processing, since we could not record hERG currents from channel mutants after replacement of the threonine residue by alanine. Whether this observation is caused by impaired trafficking due to the lack of PKC-dependent phosphorylation as this position will require further experiments.

Sanguinetti: Can someone summarize the effect of PKA- versus PKC-mediated modulation of I_{Kr} and or hERG? What is the net response if you stimulate both? I am confused about this.

Thomas: We know that PKC phosphorylation of hERG channels causes a marked shift of approximately 37 mV towards more positive potentials. In addition, PKA-dependent hERG phosphorylation causes a positive shift of 14 mV. Moreover, there is also a negative shift caused by cAMP binding, which lies around −7 mV. Now there is also 14-3-3 binding which seems to counteract the PKA phosphorylation. Thus far, we have seen several different signal transduction cascades contributing to a complex regulatory system of hERG activation, and I am sure that there are still several discoveries to come in the future. At this point, the net effect of PKC- and PKA-dependent hERG phosphorylation is not clear. What confuses us even more is the fact that apparently activation of PKC and subsequent hERG modulation requires PKA, but is independent of PKA-dependent hERG phosphorylation. We are planning to investigate the effects of α- and β-adrenergic stimulation on I_{Kr} in native guinea pig cardiomyocytes in the near future. This should provide additional insights into adrenergic regulation of hERG.

Tom, have you ever compared the half maximal activation voltages after PKA phosphorylation and after 14-3-3 binding? Does PKC-dependent phosphorylation affect the 14-3-3-effect, and is the activation shift moved towards more positive or more negative potentials?

McDonald: When we take the cells with hERG and 14-3-3 coexpressed and try to stimulate them with cAMP, we no longer see any additional shift in voltage-dependence of activation. From our back-phosphorylation experiments we think that this is because the sites are already phosphorylated and they are kept in that phosphorylated state by the 14-3-3 under basal conditions. We lose any additional effect of PKA because the phosphates are already there. We haven't added PKC stimulation in this situation. PKC is going to phosphorylate different sites, which have all been mutated by your group. You would say that the PKC regulation of hERG is not due to direct phosphorylation of the channel: is that correct?

Thomas: Right. Our mutagenesis studies show that removal of 17 out of 18 PKC-dependent phosphorylation sites in hERG does not influence the response to PKC-activation or α-adrenergic activation. However, you have to keep in mind that functional channel expression was prevented by mutation of the 18th phosphorylation site, T74. Thus, although it appears to be rather unlikely, this site might be the one responsible for PKC-dependent hERG regulation. We continue investigating the significance of the T74 site at the moment by performing additional mutagenesis experiments.

McDonald: hERG is a sticky channel not just for drugs but for other proteins. This is just one protein that we have worked out. There are others from the yeast two-hybrid that we haven't examined as closely.

Robertson: You have 14-3-3 binding, you add PKA and you don't get a shift in V_{1max}.

McDonald: There is no further shift.

Robertson: What about cAMP binding to the cyclic nucleotide binding domain: shouldn't you be getting a positive shift due to that?

McDonald: No. If we take hERG that can no longer be phosphorylated by mutagenesis, don't add extra 14-3-3 but add cAMP, we get a left shift.

Robertson: Whether 14-3-3 is there or not, you would get cAMP acting at the cyclic nucleotide binding site and a positive shift.

McDonald: There was no further shift due to cAMP when 14-3-3 was co-expressed with hERG. Thus we think that 14-3-3 binding to the channel overrides the direct effects of cAMP.

Robertson: Can we explore this a bit further? In the experiment you did to show binding of cyclic nucleotides to hERG, you labelled the cyclic nucleotide and you showed it binding to hERG.

McDonald: We used two methods to do this. One was using tritiated cAMP which we ran against immunoprecipitated hERG, and then we did the appropriate controls. We did the same with tritiated cGMP and saw no significant binding. We also took a photo-activatable [^{32}P]cAMP which with ultraviolet light exposure will covalently bind to whatever protein it is binding. Both those methods demonstrated specific binding.

Robertson: Do those experiments indicate for sure that cyclic nucleotide is binding either to hERG or to a protein that is binding to hERG? Can you rule out binding to an intermediate? You are immunoprecipitating in both cases. You are pulling down a complex of proteins. For the benefit of those who don't normally follow this, the reason this is an issue is that the hERG cyclic nucleotide binding domain doesn't have a good homology to other cyclic nucleotide binding domains that are known to bind. People who study this see a lack of key residues. It has been a big issue for a long time as to whether hERG is like cyclic nucleotide gated channels. This is why I wonder whether it is a direct interaction.

McDonald: I understand your concern, and this is a caveat to our result: we can't completely rule out that there isn't an intermediate. Maybe we are bringing down PKA itself. In our 2001 paper, however, we mutated the arginine at position 823 which is in the segment with the highest homology to nucleotide binding domains. These mutants had reduced tritiated cAMP binding as did a mutant in which we excised the segment entirely. While the same caveats exist we believe this more strongly supports direct hERG–cAMP interactions.

Robertson: It is a tough experiment to do.

McDonald: With recombinant proteins and GST it may be possible, because there is no protein kinase A in *E. coli*.

January: Are there other 14-3-3 isoforms in heart?

McDonald: Epsilon is by far and away the most common. Others may be expressed in lower abundance.

January: Do the various 14-3-3s all have the same affinities?

McDonald: They are very close: within 100-fold of each other for the same targets. They are a little promiscuous, too, in that they can heterodimerize. β can go with ε, for example.

References

Connors SP, Terrar DA 1990 The effect of forskolin on activation and de-activation of time-dependent potassium current in ventricular cells isolated from guinea-pig heart. J Physiol 429:109

Heath BM, Terrar DA 2000 Protein kinase C enhances the rapidly activating delayed rectifier potassium current, I_{Kr}, through a reduction in C-type inactivation in guinea-pig ventricular myocytes. J Physiol 522:391–402

Lei M, Brown HF, Terrar DA 2000 Modulation of delayed rectifier potassium current, iK, by isoprenaline in rabbit isolated pacemaker cells. Exp Physiol 85:27–35

General discussion II

January: I'd like to mention some work we have done on a new area. There is a long history to this: It goes back to class III anti-arrhythmic drug development days when we were involved in clinical trials of MK499. We were involved in Merck's dose ranging to identify the upper dose ranges for human administration of MK499. It was clearly a QT-prolonging drug: this was evident on electrocardiograms (ECGs). Around that time reports were appearing of I_{Kr} being rather small in many isolated cell systems, something which puzzled me. Recently, we set out to look at why this was the case. We were all trained to think about macroscopic currents in cells in terms of biophysical and biochemical properties of channels, gating and phosphorylation and so on, but obviously there is the biogenic side of channels that will regulate expression. This led us into the issue of extracellular enzymes and what they do to ion channel proteins. We opted to look at what various enzymes do to hERG channels heterologously expressed in cell lines. The enzymes we have studied are protease 14, protease 24, proteinase K (which has not been used much in the heart but has been used in other cell systems), collagenase type 2, crude hyaluronidase and trypsin. We perfuse these just like drugs. When we put in protease 24 (280 µg/ml), the hERG current disappears within 15–20 minutes without killing cells; the holding and leak currents are stable. At 37 °C, when we put protease 24 on the current is abolished in minutes. Thus, this is a rapid effect. Similar findings are seen with protease 14 and proteinase K. Collagenase and hyaluronidase have no effect. Trypsin has a minimal effect. Certain proteases will therefore abolish hERG current. If we look at this in Western blot and treat cells with protease 24 or 14, or proteinase K, the upper or mature protein band is abolished and we see the appearance of degradation products. The upper band reappears if we return the cells to culture, as does the hERG current which comes back to starting levels. In heterologous expression, KCNQ1 channels are not affected by protease 24, nor is KCNJ2. hH1a also is unaffected, as is a reconstituted L-type Ca^{2+} channel conducting barium current. In native myocytes isolated with collagenase only and held at -50 mV. We can record I_{K1}, and with depolarization we get activation of delayed rectifier currents, I_{Kr} and I_{Ks}. The tail current is diminished with enzymes, and we get the same pattern when we put E4031 on the myocytes. If we look at the rate of disappearance of tail current with enzymes and the rate of block of E4031, they are almost the same. This is work in progress, which started with bacterial and fungal derived-enzymes. We are now asking structural and human disease questions about yet another potential regulatory mechanism for hERG. This has been an interesting project.

Sanguinetti: One of the reasons I wanted you to talk about this is the issue of why I_{Kr} is different from cell to cell and lab to lab. One of the differences might be how much protease is used to isolate cells. Do you know anything about how much protease you have in there versus the contaminants in a typical lot of collagenase?

January: We study the most typical enzyme preparations. We have not looked yet at purity; there are cleaner preparations. We are going to try to test them.

Sanguinetti: I meant how much protease you are using versus the contaminants in a collagenase.

January: We have not assayed them to sort them out.

Brown: Is there a state-independent component of this effect? Do you have some sense of how much might be gating dependent?

January: If we follow the same protocol and hold continuously at -80 mV, wait five minutes and start pulsing, there is a five minute gap in the decay curve but otherwise it is unchanged. I don't think this is state dependent. There is no reason to think that these enzymes are getting into the cells: they are large proteins that should stay extracellular.

Brown: I was just wondering whether bits of the protein are popping out?

January: We have just done one set of experiments to ask whether we could see any evidence of state dependence.

Abbott: Do you see a breakdown product when you use trypsin? Or do you think it might just be an effect in which the channel protein reaches the plasma membrane intact but then trypsin clips one of the extracellular linkers?

January: If you look at the degradation band it is about half the molecule. Our antibody is recognizing the C-terminus and the clipping appears to occur in extracellular linkers. So one or more of the extracellular linkers of hERG seems to be sensitive. We have started looking at the homology with extracellular linkers of other ion channels. There is a lot of variability. These are not very specific enzymes. We focused on protease 24 because there is more known about its substrate specificity. These enzymes clip at multiple bonds and multiple amino acids, so it will take a lot of work to identify where the clipping has occurred.

Tseng: You say that I_{Ks} is not sensitive to the enzymatic digestion. A few years ago Stan Nattel said that the I_{Ks} is very sensitive to enzymatic digestion (Yue et al 1995).

January: In 1996, Nattel's group said that it was I_K that was suppressed depending on how they isolated the atrial myocytes. They had not sorted out whether it was I_{Ks} or I_{Kr}.

Mitcheson: Do you have any reason to suspect that this might happen pathophysiologically, such as during ischaemia? Is it possible that the breakdown products of cells could cause this phenomenon?

January: That is what we are interested in. Right now we have a laboratory phenomenon, but it might be a pathological mechanism. We don't have any data.

Hancox: In addition to there being quite a lot of variation in I_{Kr}, think about the situation for rabbit: for years, no one saw any I_{Ks} at all in rabbit ventricular myocytes, and then papers came out showing I_{Ks} (e.g. Salata et al 1996). This is clearly very susceptible to the conditions in different laboratories. When I worked on rabbit I didn't see much I_{Ks}, but it could be clearly seen in some papers (Salata et al 1996, Cheng et al 1999). So, I would have thought that I_{Ks} would have been particularly susceptible to being interfered with by something experimental, whether it is the enzyme, phosphorylation or something else.

January: If we make a matrix with a large number of enzymes to test and a large number of channels to test, it becomes an impossibly large number of combinations. So I don't know the answer.

Tseng: I have something to add about the pathological implications of your observations. We have done Western blot experiments on dog hearts, comparing control dogs with dogs with myocardial infarction. We probed the protein with anti-hERG antibody. In addition to the 150 kDa band as we also saw in control dogs, in dogs with myocardial infarction we saw a stronger band at 94 kDa. Perhaps this reflected more degradation of dog ERG protein by the enzyme in myocardial infarction.

Robertson: Typical degradation products are in the 65–70 kDa range.

January: Peter Schwartz could answer this better than me, but when people have myocardial infarctions, among the changes that occur in their ECGs is a degree of QT interval prolongation.

Schwartz: This is true for a certain number of people. In 1978 in *Circulation*, with Stewart Wolf, we showed that among post MI patients, by the degree of QT prolongation we could identify a subset at a much higher risk of sudden death (Schwartz & Wolf 1978). We used these data for risk stratification but, as the data were actually collected in the 1960s, we never went any further in terms of looking at mechanisms.

Wanke: I would like to describe some work we have done on investigating whether there are scorpion toxins more specific for hERGs 1, 2 and 3. We have shown that the species from a particular genus of Mexican scorpion have at least 1, 2 or 3 toxins specific for hERG1 (Gurrola et al 1999, Corona et al 2002). It had been shown (Saganich et al 2001) that in the brain most regions co-express the three types of hERG channel. The idea we had was to use hERG toxin 1 for the three types of channels. The results show that hERG1 is blocked differently at different holding potentials. At negative holding potentials the block is effective, but at depolarized potentials the block is mostly ineffective. This toxin has no effect on hERG2. Some of the amino acids that are important for the binding of hERG toxin to hERG channel differ among the three hERG channels. For hERG3 it is possible to show that if you apply the toxin there is a blocking effect, but practically there is no possibility of recovering from the block. If you apply the toxin from -70 mV it is practically not reversible. The block is less voltage dependent because it also works from a holding potential of -30 mV. There are also some differences in the IC_{50} of the blocking. It is about 48 ng/ml instead of 94 for hERG1. In conclusion, if one has a preparations of neurons from the CNS, by applying hERG1 and washing one can see the amount of hERG1 current. The current that disappears but not reappears is probably the current which originates from hERG3 channels. Applying high E-4031 concentrations or other very strong blockers one can remove all three channels. By using the combination of these observations in principle it is possible to dissect one channel from the other. We are trying to dissect the effects of the toxins using heteromultimers of hERGs 1–3 and see what happens.

Sanguinetti: Is there any difference between hERGs 1–3 with organic blockers? Is there any drug showing specificity that can't be explained by gating differences?

Wanke: No, this has been shown by Shi et al (1997).

Gosling: The slopes of your curves look profoundly different. Against hERG3 it looked considerably steeper than it did for hERG1. This has implications for the number of toxin molecules interacting with the channel. Is there a difference in stoichiometry?

Wanke: Possibly. We have to check this with heteromultimeric expression.

Brown: How pure is this toxin?

Wanke: It is pure. It is a peptide of 43 amino acids. It is also quite stable.

Hancox: Is the difference at +30 mV because of a different state-dependent blocking mechanism?

Wanke: Yes, in hERG3 there is a clear difference because the binding of the toxin is strongly voltage-dependent.

Witchel: We were able to get rid of a lot of that residual current that won't block by using very short protocols. We felt it was mostly either a closed channel block or something like that. By holding the cell at -80 or -100 mV and then just stepping for just 10 ms, we were able to get our block percentage above 95.

Tseng: How did you measure hERG current with a 10 ms pulse?

Witchel: It was small.

Hancox: If you pulse at 37 °C and +40 mV for 10 ms you can get a tail current.

References

Cheng J, Kamiya K, Liu W, Tsuji Y, Toyama J, Kodama I 1999 Heterogeneous distribution of the two components of delayed rectifier K^+ current: a potential mechanism of the proarrhythmic effects of methanesulfonanilideclass III agents. Cardiovasc Res 43:135–147

Corona M, Gurrola GB, Merino E et al 2002 A large number of novel Ergtoxin-like genes and ERG K^+-channels blocking peptides from scorpions of the genus Centruroides. FEBS Lett 532:121–126

Gurrola GB, Rosati B, Rocchetti M et al 1999 A toxin to nervous, cardiac, and endocrine ERG K^+ channels isolated from *Centruroides noxius* scorpion venom. FASEB J 13:953–962

Saganich MJ, Machado E, Rudy B 2001 Differential expression of genes encoding subthreshold-operating voltage-gated K^+ channels in brain. J Neurosci 21:4609–4624

Salata JJ, Jurkiewicz NK, Jow B et al 1996 I_K of rabbit ventricle is composed of two currents: evidence for I_{Ks}. Am J Physiol 271:H2477–2489

Schwartz PJ, Wolf S 1978 QT interval prolongation as predictor of sudden death in patients with myocardial infarction. Circulation 57:1074–1077

Shi W, Wymore RS, Wang HS et al 1997 Identification of two nervous system-specific members of the erg potassium channel gene family. J Neurosci 17:9423–9432

Yue L, Feng J, Li G-R, Nattel S 1995 Characterization of repolarizing currents in canine atrial myocytes: properties of currents and role of cell isolation methods. Circulation 92:I-753

Does hERG coassemble with a β subunit? Evidence for roles of MinK and MiRP1

Arun Anantharam*† and Geoffrey W. Abbott*‡[1]

*Division of Cardiology, Department of Medicine, †Graduate Program in Neuroscience, ‡Department of Pharmacology, Weill Medical College of Cornell University, 520 East 70th Street, New York, NY 10021, USA

Abstract. The voltage-gated potassium channel formed by hERG pore-forming alpha subunits generates the I_{Kr} cardiac potassium current, and is considered essential for human ventricular repolarization. What is not certain is whether human I_{Kr} channels contain ancillary subunits *in vivo*. Two chief contenders for this role are MinK (encoded by *KCNE1*) and MiRP1 (*KCNE2*). MinK and MiRP1 are single transmembrane domain peptides that can co-assemble with hERG in heterologous systems. MinK increases hERG currents by an unknown mechanism. MiRP1 alters hERG current density and gating, although no consensus has been reached as to the precise extent of these effects. Here we discuss key aspects of the debate surrounding the potential roles of MinK and MiRP1 in I_{Kr}: inconsistencies between reports of the effects of MiRP1 on hERG *in vitro*; association with long QT syndrome of inherited mutations in MinK and MiRP1; and a role for MiRP1 polymorphisms in acquired arrhythmia despite the apparent inability of MiRP1 to impinge upon the unique inner vestibule drug-binding site that dominates hERG pharmacology.

2005 The hERG cardiac potassium channel: structure, function, and long QT syndrome. Wiley, Chichester (Novartis Foundation Symposium 266) p 100–117

Ion channels provide the basis for cellular electrical excitability, facilitating rapid, coordinated cellular action across excitable tissues such as the nervous system and the heart. The cardiac action potential requires the concerted action of a range of ion channels. Sensing the depolarization caused by sodium and calcium ion influx, voltage-gated potassium (Kv) channels open to allow potassium ions out, repolarizing the myocardium and ending each heart-beat.

---

[1]This paper was presented at the symposium by Geoffrey W. Abbott to whom correspondence should be addressed.

Although Kv α subunits possess all the elements necessary for voltage-sensing, gating and ion selectivity, some native potassium currents are formed by a combination of α subunits and other ancillary or β subunits. Ancillary subunits do not form channels alone, but can associate with Kv α subunits to alter channel function. One class of ancillary subunits, encoded by *KCNE* genes, is characterized by a single membrane span and a noted promiscuity of interaction with α subunit partners from various subfamilies in heterologous expression systems (Abbott et al 1999, 2001, Barhanin et al 1996, Sanguinetti et al 1996, Takumi et al 1988). Here, we discuss the current evidence for and against the contribution of two *KCNE* peptides—MinK (*KCNE1*) and MiRP1 (*KCNE2*)—to the cardiac I_{Kr} current, essential for human ventricular repolarization.

MinK is essential for I_{Ks}, but what about I_{Kr}?

MinK coassembles with hERG, at least in vitro

In human ventricular myocytes, two Kv currents provide the major repolarizing force to end the cardiac action potential: I_{Ks} and I_{Kr}. The molecular correlate of I_{Ks} is widely accepted to be a Kv channel formed from four KCNQ1 α subunits and two MinK β subunits (Barhanin et al 1996, Sanguinetti et al 1996, Chen et al 2003a) (Fig. 1). MinK, discovered by expression cloning from size-fractionated rat kidney mRNA, converts the delayed rectifier channel formed by KCNQ1 α subunits alone to a slowly activating channel with fourfold higher unitary conductance, and gating and pharmacology characteristic of cardiac I_{Ks} (Barhanin et al 1996, Sanguinetti et al 1996, Takumi et al 1988). Inherited loss-of-function mutations in both KCNQ1 and MinK associate with long QT syndrome (LQTS), a form of cardiac arrhythmia caused by inefficient myocyte repolarization (Splawski et al 1997, Wang et al 1996). Thus, MinK and KCNQ1 are considered obligate partners in the I_{Ks} channel.

Reduction of the I_{Kr} current in murine AT-1 (atrial tumour) cells by MinK antisense suppression suggested MinK might also regulate the molecular correlate of I_{Kr}—the ERG α subunit (Yang et al 1995). A heterologous expression study followed, showing that MinK up-regulates hERG (the human ERG α subunit) current twofold (McDonald et al 1997). In contrast to effects on KCNQ1, MinK does not alter hERG unitary conductance or gating kinetics, but increases current density by an unknown mechanism that does not increase hERG protein density or half-life at the plasma membrane. These data introduced two important aspects of *KCNE* subunit physiology: promiscuity of partnering and versatility of function (i.e. profoundly different effects depending on which α subunit is modulated).

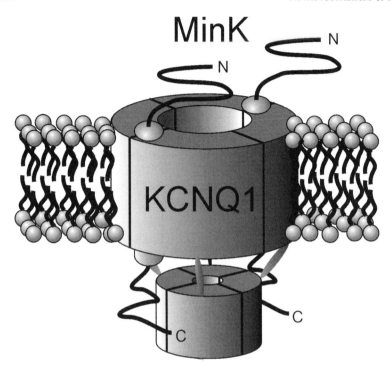

FIG. 1. Cartoon of a MinK–KCNQ1 complex. Representation of a MinK–KCNQ1 complex showing a tetramer of KCNQ1 α subunits, and two MinK subunits (N- and C-termini labelled) embedded opposite one another between α-α subunit borders. Positioning is speculative; stoichiometry is as determined by Chen and colleagues (Chen et al 2003a).

Could MinK mutations cause LQTS by disrupting I_{Kr}?

The link between MinK, hERG and I_{Kr} also raised another issue — could LQTS associated with inherited mutations in MinK be in part due to reduction in human I_{Kr}, as well as I_{Ks}? A D76N variant of MinK was one of the first two inherited mutations in MinK found to associate with human LQTS (Splawski et al 1997). D76N-MinK-KCNQ1 channels have reduced unitary conductance and impaired gating compared to wild-type, a loss-of-function effect consistent with inefficient repolarization and LQTS (Sesti & Goldstein 1998). However, the analogous rat MinK variant, D77N, also eliminates the effects of MinK on hERG (McDonald et al 1997); this loss of up-regulation could also potentially prolong the QT interval if MinK-hERG channels form I_{Kr} in some or all ventricular myocytes. The D76N mutation in MinK therefore conceivably causes LQTS by disruption of I_{Kr}.

Had the MinK-hERG interaction been discovered before the MinK–KCNQ1 interaction, the finding that inherited MinK mutations associate with LQTS and cause loss-of-function of MinK-hERG channels would doubtless have been taken as evidence that MinK-hERG channels form I_{Kr} in human heart (see MiRP1 below). In fact, little attention has been paid to the possible link between MinK mutations and I_{Kr} dysfunction except for one study in which the previously reported doubling of hERG current by MinK was repeated in both HEK cells and oocytes; inherited MinK mutations were shown to reduce this effect and the possibility of MinK mutations causing disease by I_{Kr} disruption was discussed (Bianchi et al 1999). There are also reasons other than historical bias for the greater attention to MinK-KCNQ1 than MinK-hERG. Homomeric KCNQ1 channels pass delayed rectifier potassium currents with no reported native correlate in mammalian heart, despite cardiac expression. Homomeric hERG channels pass a distinctive, rapidly inactivating, inwardly rectifying, voltage-gated potassium current that largely resembles native I_{Kr}, but with relatively subtle differences in pharmacology, unitary conductance and gating kinetics (Sanguinetti et al 1995). Coassembly of KCNQ1 with MinK produces striking effects on channel gating and is required for recapitulation of the properties of the I_{Ks} current, explaining the cardiac expression of KCNQ1 and leaving little doubt as to the molecular basis of I_{Ks} (Sanguinetti et al 1996, Barhanin et al 1996). In contrast, coassembly of hERG with MinK does not significantly alter the unitary conductance, gating or pharmacology of hERG; therefore effects of MinK on hERG are neither striking nor necessary for recapitulation of the properties of I_{Kr} (McDonald et al 1997).

Weighing the evidence for MinK regulation of human I_{Kr}

Neither the lack of interest in a link between human MinK mutations and I_{Kr} dysfunction, nor the subtlety of effects of MinK on hERG, provides evidence *against* a role for MinK in human I_{Kr}. Equally so, the inability of MinK to eliminate the relatively subtle differences between hERG and I_{Kr} does not argue against their native coassembly — a third subunit could also be required as for I_{Ks} channels, which require yotiao for β-adrenergic receptor modulation (Marx et al 2002). There is, however, supportive evidence for native coassembly of MinK and hERG. First, as mentioned, MinK antisense suppression reduces native murine I_{Kr} (Yang et al 1995); second, expression of D77N-MinK in neonatal mouse heart reduces I_{Kr} (Ohyama et al 2001); third, MinK forms native complexes with ERG in equine heart, assessed by co-immunoprecipitation (Finley et al 2002). This latter report was also the first example of native co-immunoprecipitation of MinK and KCNQ1 from any species, highlighting the previous willingness to accept the role of MinK in native I_{Ks} long before biochemical evidence of their native interaction, because of functional and genetic evidence.

The evidence for MinK contribution to I_{Kr} in some species is compelling, but for human I_{Kr} we are left with a doubling of hERG current density in a heterologous expression system. Extrapolation of evidence from cardiac tissue from other species to human cardiac physiology is often unwise, especially in the case of Kv channels, because of species-dependent differences in expression and function. Two forms of additional evidence might strengthen the hypothesis that MinK modulates human I_{Kr} *in vivo*: first, native co-immunoprecipitation from human cardiac tissue; second, association with LQTS of an inherited MinK mutation that does not affect I_{Ks} function but significantly affects I_{Kr}, as assessed by electrophysiological recording of wild-type and mutant channels in a heterologous system. As far as the authors are aware, neither has been reported and so at present we must rely upon indirect evidence supporting the possibility of MinK–hERG interaction in human heart.

The evidence for and against a role for MiRP1 in I_{Kr}

Heterologous coexpression of MiRP1 and hERG produces mixed results

A decade after the cloning of MinK, homology searches of the expressed sequence tag (EST) database revealed four related proteins named MinK-related peptides (MiRPs) and encoded by *KCNE* genes 2–5 (Piccini et al 1999, Abbott et al 1999). MiRP1 (*KCNE2*) mRNA expression was found in rat heart by Northern blot, and subsequent studies have supported the cardiac expression of MiRP1 from rodent to human (Franco et al 2001, Finley et al 2002, Yu et al 2001, Chen et al 2003b). Rat MiRP1 was found to coassemble with hERG when coexpressed in COS cells, and in *Xenopus* oocytes and CHO cells modulation by rat and human MiRP1, respectively, reduced hERG current density by 40%, increased deactivation rate and slightly shifted the voltage dependence of activation; the current density was reflected in a 40% reduction in unitary conductance upon rat MiRP1 coexpression (Abbott et al 1999).

Inconsistent effects of MiRP1 on hERG were reported in several subsequent studies. In one study MiRP1 speeded up hERG deactivation and reduced current density, but had no effects on voltage dependence (Mazhari et al 2001), whereas others reported speeded deactivation and 30% reduction in tail current density using coexpression of hERG and hMiRP1 in CHO cells and qualitatively similar effects in *Xenopus* oocytes (Weerapura et al 2002). Results from parallel recordings of native I_{Kr} led to the conclusion that MiRP1 does not redress differences between hERG and I_{Kr}, although it should be noted that human subunits in CHO cells were compared to guinea-pig I_{Kr}. Another group reported only minor effects on hERG unless relatively large concentrations of MiRP1 cRNA were co-injected in oocytes, although here MiRP1 was injected late, after hERG currents were observed (Zhang et al 2001). Isbrandt and coworkers reported hMiRP1–hERG plasma membrane co-localization in co-transfected CHO cells, and a long QT syndrome (LQTS)-associated hMiRP1 mutation (V65M) that accelerated inactivation of hMiRP1-hERG channels (Isbrandt et al 2002). Cui and colleagues showed that MiRP1 affects hERG regulation by PKA and also accentuates the effects of arrhythmia-associated mutations in the hERG cyclic nucleotide binding domain (Cui et al 2000, 2001). To summarize, MiRP1 modulates hERG, but effects vary depending upon cell-type, experimental conditions and possibly the species of the subunits used.

In an attempt to explain some of the variability of MiRP1 effects on hERG, we examined the *Xenopus* oocyte system and found that oocytes endogenously express several *KCNE*-encoded proteins, including MinK and MiRP2, both of which can modulate hERG (Anantharam et al 2003). Using RNAi knockdown of endogenous xMiRP2 (both human and *Xenopus* MiRP2 suppress hERG current in oocyte coexpression experiments) we found that endogenous xMiRP2 affects hERG current density, but that this is overcome with higher levels of hERG expression (Fig. 2a,b). We also found that at lower levels of hERG expression this affected the results of MiRP1 coexpression because hMiRP1 appeared to 'rescue' hERG from xMiRP2 suppression, thus paradoxically hMiRP1 up-regulated hERG currents under these conditions (Fig. 2c). While this does not explain all the variability seen in MiRP1-hERG coexpression experiments, endogenous MiRPs or other hERG regulators in oocytes and in mammalian cell lines may contribute to variability of observed effects.

Association of MiRP1 mutations with LQTS does not implicitly indicate a role in I_{Kr}

The heterologous studies showed that MiRP1 is capable of modulating hERG, but that did not present convincing evidence that this is the case in cardiac myocytes. A screen of DNA from patients with LQTS that had previously been found not to harbour mutations in hERG, KCNQ1 or SCN5A — other previously-linked cardiac ion channels — revealed three MiRP1 variants that did not appear in over 1000 control individuals, and a MiRP1 polymorphism, T8A, that occurred in both control and LQTS patients (Abbott et al 1999). A later study revealed another arrhythmia-associated MiRP1 mutation (Isbrandt et al 2002). When tested, all four disease-associated mutations decreased potassium flux through MiRP1-hERG channels, supportive of a possible causative link between MiRP1 mutations and pro-arrhythmic disruption of I_{Kr}. However, as discussed for MinK, because MiRP1 potentially modulates a number of cardiac ion channels — Kv4 subfamily members, KCNQ1 and even HCN pacemaker

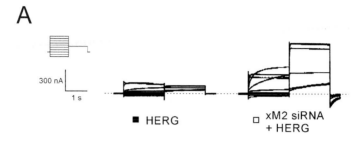
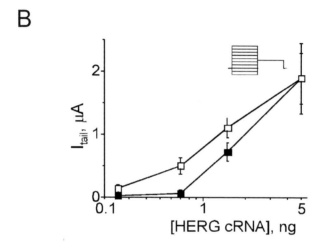
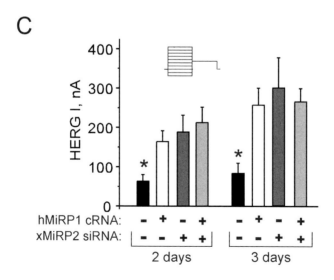

channels — even firm association of MiRP1 mutations with LQTS and disruption of MiRP1-hERG currents *in vitro* is not sufficient to directly support a role for MiRP1 in native I_{Kr}. Another possible argument against the pathophysiological significance of MiRP1-hERG current disruption by MiRP1 mutations was that effects were too subtle to cause disease. This issue was addressed by two groups using combined heterologous expression and *in silico* studies to measure and model the pro-arrhythmic effects of MiRP1 mutants on I_{Kr} — both groups concluding that the reported MiRP1 mutants were sufficient to cause arrhythmia (Lu et al 2003, Mazhari et al 2001).

Effects of MiRP1 on hERG pharmacology offer tentative support for a role in human I_{Kr}

The sensitivity of hERG to block by therapeutic drugs is a major concern for pharmaceutical companies. A number of therapeutic compounds have been removed from the market because they block hERG channels at clinical concentrations and predispose to LQTS. The pharmacology of hERG appears to be dominated by an internal vestibule containing aromatic and hydrophobic residues not present in other α subunits — these residues together with the lack of a kink in the S6 pore-lining helix ensure that the hERG internal vestibule presents a large, sticky non-specific binding site for a wide range of small molecules (Mitcheson et al 2000). MiRP1 does not appear to impinge on this binding site, because several reports indicate that MiRP1 does not alter the affinity of drugs predicted to bind in the internal vestibule — quinidine, dofetilide, vesnarinone and sotalol (Numaguchi et al 2000, Kamiya et al 2001, Weerapura et al 2002). We found that MiRP1 alters hERG block by the class III anti-arrhythmic E-4031, slightly increasing sensitivity but more importantly altering the binding kinetics from a monophasic response in hERG alone

FIG. 2. Endogenous *Xenopus* oocyte xMiRP2 affects hMiRP1-hERG co-expression analysis. (A) Exemplar current families recorded using two electrode voltage clamp at room temperature with 4 mM external KCl (oocytes pulsed at voltages between -120 and $+60$ mV followed by a -30 mV tail pulse from a holding voltage of -80 mV) for oocytes injected with 0.7 ng hERG cRNA ± 500 pg xMiRP2 siRNA (for specific RNAi silencing of endogenous xMiRP2). (B) Mean peak tail currents from oocytes injected with various hERG cRNA doses as indicated, with (open squares) or without (filled squares) 500 pg xMiRP2 siRNA as in panel A; $n = 10$–17 oocytes per point. Points were joined arbitrarily with straight lines. (C) Mean peak tail currents from oocytes injected with 0.5 ng hERG cRNA, ± 2 ng hMiRP1 cRNA, ± 500 pg xMiRP2 siRNA; $n = 5$–10 oocytes per group. Asterisks indicate significant differences from other groups on same day ($P < 0.01$, unpaired Student's *t*-test). Data from Anantharam et al (2003).

channels that reflects blockade only upon repetitive pulsing to depolarized potentials, to a biphasic response that also incorporates block in the closed state (Abbott et al 1999). This supported a role for MiRP1 in native I_{Kr}, also characterized by biphasic E-4031 block (Sanguinetti et al 1995); thus we found that MiRP1 was necessary to recapitulate one of the pharmacological properties of I_{Kr}. However, in a subsequent study no such effect of MiRP1 was found (Weerapura et al 2002) — this discrepancy remains unexplained.

In contrast to the relative lack of MiRP1 influence on the affinity of drugs that presumably block from the inside, MiRP1 may form part of an extracellular drug-binding site in MiRP1-hERG channels. Two polymorphisms have been identified in the predicted extracellular portion of MiRP1. One, T8A, is found in 1–2% of Caucasian Americans, but is absent in African-Americans. Another, Q9E, is found in 3% of African Americans but not in Caucasian Americans (we originally reported it as a rare mutation and this presumably reflects a population bias in our study) (Ackerman et al 2003, Abbott et al 1999, Sesti et al 2000). Both polymorphisms increase three- to fourfold the sensitivity of MiRP1-hERG channels to blockade by the specific drugs implicated in acquired LQTS in patients harbouring the corresponding polymorphisms: T8A with sulfamethoxazole and Q9E with clarithromycin (Sesti et al 2000, Abbott et al 1999). Both these variants lie within a predicted extracellular domain and both alter drug sensitivity, suggesting MiRP1 impingement on an extracellular drug-binding site, although it is formally possible that the variants affect intracellular drug binding by long-range steric interactions. The T8A variant disrupts MiRP1 glycosylation, likely 'deprotecting' the channel complex to increase drug affinity. Interestingly, unlike wild-type MiRP1-hERG channels, the sulfamethoxazole K_i of hERG-alone channels is within the predicted serum concentration range for patients receiving sulfamethoxazole as an antibiotic (Park et al 2003). These pharmacogenetic and pathologic effects are consistent with a role for MiRP1 in human I_{Kr}.

Co-immunoprecipitation of MiRP1 and rERG from rat heart membranes

Co-immunoprecipitation of equine MinK and ERG from native cardiac tissue provided compelling evidence for the role of MinK in I_{Kr} (Finley et al 2002). Here we show preliminary findings in which the molecular identity of rat I_{Kr} was probed using co-immunoprecipitation from rat heart membranes. Anti-MiRP1 antibody enriched rERG, as assessed by anti-ERG Western blotting (Fig. 3). Anti-MinK co-imunoprecipitation gave rERG signal equivalent to that seen with solubilized, non-immunoprecipitated rat heart membranes; anti-adenosine A1 receptor co-immunoprecipitation gave no rERG signal, suggesting specificity of the MiRP1 co-immunoprecipitation. These data are preliminary,

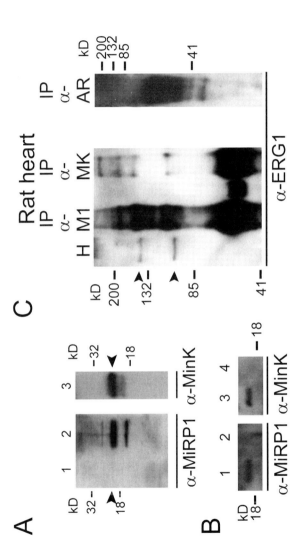

FIG. 3. rERG1 protein is enriched in anti-MiRP1 co-immunoprecipitated fraction from rat heart membranes. Western blots with numbered lines indicating MW marker size (kDa) and migration distance. Fluorography was via 1/5000 HRP-conjugated goat anti-rabbit secondary antibody (BioRad). All primary antibodies IgG purified, 1/1000. (A) 1, Untransfected CHO cell lysate probed with anti-MiRP1 IgG (in-house). 2, MiRP1-transfected CHO cell lysate probed with anti-MiRP1 IgG. 3, Rat heart membrane fraction probed using anti-MinK IgG (in-house). Arrows: mature glycosylated forms of MiRP1 (left) and MinK (right). (B) 1, Rat heart membrane fraction probed using 1/1000 anti-MiRP1 IgG. 2, Untransfected CHO cell lysate probed with 1/1000 anti-MiRP1 IgG. 3, MinK transfected CHO cell lysate probed using 1/1000 anti-MinK IgG. 4, Untransfected CHO cell lysate probed with 1/1000 anti-MinK IgG. (C) 'H', Rat heart membrane fraction probed using 1/1000 anti-ERG1 IgG (Sigma). 'IP α-M1', Anti-MiRP1 IP fraction from rat heart membrane, anti-ERG1 blot. 'IP α-MK', Anti-MinK IP fraction from rat heart membrane, anti-ERG1 blot. 'IP α-AR', Anti-adenosine A1 receptor (Santa Cruz Biotech) IP fraction from rat heart membrane, anti-ERG1 blot. Arrows: Immature and mature forms of rERG1. Enrichment of rERG was only observed with α-MiRP1 IP. In all IP lanes, band between 41 and 85 kDa markers is the immunoprecipitating antibody detected by the goat anti-rabbit secondary antibody.

but supportive of MiRP1-rERG coassembly in rat heart. Future studies will involve confirmation of these results, analysis of regional co-assembly in rat heart, and similar experiments on larger mammals.

Conclusions

Given the controversy surrounding this topic, it is important to note that the molecular composition of I_{Kr} is highly likely to vary regionally within one species, and also between species, thus hERG may well exist alone in some myocytes but with MinK, MiRP1 and/or other β subunits elsewhere. Add to this the possibility that I_{Kr} composition varies during development from neonate to adult, as could be implied by the work of Franco et al (2001), and it is apparent that attempts to correlate I_{Kr} to a rigid composition of specific subunits may be conceptually flawed. We conclude that the amassed evidence supports a role for both MinK and MiRP1 in human cardiac physiology and pathophysiology, and a role for regulation of I_{Kr} in some myocytes in some species, but that there is no conclusive evidence for or against their role in human I_{Kr}.

References

Abbott GW, Butler MH, Bendahhou S et al 2001 MiRP2 forms potassium channels in skeletal muscle with Kv3.4 and is associated with periodic paralysis. Cell 104:217–231

Abbott GW, Sesti F, Splawski I et al 1999 MiRP1 forms IKr potassium channels with HERG and is associated with cardiac arrhythmia. Cell 97:175–187

Ackerman MJ, Tester DJ, Jones GS et al 2003 Ethnic differences in cardiac potassium channel variants: implications for genetic suceptibility to sudden cardiac death and genetic testing for congenital long QT syndrome. Mayo Clin Proc 78:1479–1487

Anantharam A, Lewis A, Panaghie G et al 2003 RNA interference reveals that endogenous Xenopus MinK-related peptides govern nammalian K^+ channel function in oocyte expression studies. J Biol Chem 278:11739–11745

Barhanin J, Lesage F, Guillemare E et al 1996 K(V)LQT1 and lsK (minK) proteins associate to form the I_{Ks} cardiac potassium current. Nature 384:78–80

Bianchi L, Shen Z, Dennis AT et al 1999 Cellular dysfunction of LQT5-minK mutants: abnormalities of IKs, IKr and trafficking in long QT syndrome. Hum Mol Genet 8:1499–1507

Chen H, Kim LA, Rajan S, Xu SB, Goldstein SA 2003a Charybdotoxin binding in the IKs pore demonstrates two mink subunits in each channel complex. Neuron 40:15–23

Chen YH, Xu SJ, Bendahhou S et al 2003b KCNQ1 gain-of-function mutation in familial atrial fibrillation. Science 299:251–254

Cui J, Melman Y, Palma E, Fishman GI, McDonald TV 2000 Cyclic AMP regulates the HERG K^+ channel by dual pathways. Curr Biol 10:671–674

Cui J, Kagan A, Qin D et al 2001 Analysis of the cyclic nucleotide binding domain of the HERG potassium channel and interactions with KCNE2. J Biol Chem 276:17244–17251

Finley MR, Li Y, Hua F et al 2002 Expression and coassociation of ERG1, KCNQ1, and KCNE1 potassium channel proteins in horse heart. Am J Physiol Heart Circ Physiol 283: H126–138

Franco D, Demolombe S, Kupershmidt S et al 2001 Divergent expression of delayed rectifier K^+ channel subunits during mouse heart development. Cardiovasc Res 52:65–75

Isbrandt D, Friederich P, Solth A et al 2002 Identification and functional characterization of a novel KCNE2 (MiRP1) mutation that alters HERG channel kinetics. J Mol Med 80:524–532

Kamiya K, Mitcheson JS, Yasui K, Kodama I, Sanguinetti MC 2001 Open channel block of HERG K^+ channels by vesnarinone. Mol Pharmacol 60:244–253

Lu Y, Mahaut-Smith MP, Huang CL, Vandenberg JI 2003 Mutant MiRP1 subunits modulate HERG K+ channel gating: a mechanism for pro-arrhythmia in long QT syndrome type 6. J Physiol 551:253–262

Marx SO, Kurokawa J, Reiken S et al 2002 Requirement of a macromolecular signaling complex for beta adrenergic receptor modulation of the KCNQ1-KCNE1 potassium channel. Science 295:496–499

Mazhari R, Greenstein JL, Winslow RL, Marban E, Nuss HB 2001 Molecular interactions between two long-QT syndrome gene products, HERG and KCNE2, rationalized by in vitro and in silico analysis. Circ Res 89:33–38

McDonald TV, Yu Z, Ming Z et al 1997 A minK-HERG complex regulates the cardiac potassium current I(Kr). Nature 388:289–292

Mitcheson JS, Chen J, Lin M, Culberson C, Sanguinetti MC 2000 A structural basis for drug-induced long QT syndrome. Proc Natl Acad Sci USA 97:12329–12333

Numaguchi H, Mullins FM, Johnson JP Jr et al 2000 Probing the interaction between inactivation gating and Dd-sotalol block of HERG. Circ Res 87:1012–1018

Ohyama H, Kajita H, Omori K et al 2001 Inhibition of cardiac delayed rectifier K^+ currents by an antisense oligodeoxynucleotide against IsK (minK) and over-expression of IsK mutant D77N in neonatal mouse hearts. Pflügers Arch 442:329–335

Park KH, Kwok SM, Sharon C, Berga R, Sesti F 2003 N-glycosylation-dependent block is a novel mechanism for drug-induced cardiac arrhythmia. FASEB J 17:2308–2309

Piccini M, Vitelli F, Seri M et al 1999 KCNE1-like gene is deleted in AMME contiguous gene syndrome: identification and characterization of the human and mouse homologs. Genomics 60:251–257

Sanguinetti MC, Jiang C, Curran ME, Keating MT 1995 A mechanistic link between an inherited and an acquired cardiac arrhythmia: HERG encodes the IKr potassium channel. Cell 81:299–307

Sanguinetti MC, Curran ME, Zou A et al 1996 Coassembly of K(V)LQT1 and minK (IsK) proteins to form cardiac I_{Ks} potassium channel. Nature 384:80–83

Sesti F, Goldstein SA 1998 Single-channel characteristics of wild-type I_{Ks} channels and channels formed with two minK mutants that cause long QT syndrome. J Gen Physiol 112:651–663

Sesti F, Abbott GW, Wei J et al 2000 A common polymorphism associated with antibiotic-induced cardiac arrhythmia. Proc Natl Acad Sci USA 97:10613–10618

Splawski I, Tristani-Firouzi M, Lehmann MH, Sanguinetti MC, Keating MT 1997 Mutations in the hminK gene cause long QT syndrome and suppress IKs function. Nat Genet 17:338–340

Takumi T, Ohkubo H, Nakanishi S 1988 Cloning of a membrane protein that induces a slow voltage-gated potassium current. Science 242:1042–1045

Wang Q, Curran ME, Splawski I et al 1996 Positional cloning of a novel potassium channel gene: KVLQT1 mutations cause cardiac arrhythmias. Nat Genet 12:17–23

Weerapura M, Nattel S, Chartier D, Caballero R, Hebert TE 2002 A comparison of currents carried by HERG, with and without coexpression of MiRP1, and the native rapid delayed rectifier current. Is MiRP1 the missing link? J Physiol 540:15–27

Yang T, Kupershmidt S, Roden DM 1995 Anti-minK antisense decreases the amplitude of the rapidly activating cardiac delayed rectifier K^+ current. Circ Res 77:1246–1253

Yu H, Wu J, Potapova I et al 2001 MinK-related peptide 1: a beta subunit for the HCN ion channel subunit family enhances expression and speeds activation. Circ Res 88: E84–E87

Zhang M, Jiang M, Tseng GN 2001 mink-related peptide 1 associates with Kv4.2 and modulates its gating function: potential role as beta subunit of cardiac transient outward channel? Circ Res 88:1012–1019

DISCUSSION

Traebert: Why have you done the co-IP in the rat cardiac myocytes? I thought the I_{Kr} of hERG was expressed at very low levels in rodents and plays only a minor role in the ventricular repolarization process.

Abbott: We are going to use several different species. We had rat myocytes on hand, and we could detect hERG and MiRP1, so we thought this was a good place to start. But now we are going to look in dog and pig tissue. The antibodies should still recognize isoforms of different species based on the epitope sequences.

Traebert: What is your opinion about how the coexpression MiRP1 and hERG influences the pharmacology?

Abbott: Basically, we found differences with E-4031, but Nattel's group claimed that they didn't see this in CHO cells (Weerapura et al 2002). They used the same expression system as us. I can't explain why their results disagree with ours, although I believe they did their experiments at body temperature whereas ours were done at room temperature.

Hebert: We tried hard to replicate the results from your *Cell* article. When we used *Xenopus* oocytes we saw what you saw. When we switched to the mammalian system we could not make MiRP1 do anything to hERG in terms of its pharmacology.

Abbott: Was this with E-4031?

Hebert: Yes, and with quinidine and dofetilide.

Abbott: A number of groups have failed to observe differences in general with drugs that block inside hERG, such as quinidine. But we do see the effects of MiRP1 on drugs such as sulfamethoxazole and clarithromycin. We see significant and consistent effects with these. Frederico Sesti's group have gone on to study the mechanism of sulfamethoxazole block of MiRP1 hERG channels and the MiRP1 T8A polymorphism. The T8A polymorphism increases the sensitivity of those channels. I speculate that there is an external drug-binding site that might be important in some pathophysiological circumstances as well. The two polymorphisms our collaborators identified that associate with drug-induced arrhythmia also increased sensitivity to drug blockade of MiRP1-hERG channels and are both positioned in the extracellular portion of MiRP1, in the N-terminus. Q9E increases the sensitivity to clarithromycin block three to fourfold, and T8A increases the sensitivity to sulfamethoxazole block threefold. Both of these drugs were the precipitating drugs in the drug-induced arrhythmia cases (Abbott et al 1999, Sesti et al 2000). Federico has found that T8A disrupts a glycosylation site in MiRP1. When he studied the effects of glycosylation he came up with a hypothesis that MiRP1 glycosylation protects MiRP1-hERG channels from external drug block. If you get rid of the sugar, it leaves the channels susceptible to blockade at lower concentrations by drugs like sulfamethoxazole. In this study, which we didn't look at in the original *PNAS* study in 2000, he found that hERG channels alone have a very high sensitivity to block by sulfamethoxazole. Wild-type MiRP1 reduces the sensitivity of the channels three or fourfold, likely by virtue of the sugar group. If you look at the sensitivity to sulfamethoxazole of hERG channels alone it is in the clinical serum range, whereas with wild-type MiRP1 it is out of that range.

Hebert: The problem is that there is no consistent effect on overall pharmacology.

Abbott: Definitely. I am very surprised that it doesn't affect the internal drug binding site much, if at all.

Hebert: We are moving more towards the idea that MiRP1, MinK and these other proteins serve more of a chaperone role.

Abbott: Then you have to explain I_{Ks} and the effects on KCNQ1. Are you saying that there is a different mechanism for different α and β combinations? It is possible.

Hebert: We published another paper (Ehrlich et al 2004) which confuses me even more than the MiRP1-hERG story. We co-IPed KvLQT1 with hERG. I thought originally that if they both interacted with MinK, this might be the reason why we could co-immunoprecipitate them together. We did the experiment with a purified GST fusion protein and the C-terminus of hERG, with cells expressing KvLQT1. They still pull it down, but it doesn't seem to involve MinK in that complex. I would have been more comfortable had it done so.

Abbott: Neither MinK or MiRP1 have dramatic effects on hERG. I think this is where the problem lies. It is very easy to believe that MinK has effects on Q1, whereas the effects of MiRP1 and MinK on hERG are relatively subtle and they are inconsistent between different groups. Certain point mutants in either MinK or MiRP1 with hERG so dramatically change the current properties that it is hard to imagine them just having a chaperone role and then leaving.

Hebert: I'm not suggesting that they are leaving, but I just don't think they play a role in the active pharmacology of the channel after it gets out to the membrane.

Abbott: With MiRP2-Kv3.4 we saw a large effect on our sensitivity to BDS-II and we saw a 30-fold reduction in toxin affinity. This argues that with the bigger, bulkier blockers we can see the difference, but small molecules don't impinge on the binding site enough for us to see a difference.

Hebert: The clearest evidence that they are playing an important role is that they have been associated with long QT syndrome.

114   DISCUSSION

Tristani-Firouzi: I am not sure that is so clear. These are very small sporadic cases. I wouldn't say that there is enough evidence to say that mutations in MiRP1 cause LQTS or that they are associated with polymorphisms that are associated with acquired disease.

Abbott: The evidence is more than sporadic. If you look at genetic studies perhaps 8–10% of acquired arrhythmia patients have a MinK or MiRP1 polymorphism (that is significantly less common in the general population).

Tristani-Firouzi: That's an association, it is not a linkage.

Abbott: Yes, I agree that genetic association is weaker. It is only thought that 1–2% of inherited arrhythmias are due to MiRP1 dysfunction. Other factors may play a role.

Hancox: Can I ask about the sulfamethoxazole again? If it is acting from the outside, then presumably it has a very different kinetic profile.

Abbott: As I remember, it is quite rapid. I'll have to check the paper.

McDonald: What is the relative abundance of MinK and MiRP in the rat hearts? You have two different primary antibodies, so these will have different affinities. The same would apply for co-IP.

Abbott: Our results aren't that quantitative for those reasons. If the antibodies work the same on a Western as they do on a co-IP, then it argues that the association with MiRP1 is stronger. On the same exposure MinK and MiRP1 give similar signals on Westerns.

McDonald: If they have different affinities those two signals don't mean the same amount of protein.

Abbott: I agree, but going back to the co-IP we see a big difference.

McDonald: As anyone who works with antibodies will know, some will do very well with Westerns and not well with IPs. Let's say that a ventricular myocyte is expressing hERG, KvLQT, MiRP1 and MinK. How do these things find each other? Are they promiscuous within the *in vivo* cell?

Abbott: I'd say it is a mistake to say that I_{Kr} is one particular channel type. I'd be surprised if it didn't exist as hERG in some cells, hERG-MinK in others, and hERG-MiRP1 in others, and dynamic associations of one or more MiRPs with hERG.

McDonald: What do you mean by dynamic association?

Abbott: It can perhaps get endocytosed and then another MiRP can be put in its place, or as hERG starts being made, perhaps the profile of which MiRPs are associated with hERG in the ER changes with different cellular environments and stress levels. I think it is a fluid kind of process. We do see differences in cell type even in the same study. Last year we found that MiRP2 can form a complex with Kv2.1 in rodent brain (McCrossan et al 2003). We also found complexes with MiRP2 and Kv3.1. When we looked in PC12 cells expressing all three different proteins we found there was native association, so we knocked down MiRP2 with RNAi, and the results were consistent with the native association. We also did co-localization. Then in hippocampus MiRP2 associated with 2.1 but not 3.1. I think it is very cell-type dependent. There are a lot of unknown signals and we probably don't know all the roles of MiRPs, either.

Gosling: Is it definite that you can't get heteromultimers between hERG and KvLQT1?

Hebert: We see a functional effect when we co-express KvLQT1 with hERG. We suggested this in our paper (Ehrlich et al 2004) and frankly I was stunned that it was accepted. It is the first example of an α–α interaction, to my knowledge.

Gosling: Could a concatameric approach, attempting to force the KvLQT1 and hERG alpha subunits to form a channel, provide some insight?

Hebert: That's a good idea. I'm not a hERG guy: I'm a signalling person who occasionally publishes papers on hERG. We are starting to look at signalling molecules as dynamic complexes that have stably interacting partners and partners that are recruited to and from them. This whole hERG-MinK-MiRP-KvLQT1 story seems very similar to this. Geoff Abbott is exactly right when he says that depending on what cell type you look at, you may see a different electrophysiological picture depending on which interacting partners are there. This is the sad reality of my life now. I look at living cells and image β receptor signalling complexes. It changes depending on where we look in the cell, and where we look inside the cell with subcellular fractionation. I think it is going to be a very similar story with these. You could even start to think of the 'I_K-osome' here.

McDonald: Your data were very interesting showing MiRP2 having an almost antichaperone role. Do you have any evidence that MinK or MiRP1 plays a chaperone role with hERG? I have looked very carefully with MiRP1 and hERG and I never see any positive chaperoning effect with the addition of either MiRP1 or MinK.

Robertson: Do you mean more expression?

McDonald: More expression or more localization at the surface.

Hebert: It may be a negative chaperone as well. There are examples in the Kir family, the inward rectifiers. Kir3.3 targets itself to the lysosome. It drags the other Kir channels there too. It is a negative sort of regulator of channel activity.

Robertson: Does human MiRP2 have the same effect?

Abbott: Human and *Xenopus* MiRP2 both suppress hERG in oocytes. Neither does it in CHO cells. In mammalian cells it does not happen. Currently we are working with chimeras of MiRP1 and MiRP2. There is some reason why MiRP2 has an effect in oocytes and not CHO cells and we are trying to work out what this is.

Robertson: Could it be that there is already an abundance of MiRP2 in CHO cells?

Abbott: I don't think that is it. It is α-subunit dependent. If you look at the same MiRP2 with KvLQT1 or Kv3.4 or Kv3.1 you see effects (in CHO cells). That lack of trafficking (to the surface) effect does not happen in CHO cells with that complex. However, with other complexes and other MiRPs in CHO cells we see exactly the same effect: 100% suppression, which occurs by retention in the ER. I haven't seen this with hERG or KvLQT1 in CHO cells. Another group find a lot of suppressive effects with MiRP3 and 4, but they don't see ER retention. They see non-functional channels in the membrane (Grunnet et al 2002).

Sanguinetti: One topic that is of high interest to this group, and especially to industry, is whether we need MiRP1 or MinK co-expressed with hERG to screen drugs. If so, why?

Brown: Until we can figure out the stoichiometry exactly we can't answer that. Consider the rat heart. Is MiRP2 in the rat heart? MiRP1 is pulling down tons of hERG and there is no current at all. It is better the devil you know than the devil you don't, so we just use hERG. But we would only be too happy to use a better system. If someone shows that these subunits are present and what their proper stoichiometry is, it would be useful.

Sanguinetti: My concern is that perhaps certain drugs work at a different site on the channel and thus, may require MiRP1 to exhibit physiologically relevant block.

Abbott: MiRP1 seems to decrease sensitivity in those two cases (sulfamethoxazole and clarithromycin) and then the mutants bring it back.

Brown: I thought you said that MiRP1 produced tonic block with E-4031?

Abbott: Yes, E-4031 produces tonic block of MiRP1-hERG, but not hERG channels alone in CHO cells.

Brown: Then you can ask, does the tonic block that you see have a different concentration dependence? And what is the rate of block once you start pulsing the cells? Cocaine is a case in point. It was a drug that looked like it was a tonic blocker; it was just a very fast-activated channel blocker.

Abbott: We did it in parallel with hERG alone. It could be that the way MiRP1 affects kinetics could affect drug block. E-4031 presumably blocks from the inside. If you look at the affinity change, it wasn't that big. In oocytes it did not happen. We saw a two to threefold shift in affinity, but it didn't change the kinetics of block.

Mitcheson: What does MiRP1 do to single channel kinetics?

Abbott: We haven't looked in any detail. The single channel results recapitulate what we saw macroscopically in terms of deactivation rate. We also saw a reduction in unitary conductance from 13 pS to 8 pS. This was a nice experiment because it was done completely blind.

Traebert: In recent years we have tested a large number of compounds. Companies are anxious about being confronted by the FDA about what effects these compounds have on hERG. We have screened both for hERG and hERG-MiRP1, and have never seen a difference. Have you ever tried to express hERG and MiRP1 in a stably transfected cell line?

Abbott: No.

McDonald: We have a stable hERG-expressing cell line, which we transiently transfected with MiRP1 and MinK.

Traebert: The problem is if the cells express both they seem not to survive. If they are transfected transiently they are fine.

References

Abbott GW, Sesti F, Splawski I et al 1999 MiRP1 forms I_{Kr} potassium channels with HERG and is associated with cardiac arrhythmia. Cell 97:175–187

Ehrlich JR, Pourrier M, Weerapura M et al 2004 KvLQT1 modulates the distribution and biophysical properties of HERG. A novel alpha-subunit interaction between delayed rectifier currents. J Biol Chem 279:1233–1241

Grunnet M, Jespersen T, Rasmussen HB et al 2002 KCNE4 is an inhibitory subunit to the KCNQ1 channel. J Physiol 542:119–130

McCrossan ZA, Lewis A, Panaghie G et al 2003 MinK-related peptide 2 modulates Kv2.1 and Kv3.1 potassium channels in mammalian brain. J Neurosci 23:8077–8091

Sesti F, Abbott GW, Wei J et al 2000 A common polymorphism associated with antibiotic-induced cardiac arrhythmia. Proc Natl Acad Sci USA 97:10613–10618

Weerapura M, Nattel S, Chartier D, Caballero R, Hebert TE 2002 A comparison of currents carried by HERG, with and without coexpression of MiRP1, and the native rapid delayed rectifier current. Is MiRP1 the missing link? J Physiol 540:15–27

hERG block, QT liability and sudden cardiac death

Arthur M. Brown

MetroHealth Campus, Case Western Reserve University, Cleveland, and ChanTest, Inc., 14656 Neo Parkway, Cleveland, OH 44128, USA

Abstract. Non-cardiac drugs may prolong action potential duration (APD) and QT leading to Torsade de Pointes (TdP) and sudden cardiac death. TdP is rare and QT is used as a surrogate marker in the clinic. For non-cardiac drugs, APD/QT liability is always associated with a reduction in hERG current produced by either direct channel block or inhibition of trafficking. hERG and APD liabilities correlate better when APDs are measured in rabbit versus canine Purkinje fibres. hERG and APD/QT liabilities may be dissociated when hERG block is offset by block of calcium or sodium currents. hERG liability may be placed in context by calculating a safety margin (SM) from the IC_{50} for inhibition of hERG current measured by patch clamp divided by the effective therapeutic plasma concentration of the drug. The SM is uncertain because literature values for IC_{50} may vary by 50-fold and small differences in plasma protein binding have large effects. With quality control, the IC_{50} 95% confidence limits vary less than twofold. Ideally, hERG liability should be determined during lead optimization. Patch clamp has insufficient throughput for this purpose. A novel high-throughput screen has been developed to detect drugs that block hERG directly and/or inhibit hERG trafficking.

2005 The hERG cardiac potassium channel: structure, function, and long QT syndrome. Wiley, Chichester (Novartis Foundation Symposium 266) p 118–135

In the 1990s, several blockbuster, non-cardiac drugs including terfenadine (Seldane) (Morganroth et al 1993, Woosley et al 1993) and cisapride (Propulsid) (Rampe et al 1997, Mohammad et al 1997) were associated with prolongation of the QT interval of the electrocardiogram (ECG), polymorphic ventricular tachycardia (Torsade de Pointes, TdP) and sudden cardiac death. These non-cardiac drugs delayed cardiac repolarization, prolonged action potential duration (APD) and were subsequently shown to block the rapidly repolarizing cardiac potassium current I_{Kr}. Earlier in the 1950s and 1960s, QT prolongation and TdP were linked to a rare familial disease, hereditary long QT syndrome (HLQTS) (Jervell & Lange-Nielsen 1957, Romano et al 1963, Ward 1964). In the mid-1990s, the LQT2 form of HLQTS was linked to the human *ether-a-go-go* gene (hERG) (Curran et al 1995) and hERG was shown to produce a current that had the essential properties of I_{Kr} (Sanguinetti et al 1995). A straightforward deduction was that a non-cardiac drug associated with acquired LQTS (ALQTS) such as terfenadine might block hERG. This deduction was demonstrated by Roy et al in 1996.

Terfenadine was the first non-sedating antihistamine used to treat seasonal hay fever; cisapride was a $5HT_4$ agonist used to treat gastro-oesophageal reflux disease (GERD). The risk of TdP was very low for either drug on the order of about five cases per million patient months. Despite these very low frequencies, the risks were unacceptable for the regulatory authorities. Taking hay fever as an example, the lifetime risk to a patient is about zero. The lifetime benefit for treatment of hay fever symptoms with terfenadine is also about zero. For regulatory authorities like FDA, a risk of symptomatic relief as low as 0.0002% was not acceptable for a disease that carried no lifetime risk and a drug that carried no lifetime benefit, considering that hundreds of deaths might occur among the hundreds of millions of patients with prescriptions for terfenadine. The argument was strengthened further by the presence of me-too, second generation antihistamines without terfenadine's QT/TdP liabilities. In 1998, terfenadine was withdrawn from the US market. The FDA took a similar approach with cisapride, astemizole and sparfloxacin which were withdrawn from the market, and sertindole which did not receive approval. It is not surprising that sudden cardiac death due to non-cardiac drugs has become a major safety and financial issue for the pharmaceutical industry and the agencies that regulate it.

Measuring block of hERG current

As a practical matter, drug developers cannot wait until Phase I clinical trials to determine drug safety. Rather, this determination should be made long before an IND submission, preferably during lead development or lead optimization. Fail early, fail cheap is the mantra of the pharmaceutical industry these days. A 10% improvement in predicting QT liability might save as much as $100 million per drug in drug development costs.

To this point, hERG is the only proven molecular target for non-cardiac drugs that carry a defective repolarization liability. Drug block of hERG current expressed heterologously in cell lines is the most direct test of this propensity. The first direct test of hERG as the molecular target for block in ALQTS showed that terfenadine, the poster drug for drug-induced ALQTS, blocked hERG current transiently expressed in *Xenopus* oocytes (Roy et al 1996). Because drug access may be limited in *Xenopus* oocytes, it is preferable to express hERG in mammalian cell lines such as HEK 293 and CHO cells in which access is usually not limiting. Stable expression with a known passage number for the transfected cells is the preferred method of expression.

Gigaseal, whole-cell patch clamp recording is usually done in HEK293 cells stably transfected with hERG cDNA. Acceptance criteria include: gigaseal; stable leakage current; low access resistance; low voltage error; normal test pulse current waveform (e.g. hERG peak tail current greater than prepulse current amplitude); low rundown of test pulse current amplitude; and significant inhibition in response to positive controls.

Step-pulse or step-ramp (Fig. 1) protocols may be used, with step-ramp giving lower IC_{50}s for drug block (Kirsch et al 2004). Block is gating-dependent and maximal at conditioning potentials that fully activate hERG current (about $+20$ mV). Block is measured as the fractional reduction in tail current amplitude during the test pulse or ramp to potentials of -50 or -80 mV, respectively. Block may be strongly temperature-dependent for some drugs, e.g. erythromycin, sotalol (Kirsch et al 2004).

hERG blockers access the channel intracellularly. Access is kinetically determined and may vary significantly among drugs. Consequently, attaining steady state block at each concentration of a dose–response curve as shown in Fig. 1 may be difficult and failure may be responsible for much of the variability in IC_{50}s in the literature. Experimental data should show that steady state block has been reached at each concentration of drug (Fig. 2). Block should be distinguished from a change in access resistance or plugging of pipettes or electrodes. Vehicle and positive controls should be applied with each drug. Pulse protocols should ensure complete activation of currents and action potentials, currents should not exceed the capacity of the voltage clamp amplifier, pulse frequencies should span the physiological range, perfusion lines should be clean and exchange of drug should be complete. When the requirements of the data are satisfactory and proper acceptance criteria for the whole cell gigaseal patch clamp method are used, the 95% confidence limits for IC_{50}s of a number of non-cardiac blockers vary by factors of less than two rather than the 50- to 100-fold differences that are present in the literature (Fig. 3) (Kirsch et al 2004).

Accuracy of IC_{50}s is critical for at least two reasons: first, the IC_{50} provides an estimate of the safety margin (Redfern et al 2003) for potential QT liability described subsequently; and second standard patch clamp is used as the reference or gold standard for validating the various high throughput hERG screens that have been and are being used in earlier stages of drug discovery. The standard should not be brass, it should be gold.

At about the same time as terfenadine issues were surfacing, cisapride was raising safety flags. Rampe et al (1997) and Mohammed et al (1997), using the hERG-mammalian cell system, showed that once again hERG was the molecular target. Hundreds of drugs later, it appears that hERG is the only established target for

hERG BLOCK AND QT LIABILITY

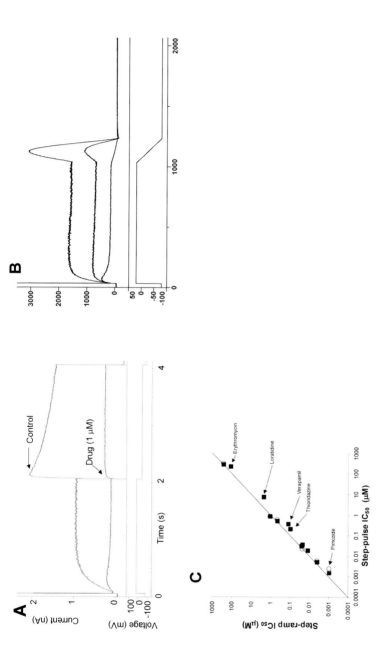

FIG. 1. Block of hERG by a non-cardiac drug. (A) hERG currents expressed in stably transfected HEK 293 cells using a conditioning prepulse to +20 mV from −80 mV for 2 s and a test potential of −50 mV done at 22 °C at 0.1 Hz. Fractional block is measured from the tail currents (see arrows). Onset and kinetics of block are observed during the prepulse. (B) E-4031 block using a step-ramp protocol with a conditioning step to +20 mV for 1 s followed by a test ramp from +20 mV to −80 mV at 0.5 V/s. Fractional block is measured from the ramp peak currents. (C) Comparison of IC_{50}s with step-ramp and step-pulse protocols.

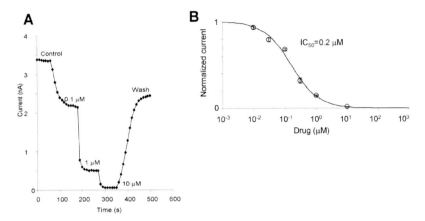

FIG. 2. Measuring steady state block of hERG. (A) Step-ramp protocol. When the on- and off-rate of drug block appear equal, i.e. when the change in successive measurements of fractional block is less than 2%, steady state has occurred. (B) Concentration–response data are fit by: % Block = $\{1-1/[1+([\text{Test}]/(\text{IC}_{50})^N]\} \cdot 100$ where [Test] is drug concentration, IC_{50} is drug concentration at half-maximal block, % block is fraction of hERG current inhibited at each concentration and N is a slope coefficient. Relationship is measured by non-linear least square fitting. 95% confidence limits are indicated by horizontal bars.

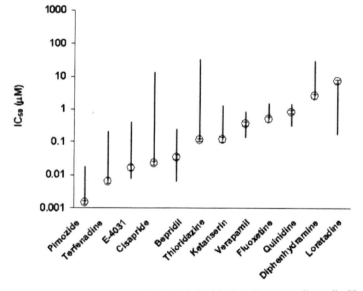

FIG. 3. Comparison of IC_{50} values for hERG/I_{Kr} blockers in mammalian cells. Vertical lines show range of high and low IC_{50}s in the literature and open circles show IC_{50}s obtained by Kirsch et al (2004) using the step-ramp protocol shown in Fig. 1.

non-cardiac drugs carrying TdP/prolonged QT liabilities. Lacerda et al (2001) showed that hERG is by far the most sensitive to drug block of the potassium channels involved in cardiac repolarization. Lacerda et al (2001) demonstrated an absence of class action effects with fexofenadine being orders of magnitude less inhibitory than terfenadine and likewise for risperidone versus sertindole. The result is significant for drug design since it indicates that hERG liability can be reduced without loss of efficacy.

Since every non-cardiac drug that impairs cardiac repolarization has been shown to carry a hERG liability, usually, but not always, (see the later discussion on trafficking inhibition) due to block of the hERG current (Brown & Rampe 2000, Ficker et al 2004), the hERG assay is now used routinely to test lead compounds for their effects on heterologously expressed hERG current. However, the relationship among reduced hERG current, QT prolongation and TdP may be drug-dependent (Roden 1993) and not all drugs with a hERG liability impair repolarization (Zhang et al 1999, Kang et al 2004). As a practical matter, the triple relationship among hERG, APD/QT and TdP operates at many diverse levels including molecular and systems pharmacology, regulatory policy and the business of drug development by the pharmaceutical industry.

The APD repolarization assay

Non-cardiac drugs linked to TdP generally prolong ventricular repolarization as a result of block of hERG. The most established preparation for measuring repolarization is the *in vitro* Purkinje fibre. The fibres are isolated from hearts and action potentials are measured using conventional intracellular recording. Action potential parameters that are measured include: APD 30, 60 and 90, AP amplitude, resting membrane potential and dV/dT max. Because drug access by superfusion and APD stability may be issues, control records must be stable for 20–25 min during control and for similar periods during exposure to drug. Drug effects should be studied over a range of stimulation rates because APD prolongation may be greater at slower rates, so-called reverse-use dependence. For some known torsadegenic drugs, e.g. terfenadine, APD prolongation in the canine cardiac Purkinje fibre does not correlate well with hERG and QT liabilities. This result has relegated the APD repolarization assay to a lesser status in some eyes (ICH Expert Working Group 2003 — S7B guidance). However, we believe that the APD repolarization assay provides important non-clinical information about mechanism, especially when there is a dissociation between hERG and QT liabilities as discussed in a subsequent paragraph.

Because species may play a role, we compared hERG liability with APD prolongation between canine and rabbit Purkinje fibres (PFs) for a series of torsadegenic and non-torsadegenic drugs (Table 1). Both assays were equally

TABLE 1 Predictive value of rabbit versus dog Purkinje fibre APD assays

Compound	Rabbit PF	Dog PF	hERG IC$_{50}$ (μM)
Torsadegenic Test Compounds			
Bepridil	Positive at 0.02 μM	Negative	0.023
Cisapride	Positive at 0.01 μM	Positive at 0.1 μM	0.026
Erythromycin	Positive at 100 μM	Positive at 100 μM	116
Haloperidol	Positive at 100 μM	Positive at 100 μM	0.03
Ketanserin	Positive at 1 μM	Positive at 0.1 μM	0.098
Pimozide	Positive at 0.01 μM	Negative	0.001
Quinidine	Positive at 1 μM	Positive at 1 μM	0.83
Sotalol	Positive at 1 μM	Positive at 50 μM	268
Terfenadine	Positive at 0.1 μM	Negative	0.004
Thioridazine	Positive at 10 μM	Positive at 10 μM	0.09
Non-Torsadegenic Test Compounds			
Amoxicillin	Negative at $\leqslant$ 2700 μM	Negative at $\leqslant$ 2700 μM	> 2700
Aspirin	Negative at $\leqslant$ 100 μM	Negative at $\leqslant$ 100 μM	> 100
Captopril	Negative at $\leqslant$ 1100 μM	Negative at $\leqslant$ 1100 μM	> 1100
Diphenhydramine	Negative at $\leqslant$ 10 μM	Negative at $\leqslant$ 10 μM	2.6
Fluoxetine	Negative at $\leqslant$ 10 μM	Negative at $\leqslant$ 10 μM	0.46
Loratadine	Negative at $\leqslant$ 100 μM	Negative at $\leqslant$ 100 μM	2.3
Pinacidil	Negative at $\leqslant$ 30 μM	Negative at $\leqslant$ 30 μM	ND
Verapamil	Mixed effect at $\leqslant$ 10 μM	Mixed effect at $\leqslant$ 1 μM	0.125
Vehicle	Negative	Negative	NE

Positive effect, statistically significant ($P < 0.05$) action potential duration (APD) prolongation versus control.
Negative effect, statistically significant ($P < 0.05$) APD shortening or no effect versus control.
Mixed effect, Significant APD$_{90}$ prolongation and significant APD$_{60}$ shortening.

selective against non-torsadegenic compounds but the rabbit PF assay had greater sensitivity to known torsadegenic compounds without any false positives for the non-torsadegenic drugs. In our series, the correlation between hERG liability and APD prolongation in rabbit PFs was 100%.

Interpreting hERG IC$_{50}$s

As noted, hERG is presently the only proven molecular target for non-cardiac drugs that carry a defective repolarization liability, and block of the hERG current expressed heterologously in cell lines is the most available, direct test of this propensity. Since many drugs will reduce hERG current at sufficiently high concentrations, the IC_{50} value should be referenced to a realistic drug concentration. A normative comparison can be achieved using the ratio of the IC_{50} for drug block of hERG (numerator) to the K_d for drug binding at its primary target or preferably, the effective therapeutic plasma concentration of the drug (denominator). The ratio is referred to as the safety margin (SM) (Redfern et al 2003). If the ratio is >100, the SM is adequate; if the ratio is <10, the SM is too low, and between 10 and 100, interpretation of the SM is indeterminate. SM calculations depend on the accuracy of the IC_{50} value for block and the measurement of free drug in the plasma. The latter is determined by the binding of drug to plasma proteins. If IC_{50} values are taken from the literature, selection must be made with caution because, as we have shown in Fig. 3, these values may be suspect. Regarding free concentrations of drug, protein binding is often in the 97–99% range and small differences in the measurement may have large effects on SM. For example, a drug with a hERG IC_{50} of 200 nM, a total plasma level of 200 nM and 99% plasma protein binding would have an acceptable SM of 100. If binding were 97%, the SM becomes 33.3 and is less satisfactory.

Drugs with a hERG liability need not impair repolarization. Drugs may block outward hERG potassium current and inward calcium or sodium currents with sufficiently similar potencies resulting in no change in APD. Prototypical drugs for such mixed ion channel effects (MICE; Hanson 2003) are the calcium channel blocker verapamil (Zhang et al 1999) and the muscarinic receptor antagonist tolterodine (Kang et al 2004). Amiodarone another drug with polypharmaceutical effects on ion channels may cause changes in QT but is not usually associated with TdP (Roden 1993). In the MICE situation, APD may not be prolonged but inhibition of inward currents may be revealed in the plateau potential providing indirect evidence for the mechanism.

Inhibition of trafficking and hERG liability

Recent reports have identified drugs that produce QT and TdP liabilities, not by direct block of hERG, but rather by inhibition of hERG trafficking. Ficker et al (2003) showed that geldanamycin blocks the ATP-ase activity of Hsp90, a cytoplasmic protein that is important for normal trafficking of hERG. As a result, the immature, core-glycosylated form of the protein is retained in the endoplasmic reticulum (ER) and surface expression and hERG currents are reduced. In ventricular myocytes, the APD is prolonged and I_{Kr} is reduced. The effects are specific as other cardiac potassium currents such as I_{Ks} and I_{Kur} are unaffected. A 17-allylamino derivative of geldanamycin is in clinical trials for cancer chemotherapy and may cause hERG block and QT/TdP liabilities.

hERG and QT/TdP liabilities appear to apply to the use arsenic trioxide (AT3) for treatment of refractory or relapsing acute promyelocytic leukaemia. AT3 treatment is often accompanied by QT prolongation and TdP has been described (Ohnishi et al 2000, Unnikrishnan et al 2001). While direct block of hERG has been reported from one lab (Drolet et al 2004), two different labs have found no evidence of direct block (Ficker et al 2004, Imredy et al 2004). Rather, it appears that AT3 blocks normal trafficking of hERG. AT3 affects hERG trafficking in a manner similar to geldanamycin, namely that the core-glycosylated, immature form of hERG is retained in the EF and does not progress to the fully glycosylated mature form. Like geldanamycin, AT3 prolongs APD and reduces I_{Kr} without affecting I_{Ks} or I_{Kur}. However, unlike geldanamycin, the mechanism by which AT3 inhibits hERG trafficking is unknown.

Pentamidine (PEN) is a drug used to treat Leishmaniasis and trypanosomiasis and is known to cause QT prolongation and TdP. Recently, we have shown that acute application of this drug does not block hERG current. However, overnight incubation is associated with reduced hERG current, absence of the fully mature protein on Western blot and inhibition of hERG trafficking. It is likely that this method of producing hERG liability is responsible for PEN's clinical effects (Fig. 4).

High-throughput hERG assays

Standard patch clamp is labour intensive and low throughput. It is used to determine concentration–response relationships and IC_{50}s for drugs that are being considered for IND submissions. Low throughput is not a bottleneck for drugs destined for the clinic because the overall numbers of drugs entering the pipeline of the entire biopharmaceutical industry are only in the low thousands per year at most and these numbers can be accommodated by the standard patch clamp hERG assay. Nor is this method cost-restrictive given the overall cost of developing an approved drug that satisfies the requirement for cardiac safety. However, the fail early, fail cheap efficiency mantra means that hundreds of thousands of compounds should be evaluated by industry much earlier during lead optimization. Here, low-throughput patch clamp will not do. The most direct approach to the problem is automation of patch clamp. Numerous methods are being implemented generally using planar arrays of chips made from materials as diverse as quartz, silicon and Sylgard. At present, none of the automated patch clamp methods has the combination of accuracy and throughput that would be desirable for earlier screening during lead development. The device with the highest throughput attains seal resistances in the low hundreds of megohms and suffers from low sensitivity as a result

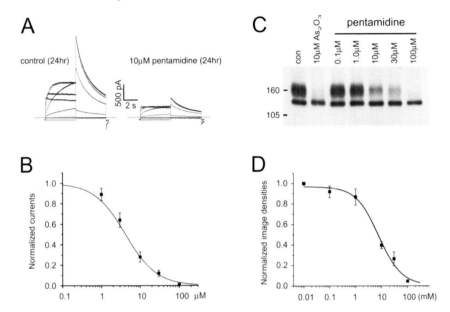

FIG. 4. Pentamidine reduces hERG current by inhibiting hERG trafficking. (A) hERG currents in HEK 293 cells recorded under control conditions and after overnight exposure to 10 μM pentamidine using depolarizing voltage steps (h.p. −80 mV). (B) Concentration-dependent reduction of hERG tail current amplitudes by overnight exposure to pentamidine. IC$_{50}$ is 5.1 μM ($n=9$–10). (C) Western blot showing effects of overnight exposure to increasing concentrations of pentamidine on hERG wild-type protein stably expressed in HEK293 cells. Pentamidine inhibits maturation and surface expression of hERG. 10 μM As$_2$O$_3$ was used as well-established inhibitor of hERG trafficking. (D) Concentration-dependent reduction of fully glycosylated 155 kDa hERG after overnight exposure to pentamidine. IC$_{50}$ is 7.8 μM ($n=3$).

(Schroeder et al 2003). Other devices which attain gigaseals have low throughput and fluidics shortcomings (Asmild et al 2003).

Alternative methods have higher throughput but do not measure functional hERG current. These methods include:

- displacement of high affinity, radioactively-labelled ligand blocker
- atomic absorption measurement of rubidium flux, and
- fluorescence detection of voltage-sensitive dyes (Tang et al 2001).

The first method only detects compounds that compete for the binding site of the labelled ligand and may miss entire classes of non-competitive compounds. Drugs that bind to remote sites may also influence ligand binding and give false positives. The second method produces depolarization with high concentrations of

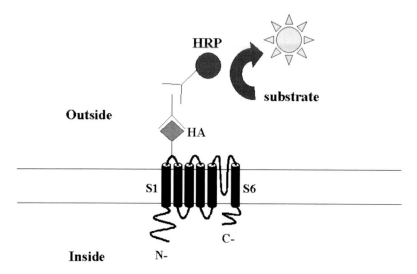

FIG. 5. hERG contains an extracellular HA epitope tag that is recognized by a specific anti-HA antibody. A chemiluminescent reaction is mediated by a secondary antibody conjugated to horseradish peroxidase (HRP).

extracellular potassium that also reduce hERG block and lower sensitivity. This method is further limited by the inactivation properties of the hERG channel and contributions from non-hERG channels to resting Rb^+ flux and the fact that the equilibrium potential of Rb^+ is continually decreasing. The third method has the first two limitations noted for the Rb^+ flux method and may be even more strongly influenced by the relative contributions of other non-potassium ion channel determinants of membrane potential.

Recently, a surface expression method called hERG-Lite™ (Fig. 5) has been introduced (Wible et al 2003). This method takes advantage of the fact that direct blockers of hERG rescue trafficking of certain hERG mutations responsible for hereditary long QT syndrome (Ficker et al 2002) or inhibit trafficking of the wild-type channel (unpublished observations). Like the other non-functional assays described above, this method has reduced sensitivity compared to standard patch clamp although the reduction is less marked for drugs with lower potencies. Within a class of drugs, the rank order of blocking potencies parallels the results obtained with patch clamp. This method also has the great advantage, in addition to detecting drugs that block hERG directly, of detecting drugs such as the chemotherapeutic arsenic trioxide and geldanamycin (Ficker et al 2003), and the anti-Leishmaniasis and trypanosomiasis drug pentamidine that produces hERG/QT liabilities and TdP by inhibition of hERG trafficking. To this point,

TABLE 2   hERG-Lite versus patch clamp

Number of compounds tested

	hERG-Lite positive	hERG-Lite negative
Patch-clamp positive	153	0
Patch-clamp negative	0	70

hERG-Lite has accurately predicted the hERG liabilities of more than 170 drugs (Table 2).

Summary

Assessment of potential QT/TdP liability is a requisite component for determination of drug safety. QT liability of non-cardiac drugs is uniformly associated with block of the cardiac hERG potassium channel and the *in vitro* hERG patch-clamp assay, when properly done, provides the most accurate, reproducible, quantitative measurement of hERG block. For some drugs, hERG block may not cause APD/QT prolongation, e.g. if it is offset by block of sodium and/or calcium currents. The APD repolarization assay in canine Purkinje fibres may give false negatives for drugs with QT/TdP liabilities but this does not appear to be the case for rabbit Purkinje fibres. The decision to go forward with a lead candidate having a hERG liability should be made with this and other considerations most importantly risk analysis, kept in mind.

Recently, it has become evident that drug-induced, acquired hERG/QT/TdP liabilities may be produced by inhibition of hERG trafficking rather than direct block of hERG current. The relative frequency of this mechanism for hERG/QT/TdP liabilities is not yet known.

While it is the reference standard at present, the hERG gigaseal patch clamp assay is too labour intensive and too low throughput to be used as a screen early in the discovery/development process. Automated patch clamp has not yet provided satisfactory sensitivity or throughput. Several indirect high-throughput screens have been used, but none are as sensitive as patch clamp and only one measures both direct block of hERG and inhibition of trafficking.

QT prolongation is a poorly understood surrogate for adverse cardiac events. The relationship between QT prolongation and TdP is not understood mechanistically and is drug-dependent. A battery of non-clinical tests must complement clinical QT measurements to provide an informed decision concerning the cardiac safety of all drugs.

Acknowledgements

I am grateful to the Study Directors, Staff Scientists, Drs Barbara Wible, Tony Lacerda and Glenn Kirsch and Mrs Dee Groynom at ChanTest, Inc. and to my colleagues Drs Eckhard Ficker and Yuri Kuryshev at the MetroHealth campus of Case Western Reserve University. This work was supported in part by NIH grant HL-36930.

References

Asmild M, Oswald N, Krzywkowski KM et al 2003 Upscaling and automation of electrophysiology: toward high throughput screening in ion channel drug discovery. Receptors Channels 9:49–58

Brown AM, Rampe D 2000 Drug-induced long QT syndrome: is hERG the root of all evil? Pharmaceutical News 7:15–20

Curran ME, Splawski I, Timothy KW, Vincent GM, Green ED, Keating MT 1995 A molecular basis for cardiac arrhythmia: HERG mutations cause long QT syndrome. Cell 80:1–20

Drolet B, Simard C, Roden DM 2004 Unusual effects of a QT-prolonging drug, arsenic trioxide, on cardiac potassium currents. Circulation 109:26–29

Ficker E, Obejero-Paz CA, Zhao S, Brown AM 2002 The binding site for channel blockers that rescue misprocessed LQT2 hERG mutations. J Biol Chem 277:4989–4998

Ficker E, Dennis AT, Wang L, Brown AM 2003 Role of the cytosolic chaperones Hsp70 and Hsp90 in maturation of the cardiac potassium channel hERG. Circ Res 92:e87–e100

Ficker E, Kuryshev Y, Dennis AT et al 2004 Mechanisms of arsenic-induced prolongation of cardiac repolarization. Mol Pharmacol 66:33–44

Hanson L 2003 Literature reports of IC50's I_{Kr} Blockade. 2003 Barnett International QT Prolongation Conference, November, Philadelphia, USA

ICH Expert Working Group 2002 Safety pharmacology studies for assessing the potential for delayed ventricular repolarization (QT interval prolongation) by human pharmaceuticals

Imredy JP, Irving WD, Clouse HK, Salata JJ 2004 Electrophysiological and pharmacological characterization of a cell line stably expressing hKCNQ1 and hKCNE1 potassium channels. 2004 Biophysical Society Annual Meeting (abstr)

Jervell A, Lange-Nielsen F 1957 Congenital deaf-mutism, functional heart disease with prolongation of the Q-T interval and sudden death. Am Heart J 54:59–68

Kang J, Chen XL, Wang H et al 2004 Cardiac ion channel effects of tolterodine. J Pharmacol Exp Ther 308:935–940

Kirsch GE, Trepakova ES, Brimecombe JC et al 2004 Variability in the measurement of hERG potassium channel inhibition: effects of temperature and stimulus pattern. J Pharmacol Toxicol Methods, accepted 50:93–101

Lacerda AE, Kramer J, Shen K-Z, Thomas D, Brown AM 2001 Comparison of block among cloned cardiac potassium channels by non-antiarrhythmic drugs. Eur Heart J Suppl 3:K23–K30

Mohammad S, Zhou Z, Gong Q, January CT 1997 Blockage of the hERG human cardiac K^+ channel by the gastrointestinal prokinetic agent cisapride. Am J Physiol 273:H2534–H2538

Morganroth J, Brown AM, Critz S et al 1993 Variability of the QT_c interval: impact on defining drug effect and low-frequency cardiac event. Am J Cardiol 72:26B–31B

Ohnishi K, Yoshida H, Shigeno K et al 2000 Prolongation of the QT interval and ventricular tachycardia in patients treated with arsenic trioxide for acute promyelocytic leukemia. Ann Intern Med 133:881–885

Rampe D, Roy ML, Dennis A, Brown AM 1997 A mechanism for the proarrhythmic effects of cisapride (Propulsid): high affinity blockade of the human cardiac potassium channel hERG. FEBS Lett 417:28–32

Redfern WS, Carlsson L, Davis AS et al 2003 Relationships between non-clinical cardiac electrophysiology, clinical QT interval prolongation and torsade de pointes for a broad range of drugs: evidence for a provisional safety margin in drug development. Cardiovasc Res 58:32–45

Roden DM 1993 Current status of class III antiarrhythmic drug therapy. Am J Cardiol 72:44B–49B

Romano C, Gemme G, Pongiglione R 1963 Aritmie cardiache rare dell'età pediatrica. Clin Pediatr 45:656–683

Roy M-L, Dumaine R, Brown AM 1996 HERG, a primary human ventricular target of the nonsedating antihistamine terfenadine. Circulation 94:817–823

Sanguinetti MC, Jiang C, Curran ME, Keating MT 1995 A mechanistic link between an inherited and an acquired cardiac arrhythmia: HERG encodes the I_{Kr} potassium channel. Cell 81:299–307

Schroeder K, Neagle B, Trezise DJ, Worley J 2003 Ionworks™ HT: a new high throughput electrophysiology measurement platform. J Biomol Screen 8:50–64

Tang W, Kang J, Wu X et al 2001 Development and evaluation of high throughput functional assay methods for HERG potassium channel. J Biomol Screen 6:325–331

Unnikrishnan D, Dutcher JP, Varshneya N et al 2001 Torsade de pointes in 3 patients with leukemia treated with arsenic trioxide. Blood 97:1514–1516

Ward OC 1964 A new familial cardiac syndrome in children. J Ir Med Assoc 54:103–106

Wible BA, Hawryluk P, Ficker E, Brown AM 2003 HERG-Lite™, a novel high throughput hERG cardiac safety test. 2003 Society for Biomolecular Screening Annual Meeting (abstr)

Woosley RL, Chen Y, Froiman JP, Gillis RA 1993 Mechanism of the cardiotoxic actions of terfenadine. J Am Med Assoc 269:1532–1536

Zhang S, Zhou Z, Gong Q, Makielski JC, January CT 1999 Mechanism of block and identification of the verapamil binding domain to HERG potassium channels. Circ Res 89:989–998

DISCUSSION

Noble: You say in your abstract that hERG liability may be dissociated from QT liability if it is offset by block of cardiac Ca^{2+} and Na^+ currents. That is totally correct. What is the prospect for a dual screen, or even a triple screen?

Brown: For the Ca^{2+} and Na^+ channels?

Noble: Why not? The problem, as Rashmi Shah's abstract indicates, is that we may be missing a lot of good drugs.

Brown: I don't like working with Na^+ channels, because they have too many subunits. The problem with the Ca^{2+} channels is that Merck now own all the patents. It's a mess. However, I agree with you in principle: we could do this.

Sanguinetti: Is there no demand?

Brown: The history is that the pharmaceutical industry picked up on hERG liability in the mid-1990s. After a while the hERG assay became accepted as some sort of preclinical predictor. Soon it was apparent that some drugs such as verapamil were carrying hERG liability, but they had no QT liability whatsoever. What the companies usually do is say that they are worried about hERG because the dossier needs to be complete. If you give them a positive hERG you can then try to explain to them that there is or isn't QT liability and provide them with a mechanistic explanation, but this is difficult because what has happened is that the pendulum has swung too far in one direction. Initially, it was very hard to persuade industry that the hERG assay was a good predictor. Now, in many cases, it has gone the other way: if a drug shows hERG liability, forget about it. What we are now trying to tell them is that this must be put into context of a risk–benefit equation.

Traebert: It is very dependent on the status of the compound. We have done lots of screens on hERG L-type Ca^{2+} channels and SCN5 Na^+ channels. In general, if you have a hERG compound that is active, you try your very best to remove this, and to look in the series of second-class compounds from the same template to find one that is not as active on hERG, but which has similar pharmacology. However, the chemist who is sitting in the lab says they want to take another compound from the series because they don't want to work with it. It is very dependent on the status of the compound. If you have a compound which is the only one of the series to go forward, but it shows an activity of about 5 μM on hERG, and an activity on the target which is below 100 nM, then it is very wise to test it also on Na^+ and Ca^{2+} channels. You have already spent a lot of money on this target and this compound, so you have to make the financial decision about whether to proceed or to kill it. If you are going to kill a compound, you have to do it as early as possible because it is burning hundreds of thousands of dollars per day, but you don't want to throw away a compound which may become a blockbuster. The problem comes when you have a patient with an inherited QT prolongation who takes your drug. If you already have a reduced hERG activity, then this overshooting of the activity of Ca^{2+} and Na^+ channel block may not be sufficient for this patient. The best case is therefore to avoid this, but sometimes you have no choice.

Gosling: I wanted to ask Buzz Brown and Luc Hondeghem about the kinetics of on and off rates in mixed ion channel effects. You may only see certain class 3 effects on removal of the molecule from hERG. The liability is actually when you remove the therapy. What are your experiences with this?

Hondeghem: I agree with that. We usually see the strongest signal upon washout. The kinetics is important: in hERG blockers there is a large subgroup that is not torsadogenic at all. One cannot say 'I have a hERG block and I must turn it off', because there are hERG blockers that can be anti-arrhythmic instead of pro-arrhythmic.

Brown: This is an important point. It is not necessarily only during the washout, either. What happens is that the use dependence may be variable. When you are trying to convert an atrial fibrillation, for example, you are looking at cycle frequencies that are in the order of 300 per minute. We have found that some drugs may be hERG blockers and their potency for steady-state block may be significant, but their use dependence is not great. On the other hand, their potency for Ca^{2+} block at the steady state is lower and their use dependence is incredibly steep. We feel that there is room for these kinds of drugs.

Shah: You suggested that the hERG IC_{50} could be correlated with therapeutic concentrations. There's a problem here: we don't know the therapeutic concentrations until we are well into phase II trials when one normally undertakes dose–response studies.

Brown: If we don't have the therapeutic concentrations, the substitute data we do have are the IC_{50}s for the primary target. This should be available at that stage.

Shah: You talked about metabolites. We don't know the full metabolite profile in humans until we are well into phase I.

Brown: This is where, if you have a concern, a primate study is the way to go. Canine studies can be misleading. Primates have a p450 system that is similar to humans.

Shah: You also talked about terodiline, a drug which was withdrawn from the market because it induced QT prolongation and TdP in human. Tolterodine is a structural analogue of terodiline. In addition to anticholinergic properties, they are both Ca^{2+} channel blockers but one prolongs the QT interval and the other doesn't. No one so far has mentioned isomers or enantiomers of QT-prolonging drugs. For terodiline, it is only the R isomer that is cardiotoxic (Hartigan-Go et al 1996). Tolterodine, which is on the market and for which we have no reports of QT prolongation in association with its clinical use, is also the R-isomer of that drug. It is important to bear in mind that although the principal chemical structure might be the same, as is probably everything else, the interactions between hERG and some chiral drugs are stereoselective. As well as the racemic mixture, both enantiomers should also be studied separately for their effect on hERG.

Brown: There is no question that the Ca^{2+} block by tolterodine is much more potent than the block by terodiline. QT prolongation is linked to EADs, which are linked to Ca^{2+} influx and this sequence may explain the relative safety of tolterodine.

Hoffmann: I would like to make a comment on the mixed cardiac ion channels, similar to the point Rashmi Shah made. There are two famous cases, verapamil and clozapin, where it works nicely and we don't see many instances of Torsade de Pointes in humans. On the other hand, our personal experience is that if you have a mixed cardiac ion channel effect in an inhibition situation, it usually doesn't work out well. There are a lot of other effects. The balanced action potential duration may have a high price in terms of Na^+ and Ca^{2+} channel effects. In our experience mixed cardiac ion channel inhibitory situations usually create problems. Examples such as verapamil are the exception, it seems.

January: Amiodarone would be the example that goes against that.

Hoffman: Amiodarone works on gene expression, too. It is a very complex drug.

Hancox: It is not just with candidates that mixed ion channel block is an issue. You can take drugs such as the selective serotonin reuptake inhibitors, and they have Ca^{2+} channel block and hERG block. At least one of them not only produces some block but also shifts the Ca^{2+} window (Witchel et al 2002). Yet they are generally very safe. I can see that if you have a candidate drug with mixed ion channel block you might think very carefully, but there are plenty of drugs out there in addition to verapamil that probably have similar profiles.

Shah: We should make a distinction between non-cardiovascular drugs and non-anti-arrhythmic drugs. We should have made this a long time ago. If you have an anti-angina (that is, non-antiarrhythmic) drug with a significant potential to prolong the QT interval, it is (these days) as good as finished — the fact that it is a cardiovascular drug does not rescue it. On the other hand, the regulatory outcome may not be as bad if the drug is an antiarrhythmic drug and therefore, an effect on ion channels is not unexpected.

Rosenbaum: It is important to mention that at the present time there is no role for drug therapy in the treatment of life-threatening ventricular arrhythmias. Until better drugs come along, drug therapy has been relegated to atrial arrhythmias.

Shah: Yes, class III antiarrhythmic drugs now in development are all for atrial arrhythmias.

Hondeghem: I would like to comment on the kinetic aspects of interaction with the hERG channel. As is well known in the clinic, some drugs may prolong the QT interval quite a bit but not be a problem; others may do hardly anything and yet cause a problem. If the drug interacts with the hERG channel in such a way that it sets off tremendous instability and beat–beat oscillations, this is when you also get the biggest problem. This has an important consequence. With certain drugs, if you don't scan the whole frequency of the spectrum you may miss it. For example, cisapride will give instability problems primarily around cycle lengths of 300 ms; if you cycle at 1000 ms the drug may not elicit much instability. Thus, to effectively uncover its problem, you need to pace at fast heart rates. Others, such as terfenadine, require you to go slow because if you go fast you won't see anything. So to do effective evaluation of a new chemical entity it is required that you study many frequencies of stimulation.

Buzz Brown, I have a question about your IC_{50}s. When you did Ca^{2+} channel blocking work you could dial in any IC_{50}, because if you did a lot of depolarizations per unit time, then your drug was tremendously potent, but if you only did a few, then the calcium channel blockers appeared much less potent. If you could dial in any IC_{50} for Ca^{2+} channel blockers, would you not have the same problem with hERG blockers?

Brown: We do the IC_{50}s at a frequency of around 1 every 5 seconds. But we always run use-dependence curves, in which we will go from 0.3 to 3 Hz. There we have observed that the hERG channel itself has use-dependence so we have to scale and normalize to the use-dependence of the hERG channel. Then we see all the tracking and kinetic effects. I wish we had more time to do this. Some of these drugs have really interesting kinetics. We have done thousands of drugs now, and in the hERG assay we have seen reverse use-dependence just twice. We know that most of these drugs show reverse use-dependence on the action potential or on the QTs. Reverse use-dependence for the hERG channel must be extremely rare.

Noble: I want to reinforce what Jules Hancox said with regard to multiple action drugs. The devil here lies in the detail, as with all problems of biological complexity. It is not a simple matter of single-action hERG blocking drugs bad, multi-action drugs good. It is the profile that will be important, and we should be trying to find the correct profiles. I am not surprised that some multiple action drugs are terrible; most will be. Incidentally, most drugs are terrible, too.

Brown: I would like to reinforce what Jules said, too. We have come a long way with the selective dopamine transmitter that has been out there forever. It is known to be perfectly safe in humans and is very potent in terms of reverting atrial fibrillation. Just as there are a lot of drugs that have been grandfathered that have a real QT liability, there are drugs out there that have a hERG liability that have never been picked up for a QT liability.

Traebert: Have you counted how many compounds have a lower IC_{50} if you patch clamp at 37 °C compared with room temperature?

Brown: We did a limited series of 14 drugs. Of these, the two that showed a big difference were erythromycin and sotalol. Erythromycin was eight times more potent at 35 °C than it was at room temperature. It is a good example illustrating the complications of drug block. Erythromycin is very slow. It is not simply a question of rate of access in terms of its block. It has to do with whether you can ever get the erythromycin at the proper concentration at the target site.

References

Hartigan-Go K, Bateman ND, Daly AK, Thomas SHL 1996 Stereoselective cardiotoxic effects of terodiline. Clin Pharmacol Ther 60:89–98

Witchel HJ, Pabbathi VK, Hofmann G, Paul AA, Hancox JC 2002 Inhibitory actions of the selective serotonin re-uptake inhibitor citalopram on HERG and ventricular L-type calcium currents. FEBS Lett 512:59–66

Structural determinants for high-affinity block of hERG potassium channels

John Mitcheson, Matthew Perry, Phillip Stansfeld, Michael C. Sanguinetti*, Harry Witchel† and Jules Hancox†

*Department of Cell Physiology and Pharmacology, University of Leicester, Maurice Shock Medical Sciences Building, University Road, Leicester LE1 9HN, UK, *Department of Physiology, Nora Eccles Harrison Cardiovascular Research & Training Institute, University of Utah, Salt Lake City, UT 84112, USA, and †University of Bristol, Cardiovascular Research Laboratories, Department of Physiology, Medical School, Bristol BS8 1TD, UK*

Abstract. Drug-induced long QT syndrome is an abnormality of cardiac action potential repolarization that can induce arrhythmias and sudden death. This unwanted side effect of some medications is most frequently associated with block of hERG channels, even though it could theoretically result from inhibition of any K^+ current with a role in repolarization. Recent studies suggest an explanation for why so many structurally diverse compounds preferentially block hERG. State dependent inhibition of hERG channel currents and slow kinetics for recovery from block suggest that many drugs bind within the inner cavity of the channel and are trapped by closure of the activation gate upon repolarization. Drug trapping studies indicate that the inner cavity of hERG is larger than other voltage-gated K^+ channels. Scanning Ala mutagenesis of S6 and pore helix domains that line the inner cavity of hERG have demonstrated that two aromatic residues (Tyr652 and Phe656) are important sites of interaction for most blockers investigated so far. These residues are unique to the EAG channel family. Ser624 and Thr623 residues at the base of the pore helices are also critical for high-affinity binding for some compounds (e.g. methanesulfonanilides) but not others (cisapride, terfenadine, propafenone).

2005 The hERG cardiac potassium channel: structure, function, and long QT syndrome. Wiley, Chichester (Novartis Foundation Symposium 266) p 136–154

*h*ERG (human ether-á-go-go related gene) encodes the pore forming, α-subunit of channels that conduct the rapid delayed rectifier potassium (K^+) current (I_{Kr}). hERG is one of several voltage-sensitive K^+ channels that help to regulate repolarization of the cardiac action potential. Unintentional block of hERG channels by quite a large number of medications can cause drug-induced long QT syndrome (LQTS), a cardiac disorder that may induce arrhythmias and sudden death due to abnormal action potential repolarization (Roden 1998). The diversity of compounds that block hERG and therefore are at risk of causing LQTS is one of the most extraordinary features of this channel. To date, it has not been possible to predict, based on chemical structure alone, whether new compounds have the potential to block hERG. Therefore it has become necessary to use relatively expensive and time consuming *in vitro* and *in vivo* screening approaches to identify compounds with the potential to induce QT prolongation as early as possible in the drug development process (Fermini & Fossa 2003). Clearly, a detailed understanding of the structural basis for the binding of compounds to hERG could facilitate the development of an *in silico* screening tool (Mitcheson et al 2000a, Mitcheson & Perry 2003).

The reason so many structurally diverse compounds blocked hERG in preference to other cardiac K^+ channels was an enigma for many years. Similarities in the pharmacological profile of hERG channel blockers suggested a common binding site and provided the first indications that hERG was structurally different from other voltage gated K^+ (K_v) channels. Block occurred from the intracellular side of the channel, the channels were required to open and recovery from block was extremely slow for many compounds. These findings led to the hypothesis that many hERG channel blockers gained access to a shared binding site within the inner cavity and were then trapped there by closure of the activation gate. In this review we will present the evidence for trapping of blockers in the cavity of K channels, highlight the differences in drug trapping that led us to propose that there are important differences in the size and structure of the inner cavity of hERG compared to other K_v channels, and present studies that have characterised the residues within the inner cavity that form the binding site for long QT compounds and provide a structural explanation for the unusual pharmacological properties of hERG channels.

Early experiments by Armstrong on open channel block of squid voltage-activated K^+ channels by charged quarternary amines (QA) provided us with a great deal of information on the nature of the activation gate and the topology of the K^+ channel pore (Armstrong 1968, 1969, 1971). Armstrong showed that QA needed to be applied from the intracellular side of the membrane and channels needed to be open before the compounds gained access to their receptor (Armstrong 1969). Furthermore, increasing the length of the alkyl groups increased blocking potency by slowing dissociation of the compound from the receptor (Armstrong 1971). These studies suggested that the activation gate was a barrier for the binding of QA to its receptor, that the gate was on the cytoplasmic side of the membrane and that the QA receptor consisted of a relatively large pocket lined by hydrophobic residues (Holmgren et al 1997). In general, QA blockers interfered with closure of the activation gate upon repolarization and caused a slowing of deactivation because the blocker prevented closure of the

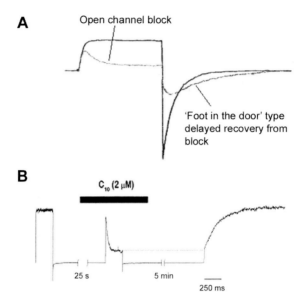

FIG. 1. (A) An example of foot in the door type recovery from block in Shaker K^+ channels (From Choi et al 1993 with permission). Currents before (dark trace) and during application of 10 μM decylethylammonium (C10) are shown superimposed. During the depolarization channels open and are then blocked (open channel block). Upon repolarization there is an apparent delayed recovery from block. The activation gate (door) is unable to close until the blocker (foot) has exited the channel. (B) An example of trapping of blocker by closure of the activation gate in the Shaker mutant I470C (From Holmgren et al 1997 with permission). Open channel block is seen in the presence of blocker so that only 20% of current is remaining at the end of the depolarisation. Upon repolarization the channels close trapping blocker inside the channel. After a long period of blocker washout a rapidly activating component of current corresponding to unblocked channels is seen, followed by a second component of current that increases slowly as channels recover from block.

activation gate. Armstrong later named this the 'foot in the door' effect (see Fig. 1). Holding the membrane at hyperpolarized potentials slowed recovery from block as though the rapid closure of the activation gate was preventing drug dissociation. This was interpreted as drug trapping, although it was not very efficient and slowed unblock rather than preventing it altogether. Subsequent studies on K^+ and Na^+ channels have shown that channel closure can occur 'silently', with no slowing of deactivation and with the blocker still bound to the channel (e.g. Strichartz 1973, Wagoner & Oxford 1990, Choquet & Korn 1992). Channels remained blocked even after drug washout (see Fig. 1). These findings were consistent with the blocker being trapped behind the activation gate within a cavity that was sufficiently large to accommodate the blocker even when the activation gate was closed (Yellen 1998).

The kinetics and state dependence of block of I_{Kr} and hERG channels by many compounds is consistent with drug trapping. Carmeliet characterised the kinetics of I_{Kr} block by dofetilide and almokalant in ventricular myocytes (Carmeliet 1992, 1993). He showed that recovery from block was slow at $-50\,\text{mV}$ and virtually absent at $-75\,\text{mV}$. This is consistent with drug trapping because at $-50\,\text{mV}$ some channel opening occurs allowing the blocker to escape, whereas at $-75\,\text{mV}$ channel open probability is very low and the blocker is unable to exit the inner cavity. More direct evidence for drug trapping in hERG channels was obtained by exploiting the unique gating properties of D540K hERG (Mitcheson et al 2000b). The drug trapping hypothesis predicts that the electrical gradient across the membrane at hyperpolarized potentials should favour recovery from block by a positively charged drug molecule if the channel is open. We tested this hypothesis with a mutant hERG channel (D540K) that opens at hyperpolarized potentials (Sanguinetti & Xu 1999).

Characterization of D540K hyperpolarization-dependent opening

The unusual gating properties of D540K hERG are shown in Fig. 2. Depolarization from a holding potential of $-90\,\text{mV}$ elicits an initial instantaneous component of current through channels that are open at the holding potential, which then rapidly inactivate. The time course of current activation is obscured by the onset of inactivation, but both activation and deactiviation are relatively fast. However, it is at hyperpolarized potentials that the greatest differences in channel behaviour compared to wild-type hERG are observed. Hyperpolarization results in a slowly activating inward current, which is sustained during long voltage pulses. The current–voltage relationship for steady state D540K hERG channel currents is similar to wild-type hERG at depolarized potentials, showing inward rectification and a peak outward current at $-20\,\text{mV}$. Between -60 and $-90\,\text{mV}$ the current amplitudes are small and the slope conductance is low indicating that at this voltage range most of the channels are in the closed state. At hyperpolarized potentials peak inward currents are large and increase linearly with voltage. Single-channel currents recorded using the cell attached configuration of the patch clamp technique show there is no difference in the single-channel amplitudes at any potential (Mitcheson et al 2000b). In wild-type channels, voltage steps to hyperpolarized potentials result in brief channel openings at the beginning of the voltage step as channels recover from inactivation and then deactivate. By comparison, D540K single-channel records are characterized by channel openings throughout the duration of the voltage step. Progressive hyperpolarization increases single-channel burst duration and open probability (Mitcheson et al 2000b). The D540K mutation destabilizes the closed state of hERG through an electrostatic repulsion with R665 on the

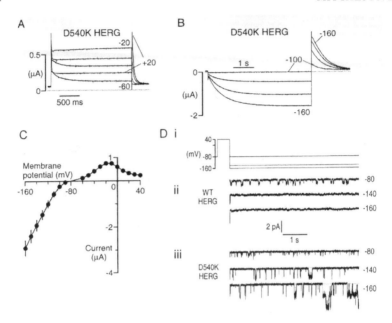

FIG. 2. D540K hERG channels open at hyperpolarized potentials because of destabilization of the closed state. (A, B) Representative whole cell current traces in response to depolarising (A) and hyperpolarizing voltage pulses (B). Tail currents were recorded at −70 mV. (C) Mean whole cell current-voltage relationship showing large inward currents at hyperpolarized potentials. (D) Comparison of wild-type (ii) and D540K hERG (iii) single channel currents in response to voltage steps from +40 to −80, −140 or −160 mV. After a brief opening upon stepping to −140 or −160 mV (as channels recover from inactivation), wild-type hERG channels remained closed (deactivated) during the remainder of the voltage step. In contrast, D540K hERG channels opened repeatedly at all potentials, with open probability being higher at the end of a step to −160 mV than at the beginning. Single channel current amplitudes were not significantly different at any potential within the range of −60 to −160 mV. From Mitcheson et al (2000b) with permission.

C-terminal end of S6 that results in hyperpolarization-dependent channel opening without altering channel conductance and K^+ selectivity (Tristani-Firouzi et al 2002).

Hyperpolarization-induced channel opening of D540K hERG facilitates recovery from MK-499 block

To test for hyperpolarization-dependent recovery from block, we applied depolarizing pulses to 0 mV repetitively until steady-state block was achieved. A concentration of 2 μM MK-499 was used for all experiments, which is sufficient to inhibit wild-type and D540K hERG currents by more than 90%. 5 s hyperpolarizing pulses to $-160\,\text{mV}$ were applied in the continued presence of MK-499 and recovery from block assessed with a depolarizing pulse to 0 mV. After 40–50 hyperpolarizing pulses the mean recovery from block of wild-type hERG currents was $5.0 \pm 0.9\%$. In contrast, the mean recovery of D540K hERG currents was $95 \pm 2.8\%$ with the same protocol (Mitcheson et al 2000b). Recovery from block could be seen during the first hyperpolarization and currents at the end of each pulse increased progressively with each hyperpolarization. Ala substitution at position Asp540 produces a channel that is unable to reopen with hyperpolarization, but has very similar voltage-dependent properties as D540K at depolarized potentials (Sanguinetti & Xu 1999). We found that D540A behaved in the same way as wild-type hERG, indicating that reopening at hyperpolarized potentials rather than an allosteric affect on drug binding due to mutation of Asp540, is critical for D540K recovery from block. As predicted for a positively charged compound, recovery of D540K channels from block by MK-499 was faster at more hyperpolarized potentials. These results provided direct evidence that slow recovery from block by methanesulfonanilides is due to trapping of the compounds in the inner cavity by closure of the activation gate, and indicated that the inner cavity is where the binding site is located (Mitcheson et al 2000b).

Trapping of aminopyridines, Ba^{2+} ions and small QA compounds has been demonstrated for a number of different K_v channels. However, these molecules are all relatively small compared to the compounds that can be trapped by I_{Kr} and hERG channels. MK-499, for example, is more than three times longer than tetraethylammonium (TEA). These findings suggest that the inner cavity of hERG is larger than many K_v channels, but the structural explanation for this is not clear. The inner helices of K^+ channels have a number of Gly or Pro residues that act as points of weakness in α helices and may provide the flexibility required for movements of the inner helices during activation gating (see Fig. 3). hERG and K_v channels have a Gly residue (G648 in hERG) in the equivalent position to the pivot point for gating of MthK and KvAP channels (Jiang et al 2002, 2003). They also have a second potential hinge point lower down the inner helices. Whereas K_v channels have a conserved Pro-X-Pro motif that creates a flexible hinge necessary for gating (Labro et al 2003), hERG has a Gly in the equivalent position to the second Pro. At present it is not clear if both hinge points are required for normal activation gating of hERG and K_v channels. The Pro-X-Pro motif of K_v channels may also limit the size of the inner cavity, but other studies suggest there is an additional more important determinant of cavity size. Shaker channels are unable to trap TEA, and repolarization results in clear 'foot in the door type' unblocking. However, mutation of a single inner cavity residue from Ile to the smaller Cys alters vestibule size so that the channel not only traps TEA, but also the much longer QA

	K⁺ signature sequence	648 652 656 Inner helices
hERG	LTSVGFGNVSPNTNSEKIFSICVML	IGSLMYASIFGNVSAIIQRLY
hEAG	LTSVGFGNIAPSTDIEKIFAVAIMM	IGSLLYATIFGNVTTIFQQMY
HCN	MLCIGYGRQAPESMTDIWLTMLSMI	VGATCYAMFIGHATALIQSLD
hKv1.1	MTTVGYGDMYPVTIGGKIVGSLCAI	AGVLTIALPVPVIVSNFNYFY
Shaker	MTTVGYGDMTPVGFWGKIVGSLCVI	AGVLTIALPVPVIVSNFNYFY
KcsA	ATTVGYGDLYPVTLWGRCVAVVVMV	AGITSFGLVTAALATWFVGRE
KvAP	ATTVGYGDVVPATPIGKVIGIAVML	TGISALTLLIGTVSNMFQKIL
MthK	IATVGYGDYSPSTPLGMYFTVTLIV	LGIGTFAVAVERLLEFLINRE
KirBac1.1	LATVGYGDMHPQTVYAHAIATLEIF	VGMSGIALSTGLVFARFARPR

FIG. 3. Alignment of K⁺ signature and inner helix sequences from selected K⁺ and hyperpolarization-activated cation channels. Position of Gly or Pro hinges that provide flexibility for movement of C-terminal ends of inner helices during gating. Shaker and mammalian K$_V$ channels have a Pro-X-Pro motif (black box). Mutation of first Pro makes channels non-functional, whereas mutation of second Pro slows activation and shifts it to more depolarised potentials (Labro et al 2003). The positions of S6 and pore helix residues important for drug binding in hERG channels are indicated by arrows or by underlining in grey. Alignments were generated using Clustal W (European Bioinformatics Institute).

compound decyltriethylammonium (Holmgren et al 1997). This suggests that the Pro-X-Pro motif alone is not sufficient to explain the inability of K$_V$ channels to trap larger molecules and that the length and volume of residues on the inner helices and how they point into the inner cavity may also be critical.

Alanine scanning mutagenesis of the inner helices identified several residues that are important for drug binding

Initial attempts to locate the hERG channel drug binding site used the 100-fold lower affinity of EAG channels for dofetilide to identify domains and single residues that were responsible for the different pharmacological properties of these closely related channels. Studies on EAG/hERG channel chimeras identified the pore of the channel (S5–S6) as the location of the binding site and showed that an intact C-type inactivation gating process was required for high-affinity block (Ficker et al 1998). Subsequent mutagenesis studies identified key residues for block by several long QT compounds on S6 and at the base of the pore helices. Lees Miller *et al* (2000) used data on drug binding sites identified in K$_V$ and Na⁺ channels together with S6 sequence alignments to identify putative drug binding sites in hERG (Lees-Miller et al 2000). They determined that the mutation F656V dramatically reduced the sensitivity of hERG to dofetilide and quinidine. In our own study we hypothesized that a number of residues that face into the inner cavity would form the drug-binding site (Mitcheson et al 2000a). Using the KcsA crystal structure as a template for the pore of K^+ channels, we identified S6 and pore helix residues as the most likely to line the inner cavity. Ala scanning mutagenesis was used to investigate the role of individual residues in block of hERG channels by MK-499. Since then we have looked at a number of different compounds and the results from a study on clofilium and ibutilide are shown in Fig. 4. Each bar represents the current remaining after steady state block with 300 nM of each drug. Four mutations on S6 (G648A, Y652A, F656A, V656A) and three mutations on the pore helices (T623A, S624A, V625A) considerably reduced block by both compounds. The S6 residues G648, Y652 and F656 are spaced four residues apart and based on homology with K^+ channel crystal structures are predicted to face into the inner cavity, separated by slightly more than a single turn of the S6 α helix. V659A, the fourth S6 mutant with reduced sensitivity to ibutilide and clofilium, deactivates very slowly and therefore does not close between depolarizations. Homology models of hERG based on KcsA or MthK show that V659 is facing away from the central pore of the channel. Therefore the decreased drug block of V659A hERG channels is probably due to a reduction of drug trapping rather than disruption of a direct interaction with drug molecules.

Substitution of Gly648 for Ala has a dramatic effect on binding affinity for some compounds (MK-499, ibutilide and clofilium) but little effect for others (terfenadine, cisapride and propafenone). Gly648 is at a position equivalent to the Gly hinge in MthK that is proposed to be the pivot point for movements of the inner helices during activation gating. Mutation of such a critical residue to Ala, which would reduce S6 flexibility, might be expected to dramatically alter the voltage and time-dependent kinetics of activation. However, this is not the case for G648A hERG. In fact, the mid-point for G648A hERG activation is -33 ± 2.5 mV compared to -16.7 mV for wild-type hERG (Mitcheson et al 2000a), suggesting it is not the only hinge point for activation gating. The negative shift in the voltage dependence of activation and the observation that block by some compounds is normal, indicate that an inability of G648A hERG channels to open preventing access of drugs to the inner cavity is not the mechanism responsible for reduced block by MK-499, clofilium and ibutilide. The voltage dependence of G648A hERG inactivation is shifted to negative potentials. Enhanced inactivation is not expected to reduce drug potency (see Sanguinetti et al 2005, this volume). Such a conserved change in amino acid is also unlikely to have a direct detrimental effect on drug interactions. Therefore the most likely explanation is that G648A exerts its effect on drug binding by sterically hindering drug interactions at neighbouring positions.

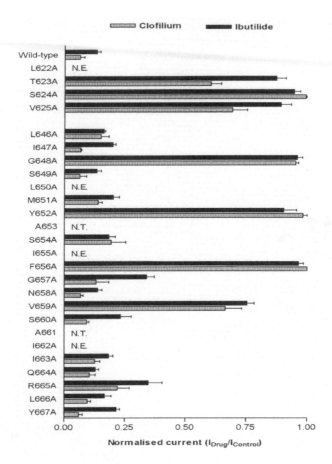

FIG. 4. Ala-scanning mutagenesis of pore helix and S6 domains of hERG. Inhibition of Ala mutants by clofilium and ibutilide was determined by calculating the percentage reduction in currents after reaching steady-state block following repetitive 5 s depolarizations to 0 mV. The concentration of drug used was 300 nM in all experiments, which is approximately 10 times the IC_{50} for both compounds. Large bars indicate mutants that are insensitive to block. NT corresponds to residues not tested; NE refers to channels that didn't give functional expression. Bottom panel, structures of ibutilide and clofilium. From Perry et al (2004) with permission.

Mutation of either Tyr652 or Phe656 drastically reduces the potency of channel block by nearly every compound tested so far (Ridley et al 2004, Kamiya et al 2001, Mitcheson et al 2000a, Lees-Miller et al 2000, Sanchez-Chapula et al 2002, 2003). For example mutation of Tyr652 results in a 94-fold increase in IC_{50} for MK 499, a ~500-fold increase for chloroquine and a 125-fold increase for quinidine (Sanchez-Chapula et al 2002, 2003, Mitcheson et al 2000a). The shifts in IC_{50} are even greater for F656A, with a 650-fold increase in IC_{50} for MK-499 compared to wild-type hERG (Mitcheson et al 2000a). So far only block by fluvoxamine appears to be insensitive to mutation in both of these residues (Milnes et al 2003). These findings indicate that Tyr652 and Phe656 are particularly important determinants of binding to hERG for a variety of structurally diverse compounds. Pharmacophore models indicate that the important features of most hERG channel blockers are a basic nitrogen, that is commonly protonated at physiological pH and one or more aromatic groups arranged around the central charged nitrogen (see Recanatini et al 2005, this volume). Thus, Tyr652 and Phe656 furnish hERG channels with the ability to form hydrophobic, π-stacking and cation-π interactions with drugs (see Sanguinetti et al 2005, this volume). Other K_v channels have aliphatic Ile or Val residues in the equivalent positions (Hanner et al 2001, Decher et al 2004). The presence of Y652 and F656 provides a structural explanation for the diversity of high affinity blockers of hERG.

Pore helix residues are important sites of interaction for some drugs

A comparison of binding sites indicates that Tyr652 and Phe656 are important for nearly all high affinity blockers, whereas there is much more variation with respect to residues closer to the selectivity filter. S624A has a relatively minor effect on MK-499 block, but a major effect on clofilium, ibutilide and vesnarinone block. Clofilium and ibutilide are structurally similar compounds. Clofilium is a quaternary whereas ibutilide is a tertiary amine and the other difference is in the polar group attached to the aromatic ring. Both compounds have a similar IC_{50} for hERG and Ala scanning mutagenesis reveals similar interactions with the S6 amino acids (Perry et al 2004). However, there are clear differences in the strength of interactions at the pore helix residues. For clofilium, mutation of Ser624 has a much greater effect on block than Thr623 and Val625, whereas all three pore helix residues are important for ibutilide binding. This is likely to reflect differences in specific interactions between the pore helix residues and the halogen and methanesulfonamide groups of clofilium and ibutilide, since this is where the compounds differ most significantly.

Further evidence for the importance of Ser624 in drug binding comes from comparing the recovery from block kinetics of wild-type and mutant hERG channels. Clofilium recovery from block in D540K and wild-type hERG channels is very slow. In the case of D540K this is surprising because as we have previously showed hyperpolarization-induced channel opening of D540K facilitates rapid recovery of drugs trapped by closure of the activation gate. The slow dissociation of clofilium from D540K channels therefore suggests that something other than trapping by the activation gate was preventing unblock. Following a washout period, V625A and T623A showed a similar slow recovery from block to wild-type hERG channels. In contrast, rapid recovery from block was observed for S624A hERG channels after the washout period. These results suggest a highly specific interaction with Ser624 is important for drug binding to clofilium at depolarised potentials and is also responsible for the remarkably slow recovery from clofilium block of wild-type hERG channels (Perry et al 2004). The intracellular TEA binding site for shaker is Thr441 (Choi et al 1993) and is located at a position equivalent to Thr623 in hERG. This site is important for block of Kv1.5 channels by bupivacaine and benzocaine, and the neighbouring residue, which is equivalent to Ser624 in hERG, is important for block of Kv1.5 by S0100176 (Franqueza et al 1997, Caballero et al 2002, Decher et al 2004). Thus interactions with conserved polar residues adjacent to the selectivity filter are a common feature of binding sites in K^+ channels.

Propafenone block of hERG channels

Closed channel homology models of hERG based on KcsA crystal structures suggest that the inner cavity is lined by relatively few residues, which include Thr623, Ser624, Gly648, Ser649, Tyr652 and Phe656 (see Fig. 5). Although most hERG channel blockers require several of these residues for normal binding, propafenone is unusual because it only interacts with Phe656 (Witchel et al 2004). Propafenone shows use and voltage dependent block consistent with open state channel block (Mergenthaler et al 2001, Paul et al 2002, Arias et al 2003). Recovery from block is slow and most studies show little effect of propafenone on channel deactivation, indicating that like many other hERG channel blockers propafenone is trapped within the inner cavity by closure of the activation gate. Untrapping is facilitated by hyperpolarization-dependent D540K channel opening. Inspection of docking to open and closed hERG channel homology models, based on KcsA and MthK crystal structures, respectively, suggest that the mutagenesis data can readily be explained in the open state. There was a greater preference for interactions involving Phe656 than Tyr652. Phe656 residues were fully accessible to propafanone and both aromatic rings were able to make simultaneous π-stacking interactions with the aromatic rings of Phe656. In contrast, propafenone was unable to favourably interact with Phe656 residues in the closed channel models. The lowest energy conformations always involved movement of propafenone further into the inner cavity. In the closed channel

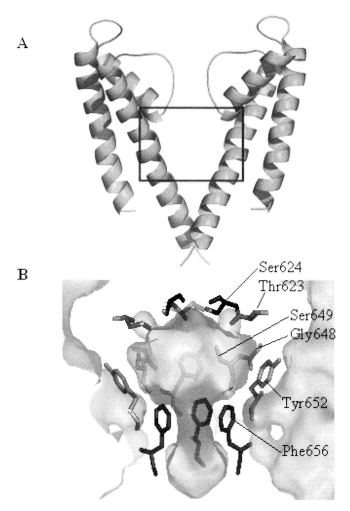

FIG. 5. Structure of inner cavity of hERG. (A) Side view of S5-S6 domains of closed channel homology model of hERG based on KcsA crystal structure (Doyle et al 1998). Only two subunits of the tetrameric channel are shown. (B) 3D representation of molecular surface of inner cavity and region of model indicated by box in panel A. The residues that contribute to drug binding are shown in stick form. Ser649 is shown because it lines the inner cavity, although there is no evidence that the side chain contributes to the drug binding site. The figure in panel B was generated using PyMOL (DeLano scientific LLC).

model there was insufficient space at the cytoplasmic end of the cavity for propafenone to make favourable interactions with Phe656 residues. The results of propafenone docking to the closed model are not consistent with the experimental data. Propafenone does not interfere with channel deactivation indicating that the channels can close around propafenone without an energetic consequence. The slow recovery from block even at hyperpolarized potentials suggests that conformational changes prevent positively charged propafenone molecules from exiting the channel (Witchel et al 2004). The structural basis for this remains unclear, but our results suggest that the inner cavity of hERG is larger at the level of Phe656 than closed state models, based upon KcsA, indicate. It is also possible that flexibility of the inner helices at a Gly residue immediately downstream of Phe656 permits channel closure without interfering with propafenone binding at position Phe656. Further studies to investigate the role of Gly residues in hERG channel gating and drug trapping are required.

In summary, recent studies have provided direct evidence that hERG channel blockers bind within the inner cavity, and are trapped there by closure of the activation gate upon repolarization. Drug trapping increases the affinity of drugs for hERG by slowing the rate of drug dissociation from its binding site. The inner cavity of hERG is larger than other K_v channels and is lined by aromatic and polar residues that permit hydrophobic, electrostatic and polar interactions with a wide range of compounds.

References

Arias C, Gonzalez T, Moreno I et al 2003 Effects of propafenone and its main metabolite, 5-hydroxypropafenone, on HERG channels. Cardiovasc Res 57:660–669

Armstrong CM 1968 Induced inactivation of the potassium permeability of squid axon membranes. Nature 219:1262–1263

Armstrong CM 1969 Inactivation of the potassium channel conductance and related phenomena caused by quarternary ammonium ion injection in squid axons. J Gen Physiol 54:553–575

Armstrong CM 1971 Interaction of tetraethylammonium ion derivatives with the potassium channels of giant axons. J Gen Physiol 58:413–437

Caballero R, Moreno I, Gonzalez T et al 2002 Putative binding sites for benzocaine on a human cardiac cloned channel (Kv1.5). Cardiovasc Res 56:104–117

Carmeliet E 1992 Voltage- and time-dependent block of the delayed K+ current in cardiac myocytes by dofetilide. J Pharmacol Exp Ther 262:809–817

Carmeliet E 1993 Use-dependent block and use-dependent unblock of the delayed rectifier K+ current by almokalant in rabbit ventricular myocytes. Circ Res 73:857–868

Choi KL, Mossman C, Aube J, Yellen G 1993 The internal quaternary ammonium receptor site of Shaker potassium channels. Neuron 10:533–541

Choquet D, Korn H 1992 Mechanism of 4-aminopyridine action on voltage-gated potassium channels in lymphocytes. J Gen Physiol 99:217–240

Decher N, Pirard B, Bundis F et al 2004 Molecular basis for Kv1.5 channel block: conservation of drug binding sites among voltage-gated K^+ channels. J Biol Chem 279:394–400

Doyle DA, Morais Cabral J, Pfuetzner RA et al 1998 The structure of the potassium channel: molecular basis of K^+ conduction and selectivity. Science 280:69–77

Fermini B, Fossa AA 2003 The impact of drug-induced QT interval prolongation on drug discovery and development. Nat Rev Drug Discov 2:439–447

Ficker E, Jarolimek W, Kiehn J, Baumann A, Brown AM 1998 Molecular determinants of dofetilide block of HERG K^+ channels. Circ Res 82:386–395

Franqueza L, Longobardo M, Vicente J et al 1997 Molecular determinants of stereoselective bupivacaine block of hKv1.5 channels. Circ Res 81:1053–1064

Hanner M, Green B, Gao YD et al 2001 Binding of correolide to the K(v)1.3 potassium channel: characterization of the binding domain by site-directed mutagenesis. Biochemistry 40:11687–11697

Holmgren M, Smith PL, Yellen G 1997 Trapping of organic blockers by closing of voltage-dependent K^+ channels: evidence for a trap door mechanism of activation gating. J Gen Physiol 109:527–535

Jiang Y, Lee A, Chen J et al 2003 X-ray structure of a voltage-dependent K+ channel. Nature 423:33–41

Jiang YX, Lee A, Chen JY et al 2002 The open pore conformation of potassium channels. Nature 417:523–526

Kamiya K, Mitcheson JS, Yasui K, Kodama I, Sanguinetti MC 2001 Open channel block of HERG K^+ channels by vesnarinone. Mol Pharmacol 60:244–253

Labro AJ, Raes AL, Bellens I, Ottschytsch N, Snyders DJ 2003 Gating of shaker-type channels requires the flexibility of S6 caused by prolines. J Biol Chem 278:50724–50731

Lees-Miller JP, Duan Y, Teng GQ, Duff HJ 2000 Molecular determinant of high-affinity dofetilide binding to HERG1 expressed in Xenopus oocytes: involvement of S6 sites. Mol Pharmacol 57:367–374

Mergenthaler J, Haverkamp W, Huttenhofer A et al 2001 Blocking effects of the antiarrhythmic drug propafenone on the HERG potassium channel. Naunyn Schmiedebergs Arch Pharmacol 363:472–480

Milnes JT, Crociani O, Arcangeli A, Hancox JC, Witchel HJ 2003 Blockade of HERG potassium currents by fluvoxamine: incomplete attenuation by S6 mutations at F656 or Y652. Br J Pharmacol 139:887–898

Mitcheson JS, Perry MD 2003 Molecular determinants of high-affinity drug binding to HERG channels. Curr Opin Drug Discov Devel 6:667–674

Mitcheson JS, Chen J, Lin M, Culberson C, Sanguinetti MC 2000a A structural basis for drug-induced long QT syndrome. Proc Natl Acad Sci USA 97:12329–12333

Mitcheson JS, Chen J, Sanguinetti MC 2000b Trapping of a methanesulfonanilide by closure of the HERG potassium channel activation gate. J Gen Physiol 115:229–240

Paul AA, Witchel HJ, Hancox JC 2002 Inhibition of the current of heterologously expressed HERG potassium channels by flecainide and comparison with quinidine, propafenone and lignocaine. Br J Pharmacol 136:717–729

Perry M, de Groot MJ, Helliwell R et al 2004 Structural determinants of HERG channel block by clofilium and ibutilide. Mol Pharmacol 66:240–249

Recanatini M, Cavalli A, Masetti M 2005 *In silico* modelling — pharmacophores and hERG channel models. In: the hERG cardiac potassium channel: structure, function and long QT syndrome (Novartis Found Symp 266). Wiley, Chichester, p 171–185

Ridley JM, Dooley PC, Milnes JT, Witchel HJ, Hancox JC 2004 Lidoflazine is a high affinity blocker of the HERG K^+ channel. J Mol Cell Cardiol 36:701–705

Roden DM 1998 Mechanisms and management of proarrhythmia. Am J Cardiol 82:49I–57I

Sanchez-Chapula JA, Ferrer T, Navarro-Polanco RA, Sanguinetti MC 2003 Voltage-dependent profile of human ether-a-go-go-related gene channel block is influenced by a single residue in the S6 transmembrane domain. Mol Pharmacol 63:1051–1058

Sanchez-Chapula JA, Navarro-Polanco RA, Culberson C, Chen J, Sanguinetti MC 2002 Molecular determinants of voltage-dependent human ether-a-go-go related gene (HERG) K^+ channel block. J Biol Chem 277:23587–23595

Sanguinetti MC, Xu QP 1999 Mutations of the S4-S5 linker alter activation properties of HERG potassium channels expressed in Xenopus oocytes. J Physiol 514:667–675

Sanguinetti MC, Chen J, Fernandez D, Kamiya K, Mitcheson J, Sanchez-Chapula JA 2005 Physicochemical basis for binding and voltage-dependent block of hERG channels by structurally diverse drugs. In: The hERG cardiac potassium channel: structure, function and long QT syndrome (Novartis Found Symp 266). Wiley, Chichester, p 159–170

Strichartz GR 1973 The inhibition of sodium currents in myelinated nerve by quaternary derivatives of lidocaine. J Gen Physiol 62:37–57

Tristani-Firouzi M, Chen J, Sanguinetti MC 2002 Interactions between S4-S5 linker and S6 transmembrane domain modulate gating of HERG K$^+$ channels. J Biol Chem 277:18994–19000

Wagoner PK, Oxford GS 1990 Aminopyridines block an inactivating potassium current having slow recovery kinetics. Biophys J 58:1481–1489

Witchel HJ, Dempsey CE, Sessions RB et al 2004 The low potency, voltage-dependent HERG blocker propafenone—molecular determinants and drug trapping. Mol Pharmacol 66:1201–1212

Yellen G 1998 The moving parts of voltage-gated ion channels. Q Rev Biophys 31:239–295

DISCUSSION

Hondeghem: There is a very substantive difference between ibutilide and clofilium. One is quaternary, the other is tertiary. So perhaps the neutral compound may perhaps diffuse partly through the lipid phase and get out anyway, while the charged one may not be lipid soluble enough? If this is the case, this should be strongly pH dependent as has been shown for Na$^+$ and Ca^{2+} channel blockers: it is possible to modulate the exit rate by changing pH. Have you tried this?

Mitcheson: No. However, what we do have is a close analogue of clofilium, which is a tertiary amine and we see exactly the same phenomenon. At first, we thought that the tertiary–quaternary difference in that nitrogen was going to be important, but it appears that this is not in this case (our unpublished work).

Recanatini: There is another difference between ibutilide and clofilium. Ibutilide is a chiral molecule that exists in two enantiomeric forms. The same holds for MK499. In your experiments, which enantiomer did you use?

Mitcheson: The preparation of ibutilide that we use is racemic. We are hoping to address this by studying the individual enantiomers. I agree with you that this could be important.

Shah: Does this structural analogue of clofilium you are looking at still have a chlorophenyl ring?

Mitcheson: Yes.

Shah: Where is, then, the modification to the structure of this analogue?

Mitcheson: It is at the central nitrogen. As opposed to having the two ethyl groups like clofilium (bottom panel of Fig. 4) it only has the one and depending on pH will be charged or uncharged.

Brown: When you substitute for the polar residues in the pore helix, have you ever checked the unit conductance? What if some of this effect is that they are just not focusing K^+, and so the apparent block is greater but it is really indirectly due to the fact that the K^+ concentration hasn't gone down?

Mitcheson: We see a lot of variability in the time course of currents when we mutate the pore helix positions. For example, Ser624 gates almost exactly like the wild-type channel. This suggests that mutating this position is not really having much effect on conductance.

Brown: That is gating. What about unit conductance?

Mitcheson: We don't know that. All I can say is that there is quite a lot of variability in the effect the pore helix mutations have on drug block. So for a lot of compounds mutating S624 has no effect on drug sensitivity, and for others, like clofilium it does. This suggests that specific differences in the nature of channel-drug interactions in this part of the binding domain are important.

Brown: I was wondering about the templates you used to build the models. You are using the closed KcsA for the trapping and the open model in the other case, but you didn't show what the D540K mutation is doing to the structure. What does this mutation, which removes the trapping, do?

Mitcheson: These models were done by Chris Dempsey at the University of Bristol (Witchel et al 2004). He used MthK as the template for the open model and KcsA as the template for the closed model. The D540K mutation should have no effect on the structure of the inner cavity. D540 interacts with R665 at the C-terminal end of S6. In the wild-type channel the interaction between the D540 and R665 stabilizes the closed state. When you reverse the charge at the D540 residue it destabilizes this closed state, causing an increased probability of the open state at hyperpolarized potentials without changing single channel conductance.

Sanguinetti: Buzz Brown, are you asking where is D540 located?

Mitcheson: On the S4–S5 linker.

Brown: Is the D540 position the same in the open channel as it is in the closed channel?

Mitcheson: We can't say for sure. It is close to the C-terminal end of S4 and so may be repositioned by changes in membrane potential.

Brown: For the Y652A, the tail current was significantly below the steady-state current. This is pretty unusual. Is there something going on with the inactivation?

Mitcheson: You are referring to the fact that there is a lot more outward current at 0 mV relative to the tail current. There might be small differences in the amount of inactivation compared to wild-type hERG, but they are relatively minor.

Netzer: Have you ever paused during pulsing and washed out your compound? We have seen some interesting results with compounds that have more-or-less no recovery. If you make a break with the stimulation but continue superfusing the culture, we observe a recovery from the block. However, during continuous pulsing there is an onset of block again, even if there should be no compound available.

Mitcheson: I think this may depend on how quickly drugs dissociate from the channel. We have only looked at a limited number of compounds with this type of protocol. When we wash out compounds we pause for one or two minutes before we start pulsing and find that there is very little recovery from block. For some compounds that dissociate from the channel more rapidly, some recovery from block is seen. When you continue with the pulsing you then observe resumed onset of block once you open the channels.

Netzer: There should be more-or-less no compound available.

Mitcheson: I think that some may be left in the cytoplasm.

Robertson: What about other EAG family members in drug block. Have you looked?

Mitcheson: No. We haven't done any mutagenesis work on any other EAG related channels.

Robertson: Have you looked at drug block of those channels?

Mitcheson: We haven't but others have. In general, there is a much lower affinity.

Ficker: I was puzzled by the Lily compound (LY97241) you referred to which blocks EAG with high affinity. In a more recent paper (Gessner et al 2004) this drug molecule has been oriented upside-down compared to what you showed. Similarly the authors dock clofilium so that its long lipophilic tail projects into a hydrophobic pore-helix pocket contrary to what you showed. I like this proposal because it may explain why washout of these compounds is so slow. However, I don't think that the orientation of these compounds is different in the inner cavity of hERG and EAG.

Mitcheson: I am aware of that study. Our data don't fit well with that idea. I am not sure that the data are particularly strong support for the idea of the molecule interpolating between the S6 and pore helix.

Ficker: It is puzzling that the washout is so slow for clofilium.

Mitcheson: We are proposing that this is due to an interaction of the clofilium molecule with Ser624, which keeps the drug molecule bound to the channel. It is not occurring for dofetilide and ibutilide. This is down to the complexity of the interactions in this region.

Robertson: So why is EAG not sensitive to the other compounds?

Mitcheson: That's a good question. We have been battling with this problem for a while. EAG has the same amino acids that we have identified, the tyrosine and the phenylalanine, in the same positions. EAG doesn't inactivate, though, whereas hERG does. One proposal is that C-type inactivation is required for high affinity binding. There may be differences in the orientation and positioning of those residues relative to the inner cavity that are different in channels that inactivate compared to those that don't.

Recanatini: I have a comment on the observation of the particular binding of clofilium. Clofilium is highly lipophilic but is a quaternary ammonium compound. It carries two chemical features which combine in strengthening the binding to the inner cavity of hERG, which is quite lipophilic. Moreover, this molecule has a high conformational flexibility. We just picture it in one conformation, which can be oriented upwards or downwards, but it can get into the cavity in different conformations. When it is in, it can also adapt to the inner cavity and improve the binding interaction in a dynamic way.

Ficker: The data in EAG presented by Gessner et al (2004) were a little more complicated. According to these authors the pore helix forms a lipophilic pocket and mutations in this pocket affect clofilium binding. This is quite distinct from whatever other groups have mutated. The long lipophilic tail of clofilium somehow squeezes into the crevice of this pocket located between S6 and the pore helix. My question was whether it is likely that the same molecule has completely different binding modes for hERG and EAG?

Sanguinetti: Maurizio Recanatini can comment on the limitations of these models. I think they are severe. We like to show models and they help us to visualize the mutagenesis results, but knowing the real open state of any channel and the drug conformation is not possible. We have crystal structures in free solutions, but we don't know how they are folding within this lipophilic environment. There are a lot of unknowns. What you are bringing up is the extreme example in which the drug can be completely flipped in one model compared to another.

Hancox: Is the tertiary analogue of clofilium closer to ibutilide in its behaviour, or more to clofilium itself?

Mitcheson: It is actually very like clofilium. It has a similar potency, but then so does ibutilide. As far as the recovery from block is concerned, it is very slow, similar to that seen with clofilium.

Hancox: This suggests that the charge interaction isn't so important.

Mitcheson: It suggests that interaction is not important for the slow rates of recovery from block. It may still be important for other drug-channel interactions.

Brown: What about antibiotics? Where is erythromycin in all this?

Mitcheson: We haven't looked at this, but we have thought about it. The problem is that it is a very low potency blocker and we can't study block in oocytes. This is quite difficult.

Brown: I don't think it is hard to do, even in oocytes if you make a macropatch.

Sanguinetti: Are you asking where the binding site is?

Brown: I'm asking can it as a structure fit into any of these molecules?

Sanguinetti: Almost anything can fit into MthK, and for KcsA there are limits. The channel couldn't close around it.

Noble: I have a more general question prompted by this fascinating discussion. I am amazed at the detail at which we now seem to understand the difference between hERG and other K^+ channels. While I take your point about the limitations of the models, I do have a challenge that fascinates me. hERG is clearly one of nature's disasters. While we can't blame evolution for not anticipating the coming of the pharmaceutical industry, we can ask the following question. Do we yet know enough about the molecular biology and the molecular genetics to ask the question what would have to be done to put this right? Clearly, not all K^+ channels are as sensitive to drugs as hERG. It is possible to build K^+ channels without having this disadvantage. Are there any mutations that influence the IC_{50}s of drugs in relation to hERG? This would be one clue. Another would be simply by considering the molecular biology of drug interaction — could we design a better hERG channel? But the first question is, are there any naturally occurring mutations in hERG that affect the IC_{50}s of drugs?

Sanguinetti: I don't think there are.

Mitcheson: As far as I know that is true.

Sanguinetti: There are no naturally occurring mutations in the key drug binding residues that have been identified by alanine scanning mutagenesis. You might expect that there would be mutations that enhance the sensitivity of the channel to drugs, which would explain some cases of drug-induced LQTS, but none have been reported so far.

References

Gessner G, Zacharias M, Bechstedt S, Schonherr R, Heinemann SH 2004 Molecular determinants for high-affinity block of human EAG potassium channels by antiarrhythmic agents. Mol Pharmacol 65:1120–1129

Witchel HJ, Dempsey CE, Sessions RB et al 2004 The low potency, voltage-dependent HERG blocker propafenone — molecular determinants and drug trapping. Mol Pharmacol 66:1201–1212

General discussion III

Arcangeli: I have a comment regarding MinK versus MiRP. When you use cell lines that stably express hERG channels, you must take into account that the cells can express MinK versus MiRP. For example, we use HEK cells which do not express MinK but express MiRP. The endogenous MiRP co-precipitates with hERG in these cells.

Abbott: I didn't realize this.

Arcangeli: When you test a drug in these cells you have to take into account that you can have hERG and MiRP together. Another interesting feature is that adherent cells are quite different in their assembly of hERG and MiRP, as compared with suspended cells. There are many features that we need to take into account.

Abbott: How much MiRP was there? Was there any saturating effect? Did you look at low levels of hERG versus high levels of hERG, for example?

Arcangeli: The levels of hERG were high because these were stable transfectants.

Abbott: We did a dose–response with the oocytes and hERG (Anantharam et al 2003). I'd be interested in knowing the relative levels of MiRP1 and hERG. It could be that a lot of drug testing is done with a mixture of hERG and hERG–MiRP1.

Robertson: Have you looked for MiRP2?

Arcangeli: No, I haven't.

Abbott: As an aside, we couldn't find *Xenopus* MiRP1. It doesn't mean that it isn't there.

Brown: You are using HEK cells. Does this apply to CHO cells also?

Arcangeli: We haven't looked at CHO cells.

Abbott: I think Steve Goldstein did RT-PCR in CHO and found MiRP1 was there (unpublished study).

Sanguinetti: We need to search a variety of heterologous expression systems to look at differential expression of MinK, MiRP and hERG.

Brown: One of the important things in tissue culture is to start with the ATCC line as passage 1, so you can track where you are in the generations.

Sanguinetti: What changes with time?

Brown: For hERG there haven't been changes with time, but some of the G proteins we have looked at have changed substantially.

January: We've got a stable hERG line that is now seven years old. The current density is still the same. This is in HEK293 cells. Anyone who has used these cells will know that there are many parent cell lines, so one is not the equivalent of the next. As a vehicle for expressing hERG it has been great.

Sanguinetti: I don't think this has been the general experience. I have heard that stably expressed cell lines do vary with time.

Hebert: We have a CHO line that is stable for hERG. It tends to die as soon as I give it to someone. It survived 10–20 generations and progressively declined after that.

January: We have passaged our line around 140 times and it has remained stable. You could of course kill it depending on how carefully these things are handled. There is a certain degree of mystery attached to culturing cell lines.

Netzer: It is also important that people who start to work with K^+ channel cell lines do not take a wild-type cell line, measure it and say it doesn't have any K^+ channels. If you let them grow dense, you will get tremendous endogenous currents, and if you treat them badly then you have a huge mess in your cell culture system. We tested more than 40 different cell lines before we found one line of CHO cells that we wanted to use as future controls. Most of the companies using these cells take a fresh batch out every few months in standard operation procedures (SOPs), but others use them for three years and have 100% positive cells.

Brown: We use somewhere between 30 and 50 passages. This is about the most we can do.

Traebert: We are linked to our SOPs. We can use them as long as they make current. But in our experience we thaw a new batch of cells between passage 20 and 30.

Brown: What I am saying is that in our experience, up to around 30 or 40 passages things are very stable.

Gosling: Would someone like to comment on the metabolic capabilities of HEKs, CHOs and other cells and what impact this can have on hERG screening?

Netzer: This is a subject we have discussed a lot. If the company insists on measuring with HEK cells we bite our lips and do it. But for scientific reasons we prefer to use CHO cells. One reason is the background K^+ current in HEK cells. Another reason is that from the metabolic perspective, a CHO cell is more comparable to a human cardiomyocyte. HEK cells are built to pump, metabolize and be active. CHO cells are more or less neutral. We have seen in the past that compounds are sometimes more active in HEK cells and not as active in CHO cells.

Hoffmann: What do we know about transporters in HEK and CHO cells?

Brown: One of the aspects of the erythromycin story is probably related to transport. We make sporadic attempts to investigate whether transporters are playing a role. We have not done this systematically.

Netzer: The company we work for compared the data we provided them with the data they obtained in their *in vivo* measurements.

Hoffmann: Is there any other experience from other groups that there may be transporters in HEK or CHO cells?

Netzer: There are many transporters in HEK cells. In CHO cells there are also transporters, but far fewer. HEK cells were described to me by a colleague as 'pump wonders'.

Brown: Leaving aside the problems of what is in the bath and what is in the cell, if you still insist on getting to a steady state before you block, you will have problems. For example if the drug is pumping it out, you are not going to get a steady state. The drug is eventually going to overwhelm the transport system, but this will take a long time. Another issue is that HEK cells are easier to patch than CHO cells. For gating purposes there is no advantage to using CHO cells. When you are looking at the tail current, the contaminating currents of the CHO cells aren't in the picture.

Gosling: With the planar patch system it seems to be the inverse: the CHO cells are easier to patch.

Brown: That has to do with clumping: when you float HEK cells into the plane of patching they clump.

Gosling: If that is the direction screening is going, then please start generating CHO lines and not HEKs.

January: As a matter of practice you can select HEK cells that don't clump.

Gosling: How?

January: You look for how well they clump when they are freed from the culture dishes. If you are used to making stable cell lines you will know that you have to pick the right cells to start with.

Netzer: A major problem is that if you have a channel in it, the cell can behave totally differently. Everyone who has tried to put the same cell line on an automated system will know what I am talking about. If you take a CHO line and put Kv1.5, Kv1.3, KCNQ2 and 3, and hERG in, you will have to develop many different protocols for the cell culture and the measurements.

Traebert: We have found that HEK cells transfected with hERG and Na^+ channels behave completely differently. The Na^+ channel cell-line clumps and cells grow very slowly whereas the hERG cell-line grows very nicely.

Hancox: The difference between using HEK/CHO on the one hand and oocytes on the other, to some extent depends on the purpose of the work. Oocytes are better in some respects for mutagenesis work, but the cell lines are more similar to the native system. One of the problems we have with slow-acting drugs is rundown. Does anyone have any magic solution to run-down in CHOs and HEKs?

Netzer: There are lots of day-to-day variances with the cells and run-down. We have seen that there is a bifunctional exponential function behind the run-down. This is what we routinely do every time we perform a biexponential run-down correction, however the non-corrected numbers are also stored. If the run-down is too large we just stop working.

Gosling: With regard to the issue of measuring bath concentration, why do people think this is such a big issue? One strategy used by pharmaceutical companies to optimize potency, selectivity or duration is actually to generate molecules which will be transported into membranes. Thus bath concentration may not be an indication of the concentration in the cell membrane at all.

Brown: Yes. Take terfenadine as an example. The tissue concentration in the heart is 10 times the plasma concentration, which is not the free concentration.

Gosling: So why take a bath measurement?

Brown: You have to start somewhere. The problem is that if you don't do that, most of your drug can be in the perfusion system. At least this way you know what is in the bath. I agree that you don't know what is in the tissue, and it is very hard to know where the drug is in the tissue.

Gosling: John Mitcheson has been questioned about the reversibility of clofilium block. If you are starting to modify the membrane permeability of a compound, its ability to partition into different components in the cell cytoplasm could also be quite profound, particularly with oocytes. Your ability to wash them out might therefore be compromised.

Brown: Or wash them in. With erythromycin, when it is applied extracellularly it doesn't get to where it ought to go, certainly at room temperature.

Reference

Anantharam A, Lewis A, Panaghie G et al RNA interference reveals that endogenous Xenopus MinK-related peptides govern mammalian K^+ channel function in oocyte expression studies. J Biol Chem 278:11739–11745

Physicochemical basis for binding and voltage-dependent block of hERG channels by structurally diverse drugs

Michael C. Sanguinetti, Jun Chen*, David Fernandez, Kaichiro Kamiya†, John Mitcheson‡ and José A. Sanchez-Chapula§

Department of Physiology, Nora Eccles Harrison Cardiovascular Research & Training Institute, University of Utah, Salt Lake City, UT, *Abbott Laboratories, Abbott Park, IL, USA, †Nagoya University, Nagoya, Japan, ‡University of Leicester, Leicester, UK, and §Universidad de Colima, Colima, Mexico

> Abstract. Blockade of hERG K^+ channels in the heart is an unintentional side effect of many drugs and can induce cardiac arrhythmia and sudden death. For this reason, most pharmaceutical companies screen compounds for hERG channel activity early in the drug discovery/development process. A detailed understanding of the drug binding site(s) on the hERG channel could enable rational design of future medications devoid of this unwanted side effect. Towards this goal, we have used site-directed mutagenesis to identify several residues of the hERG channel that comprise a common drug binding site. The initial Ala-scan identified several residues located in the S6 domain (Tyr652, Phe656) and the base of the pore helix (Thr623, Ser624, Val625) as important sites of interaction. Here, we review studies that refine our understanding of the physicochemical basis of interaction by structurally diverse drugs with aromatic residues in the S6 domain. Our findings suggest that the position of Tyr652 and Phe656 in hERG is optimal for interaction with multiple drugs, Tyr652 is an important determinant of voltage-dependent block, and the hydrophobic surface area of residue 656 and aromaticity of residue 652 are the physicochemical features required for high-affinity block by MK-499, cisapride and terfenadine.
>
> *2005 The hERG cardiac potassium channel: structure, function, and long QT syndrome. Wiley, Chichester (Novartis Foundation Symposium 266) p 159–170*

Long QT syndrome (LQTS) is a disorder of ventricular repolarization that predisposes affected individuals to ventricular arrhythmia and sudden death. The inherited form of the disorder is usually caused by mutations in cardiac K^+ or Na^+ ion channels (Keating & Sanguinetti 2001). Acquired LQTS is more common and can be induced as an unintended and rare side effect of drug treatment. In the past few years, several commonly used drugs (e.g. terfenadine, cisapride, sertindole, thioridazine, grepafloxacin) were withdrawn from the market, or their approved use severely restricted when it was discovered that these drugs caused arrhythmia, albeit very infrequently, or were associated with unexplained sudden death (Pearlstein et al 2003).

By far the most common cause of drug-induced LQTS is block of hERG channels that conduct I_{Kr}, the rapid delayed rectifier K$^+$ current (Sanguinetti et al 1995, Trudeau et al 1995). Reduction in I_{Kr} by drugs prolongs action potentials of cardiomyocytes, lengthens the QT interval and increases the risk of ventricular fibrillation and sudden death. An understanding of the physicochemical features of the drug binding site would complement pharmacophore models (Cavalli et al 2002, Ekins et al 2002) of hERG blockers and potentially define the molecular basis of the receptor fields predicted by these models. Alanine-scanning mutagenesis identified three residues located at the base of the pore helix and two aromatic residues, Tyr652 and Phe656, located in the S6 domain, that are critical for high affinity binding of several drugs (Mitcheson et al 2005, this volume). Most Kv channels have an Ile or Val residue in the positions equivalent to Tyr652 or Phe656 of hERG (Fig. 1), indirectly implying that the aromatic residues in S6 are critical components of a high affinity-binding site and explaining why hERG is so sensitive to structurally diverse drugs. The presence of two aromatic residues per subunit (eight per channel) provides several potential sites for π-stacking interactions with an aromatic moiety of a drug. In addition, the S6 domains of most K_v channels have a Pro-Val(Ile)-Pro motif that induces a kink in the α-helix and perhaps reduces the volume of the central cavity compared to channels like hERG that lack these Pro residues. The volume of the central cavity of channels in the closed state might limit the size of drugs that can be trapped by closure of the activation gate (Mitcheson et al 2005, this volume). Here, we review our recent findings that further support the importance of Tyr652 and Phe656 in hERG as residues that mediate interaction with multiple drugs.

Altered drug-sensitivity can be induced by repositioning aromatic residues along the S6 domain of EAG and hERG channels

EAG channels are structurally related to hERG, including the presence of Tyr and Phe residues in the S6 domain in positions homologous to Tyr652 and Phe656 of hERG. Despite structural similarities, EAG channels are normally insensitive to hERG blockers. However, EAG channels can acquire drug sensitivity by introduction of mutations that induce C-type inactivation (Ficker et al 2001). This finding suggests that C-type inactivation of hERG is a key determinant of high-affinity drug binding. However, the link between inactivation and drug-sensitivity has been confused by other findings. For example, some mutations in

```
Channel                    S6 domain
              636              652                 663
hERG      SEKIFSICVMLIGSLMYAS IFG NVSAII
hEAG      IEKIFAVAIMMIGSLLYAT IFG NVTTIF
Kv1.5     GGKIVGSLCAIAGVLTIAL PVP VIVSNF
Kv2.1     LGKIVGGLCCIAGVLVIAL PIP IIVNNF
Kv3.1     SGMLVGALCALAGVLTIAM PVP VIVNNF
Kv4.2     AGKIFGSICSLSGVLVIAL PVP VIVSNF
```

FIG. 1. Sequence alignment of the S6 domain for several human voltage-gated K^+ channel subunits. Note that most channels have a Val and Ile in the position equivalent to Tyr652 and Phe656 of hERG. Amino acid numbering refers to hERG.

hERG remove inactivation, but do not remove high-sensitivity to block by methanesulfonanilides (Wang et al 1997, Mitcheson et al 2000a), while other mutations increase inactivation, but greatly reduce drug sensitivity (Mitcheson et al 2000a). G628C/S631C hERG channels do not inactivate (Smith et al 1996), yet compared to wild-type hERG are only about 10-fold less sensitive to block by MK-499 (Mitcheson et al 2000b) and equally sensitive to disopyramide (Paul et al 2001).

An alternative explanation for why EAG channels are rather insensitive to many hERG blockers is that orientation of the Tyr and Phe residues in the S6 domain of hERG and EAG channels are not equivalent. Channel inactivation might cause a subtle, but important repositioning of Tyr652 and Phe656 in an orientation optimal for drug binding in hERG, whereas gating of non-inactivating EAG may not be accompanied by the same reorientation. To test this idea, we shifted the position of the Tyr or Phe in hERG or the equivalent aromatic residues in EAG in either the N-terminal ('up') or C-terminal ('down') direction, and tested the sensitivity of the resultant mutant channels to block by cisapride (Chen et al 2002). To construct a hERG channel with Tyr652 moved up one position, two mutations were introduced: Tyr652 was replaced with an Ala and Met651 was mutated to a Tyr (M651Y/Y652A). The IC_{50} for block of wild-type hERG by cisapride was 102 nM, whereas this double mutant channel was only partially blocked by 10 μM cisapride. Moving Phe656 'up' (I655F/F656A) also decreased sensitivity to cisapride. Tyr-down subunits did not functionally express. Phe-down hERG channels expressed poorly, were constitutively open and insensitive to block by cisapride. Thus, as predicted, repositioning of the Tyr or Phe residues in the S6 of hERG reduced block by cisapride. However, the mutations also significantly altered hERG channel gating, confounding a clear interpretation of the findings.

Similar experiments were performed on EAG channels that are 120-fold less sensitive to block by cisapride ($IC_{50} = 11.9\,\mu M$) compared to wild-type hERG. In EAG, Tyr481 and Phe485 are homologous to Tyr652 and Phe656 of hERG. Of all the combinations examined, repositioning aromatic residues in the C-terminal direction increased sensitivity to block by cisapride the most. The IC_{50} was decreased by 43-fold by moving Tyr down and reduced 10-fold by moving Phe down (Chen et al 2002). However, these mutations also induced inactivation, confusing the interpretation of whether inactivation or repositioning of the aromatic residues was responsible for increased sensitivity to block by cisapride. Therefore, additional S6 mutations of EAG were explored in an attempt to introduce drug sensitivity without simultaneously inducing inactivation. One combination of mutations was found that had the sought-after properties. Combining the Tyr-down with another point mutation (A478G) yielded a non-inactivating channel that was 36-times more sensitive to cisapride than wild-type channels (Chen et al 2002). Thus, although inactivation can facilitate block of hERG or EAG channels, it is not sufficient or required for block.

Physicochemical basis of voltage-dependent block of hERG by quinolines

hERG channels must open before block by MK-499, cisapride or terfenadine can occur, implying that drug access to the binding site inside the central cavity is denied when the activation gate is closed. This interpretation of state-dependent block was first proposed for blockers of squid axon K^+ channels (Armstrong 1966) and later formulated as the guarded receptor model (Starmer 1987). Despite the requirement for channel opening, MK-499 does not exhibit any obvious voltage-dependence, probably because the off (unbinding) rate is very slow. In contrast, low affinity blockers like chloroquine (Sanchez-Chapula et al 2002), quinidine (Furukawa et al 1989, Balser et al 1991, Lees-Miller et al 2000), mefloquine (Kang et al 2001a) and sparfloxacin (Kang et al 2001b) exhibit increased block with increasing membrane depolarization, and channels rapidly recover from block at a holding potential of $-80\,mV$.

Mutation of Tyr652 has dramatic effects on the voltage dependence of hERG channel block by quinolines. Mutation of Tyr652 to Ala reverses the voltage-dependence of block such that block is *diminished* with increasingly positive test potentials. A more subtle mutation, substitution of a phenyl with a benzyl moiety (Y652F), eliminated voltage-dependent block. In other words, block was essentially equivalent for all test potentials. Considering the evident importance of the -OH, we determined the effect of chloroquine on Y652T and Y652E hERG channels. Both mutations reduced the potency and reversed the voltage dependence of block by chloroquine similarly to Y652A hERG. Thus, the voltage-dependent block of wild-type hERG cannot be explained by H-bonding of the drug with an -OH or -COOH group of non-aromatic amino acids. The effects of these Tyr652 mutations were surprising because an increase in membrane depolarization should favour movement of a positively charged drug further into the central cavity and would be predicted to cause no change or an increase in block as was observed for wild-type channels. It is likely that the charged form of chloroquine is responsible for channel block because the alkylammonium ions have pK_a values of 8.4 and 10.8, and thus, the drug would be >99% charged at normal intracellular pH. However, a charged N is not required to observe a Tyr652 mutation-induced change in the voltage dependence of hERG block by a compound. Vesnarinone, an uncharged drug, normally exhibits little or no voltage-dependent block of hERG. However, block of Y652A channels by vesnarinone is voltage-dependent like the quinolines, with greater decreases in current magnitude measured at less depolarized membrane potentials.

While the structural basis of the altered voltage-dependent profile for block induced by mutation of Tyr652 is uncertain, our results suggest a possible model. A requirement for initial drug docking with F656 could explain why the voltage dependence for the onset of channel block was nearly the same (~17 mV/e-fold change in τ) regardless of whether the fractional block was a positive (wild-type), negative (Y652A/E/T) or independent (Y652F) function of transmembrane voltage. If docking to F656 was prevented, for example by mutation of the residue to Ala, then channel block would be expected to be drastically reduced, as was observed. Increased block of wild-type hERG in response to greater membrane depolarization could be explained by an enhanced interaction of drug with an additional residue, i.e. Tyr-652. Assuming a positively charged drug like chloroquine or quinidine would move further into the central cavity towards the selectivity filter and away from the F656 residues, then mutations of Tyr-652 to non-aromatic residues could preclude significant drug interaction at depolarized potentials and result in the observed reversed voltage-dependence of block. Mutation of Y652 to Phe would still allow π-stacking interactions with a drug. Whatever the exact mechanisms, our findings suggest a key structural role for Y652 in determining voltage-dependent block of hERG by chemically diverse compounds.

Physicochemical basis of drug interaction with Tyr652 and Phe656

Pharmacophore models suggest hERG channel blockers have a few features in common, including a basic nitrogen surrounded by several hydrophobic centres of mass, usually one or more aromatic rings (Cavalli et al 2002, Ekins et al 2002). The basic N could form a cation-π interaction with Phe656 or Tyr652 of hERG.

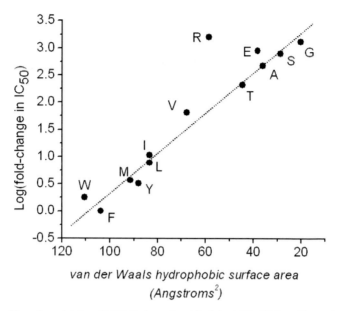

FIG. 2. Altered sensitivity of hERG channel to block by MK-499 is well correlated with the 2D van der Waals hydrophobic surface area of residue 656 (Fernandez et al 2004).

These same residues could also π-stack with aromatic groups of the drug. We investigated the relative importance of π-stacking versus cation-π interactions by mutating Tyr652 or Phe656 to many different residues and determining the sensitivity of the resulting mutant channels to block by MK-499, cisapride and terfenadine (Fernandez et al 2004).

Mutation of Phe656 to other aromatic amino acids (Trp or Tyr) only slightly altered block by MK-499, whereas the IC_{50} was increased by a factor of 3.7, 7.8 and 11 when Phe656 was mutated to Met, Leu or Ile, respectively. In contrast, the IC_{50} for block by MK-499 was increased by >1000-fold when Phe656 was mutated to Gly, Glu or Arg, but these mutations also disrupted channel gating. For MK-499, the logarithmic-fold change in IC_{50} was well correlated with several descriptors of hydrophobicity and was best correlated with the 2D van der Waals hydrophobic surface area, VHSA (Fig. 2). A similar correlation between VHSA of the residue at position 656 and potency was found for cisapride and terfenadine (Fernandez et al 2004). When the mutant channels with altered gating (Phe656 to Ala, Gly, Glu or Arg) were removed from the analysis, the correlation between VHSA and experimentally determined IC_{50} values was improved. These findings indicate that hydrophobic volume and not aromaticity is the most important physicochemical feature of Phe656 for interaction with these three drugs, and that cation-π and π-stacking interactions with Phe656 are probably not as important as we initially hypothesized (Mitcheson et al 2000a, Sanchez-Chapula et al 2002).

The importance of aromaticity vs. hydrophobicity for Tyr652 was also examined (Fernandez et al 2004). Mutation of Tyr652 to eight other residues (Phe, Trp, Ala, Val, Glu, Gln, Ile, Thr) yielded channels with relatively normal biophysical properties. The mutations Y652F or Y652W reduced or did not significantly change the sensitivity to MK-499, cisapride or terfenadine, indicating that a phenol side-chain is not an essential feature of this residue for drug sensitivity. Mutation to Thr greatly reduced sensitivity to all three drugs, confirming the lack of importance of the -OH of Tyr652. Mutation of Tyr652 to other hydrophobic (Ala, Val, Ile) residues greatly reduced drug block. Thus, aromaticity of residue 652, not merely hydrophobicity is the important determinant of interaction with these three drugs.

In summary, the three studies reviewed here provide a refined molecular understanding of the hERG binding site. First, the positioning of the side groups of Tyr652 and Phe656 with respect to the central cavity may be channel state-dependent and influence drug sensitivity. Second, Tyr652 is the key residue in determining the voltage-dependent block of quinolines. Third, the potency for block of hERG channels by structurally diverse drugs was well correlated with the hydrophobic surface area for the side chain of residue 656. In contrast, an aromatic side group at residue 652 was essential for high affinity block, suggesting the possible importance of a more specific interaction (cation-π or π-stacking) between Tyr652 of hERG and a drug.

References

Armstrong CM 1966 Time course of TEA(+)-induced anomalous rectification in squid giant axons. J Gen Physiol 50:491–503
Balser JR, Bennett PB, Hondeghem LM, Roden DM 1991 Suppression of time-dependent outward current in guinea pig ventricular myocytes: actions of quinidine and amiodarone. Circ Res 69:519–529
Cavalli A, Poluzzi E, De Ponti F, Recanatini M 2002 Toward a pharmacophore for drugs inducing the long QT syndrome: insights from a CoMFA study of hERG K^+ channel blockers. J Med Chem 45:3844–3853
Chen J, Seebohm G, Sanguinetti MC 2002 Position of aromatic residues in the S6 domain, not inactivation, dictates cisapride sensitivity of HERG and eag potassium channels. Proc Natl Acad Sci USA 99:12461–12466
Ekins S, Crumb W J, Sarazan RD, Wikel JH, Wrighton SA 2002 Three-dimensional quantitative structure-activity relationship for inhibition of human ether-a-go-go-related gene potassium channel. J Pharmacol Exp Ther 301:427–434
Fernandez D, Ghanta A, Kauffman GW, Sanguinetti MC 2004 Physicochemical features of the hERG channel drug binding site. J Biol Chem 279:10120–10127
Ficker E, Jarolimek W, Brown AM 2001 Molecular determinants of inactivation and dofetilide block in ether a-go-go (EAG) channels and EAG-related K^+ channels. Mol Pharmacol 60:1343–1348

Furukawa T, Tsujimura Y, Kitamura K, Tanaka H, Habuchi Y 1989 Time- and voltage-dependent block of the delayed K^+ current by quinidine in rabbit sinoatrial and atrioventricular nodes. J Pharmacol Exp Ther 251:756–763

Kang J, Chen XL, Wang L, Rampe D 2001a Interactions of the antimalarial drug mefloquine with the human cardiac potassium channels KvLQT1/mink and HERG. J Pharmacol Exp Ther 299:290–296

Kang J, Wang L, Chen XL, Triggle DJ, Rampe D 2001b Interactions of a series of fluoroquinolone antibacterial drugs with the human cardiac K+ channel HERG. Mol Pharmacol 59:122–126

Keating MT, Sanguinetti MC 2001 Molecular and cellular mechanisms of cardiac arrhythmias. Cell 104:569–580

Lees-Miller JP, Duan Y, Teng GQ, Duff HJ 2000 Molecular determinant of high-affinity dofetilide binding to HERG1 expressed in Xenopus oocytes: involvement of S6 sites. Mol Pharmacol 57:367–374

Mitcheson JS, Chen J, Lin M, Culberson C, Sanguinetti MC 2000a A structural basis for drug-induced long QT syndrome. Proc Natl Acad Sci USA 97:12329–12333

Mitcheson JS, Chen J, Sanguinetti MC 2000b Mechanism of methanesulfonanilide block of the HERG potassium channel. Jpn J Electrocardiol (suppl) III:67–70

Mitcheson JS, Perry M, Stansfeld P, Sanguinetti S, Witchel H, Hancox J 2005 Structural determinants for high affinity block of hERG potassium channels. In: The hERG cardiac potassium channel: structure, function and long QT syndrome (Novartis Found Symp 266). Wiley, Chichester, p 136–154

Paul AA, Witchel HJ, Hancox JC 2001 Inhibition of HERG potassium channel current by the class 1a antiarrhythmic agent disopyramide. Biochem Biophys Res Commun 280:1243–1250

Pearlstein R, Vaz R, Rampe D 2003 Understanding the structure-activity relationship of the human ether-a-go-go-related gene cardiac K^+ channel. A model for bad behavior. J Med Chem 46:2017–2722

Sanchez-Chapula JA, Navarro-Polanco RA, Culberson C, Chen J, Sanguinetti MC 2002 Molecular determinants of voltage dependent HERG K^+ channel block. J Biol Chem 277:23587–23595

Sanguinetti MC, Jiang C, Curran ME, Keating MT 1995 A mechanistic link between an inherited and an acquired cardiac arrhythmia: *HERG* encodes the I_{Kr} potassium channel. Cell 81:299–307

Smith PL, Baukrowitz T, Yellen G 1996 The inward rectification mechanism of the HERG cardiac potassium channel. Nature 379:833–836

Starmer CF 1987 Theoretical characterization of ion channel blockade. Competitive binding to periodically accessible receptors. Biophys J 52:405–412

Trudeau M, Warmke JW, Ganetzky B, Robertson GA 1995 HERG, A human inward rectifier in the voltage-gated potassium channel family. Science 269:92–95

Wang S, Morales MJ, Liu S, Strauss HC, Rasmusson RL 1997 Modulation of HERG affinity for E-4031 by [K+]o and C-type inactivation. FEBS Lett 417:43–47

DISCUSSION

Robertson: What is it about the glycine in combination with that mutation?

Sanguinetti: We don't know, but it is exciting that we can introduce or remove inactivation completely with a single mutation in S6 in a position that was thought not to be important for inactivation. This opens up possibilities for all sorts of structure–function studies on channel activation. But I don't know what is occurring there specifically.

Noble: You are partially answering the question I posed earlier, where I asked whether it might be possible to engineer an improved hERG channel, because you are showing that with site-specific mutagenesis you can greatly alter the IC_{50}s. This leads me to a functional question: can one with any of these sites alter the IC_{50} without seriously altering the functionality of the channel?

Sanguinetti: Yes.

Noble: So the answer to my question is that there is a serious possibility of putting nature right!

Sanguinetti: Yes; however, I'd rather have safer drugs than begin doing gene therapy. But you are right, that it is a possibility.

Traebert: The pharmacophore you have shown of the tryalkylated nitrogen is a good, classical approach and the best example for this is TEA which is known to block nearly all K^+ channels. But there are an increasing number of secondary amides which block the hERG channel. There also may be some molecules devoid of central nitrogen that do this. Have you any explanation for how they could interact with the hERG α subunit?

Sanguinetti: We think it is mainly hydrophobic interactions.

Recanatini: With regard to this question, your work on mutagenesis of F656 is quite remarkable. From the QSAR point of view, it is very important that you find these linear relationships between molecular descriptors and the potency of one drug against many different targets (the hERG mutants). This is the opposite of what we obtain when we study several drugs against one target. This is described as a linear free-energy relationship in physical organic chemistry. When you find this sort of empirical linear relationship it means that two systems (i.e. the protein target, and the model system where you determine the descriptor) respond linearly to the same perturbation (the variation of molecular fragments, i.e. in this case, the different amino acid side chains). I did some calculation on your data using another parameter, an experimentally determined constant that accounts for lipophilicity of the amino acid side chains, and found a relationship quite similar to the one you obtained. So when you see these linear relationships with a lipophilic parameter, you can infer that the main effect driving the interaction (the binding inside the cavity) is a hydrophobic one. Of course, for different drugs perhaps you could subtly distinguish between hydrophobic and other interactions, but I think from the data you showed that the main effect arising is the hydrophobic one. This may explain why other non-charged molecules can bind.

Sanguinetti: I just want to make clear that we think that the cation-π interactions are with tyrosine 652. This doesn't rule out cation-π occurring at Phe656. I showed data for just a few compounds, and others could be different. I want to stress that we don't think that every drug interacts with these two sites. I should also mention that Greg Kaufmann (Pfizer) used the molecular operating environment (MOE) to look at 146 different physicochemical (PC) descriptors and identify the best PC descriptors at that position relative to drug sensitivity. The only descriptors that had high scores were hydrophobic measures, and the best one turned out to be VHSA, but any hydrophobic measure was significant.

Recanatini: We usually interpret this kind of relationship to mean that the main effect is an entropic effect driven by hydrophobicity. This may help putting together the drug and the target. Then when they are close, specific interactions can strengthen the binding. These work in combination.

Sanchez-Chapula: My more recent results using Drk1 and quinidine in which we mutated the Y652 are relevant here. I find that that with Drk1 it looks like the interaction with that residue is cation-π. If I change Y652 to an aliphatic residue the potency is increased. With quinidine it looks like there is a very similar potency with tryptophan, tyrosine and phenylalanine. It also seems that the potency is decreased as the hydrophobicity of the residues is decreased. It is different for quinidine: at first I thought that it would be a similar mechanism for Drk1 and quinidine, but this doesn't seem to be the case.

Sanguinetti: When we did the 'tyrosine-down' experiments in EAG, the results I showed you for cisapride indicated that we could introduce drug sensitivity into EAG. However, we could not reconstitute sensitivity to MK-499. This is frustrating and suggests that each drug will be different. Thus, we can't simply say that it is just the position of tyrosine that determines sensitivity.

Sanchez-Chapula: With the Y652A mutant there is a blocking effect at negative potentials such as $-50\,\mathrm{mV}$. Then, we apply a pulse to $+20\,\mathrm{mV}$ and see a rapid block and then a slower unblock. When we repolarized there was a fast reblock. What happened with the drug? It is inside the cavity.

Sanguinetti: The point here is that with the Y652A mutation, if you pulse to $+40$ there appears to be no change of current during the activating pulse. But if you ratio the two currents (before and after drug), you can see there is a rapid onset of block and an extremely rapid recovery from block in the first few milliseconds. The question is: is the drug still in the central cavity but not blocking conduction of K^+ ions, or is it leaving the inner vestibule and then coming back in? We don't know. It could be either mechanism. The problem is that when we do modelling and the drug is bound in a position anywhere near tyrosine or phenylalanine (especially if it is interacting with the base of the pore helix) it is masked. The drug is interfering with the position of a K^+ which is proposed to be right in the centre of the inner cavity, coordinated by dipole moments of the pore helices. This is confusing to us. If the drug is there, it should be preventing the central K^+ ion from being coordinated, unless the channel opens widely and the drug is opposed against the S6. Then there might be enough space in there.

Hondeghem: If the drug is neutral it could also do the converse, and diffuse into the membrane. Take lidocaine, as an example. Its inactivated state interaction with the sodium channel is quite fast, and it is probably simply interacting with the receptor though the lipid phase of the membrane (even though the channel is 'closed').

Sanguinetti: An obvious answer would be that the drug just sneaks behind S6 temporarily, leaving the central cavity perfectly able to conduct K^+ ions, but we don't know what is going on.

Recanatini: Perhaps I missed this, but in your paper on the structural basis of hERG binding by drugs (Mitcheson et al 2000), when you made a model of MK-499 binding, was there any reason why you oriented the molecule the way you did, with the methylsulfonamide moiety on top? We don't know how that molecule can bind unless we do some virtual docking experiments. Do you have mutagenesis data that show that the methylsulfonamide group might bind with some specific residue?

Sanguinetti: Chris Culberson from Merck did this work using a docking program called FLOG (flexible ligands oriented on a grid). The docking shown was the energetically most favourable position that was consistent with our data. We can flip the drug and it still fits in the central cavity, even in the closed state. The reason the drug was placed in this orientation was in part because of the glycine 648 mutation. If this residue is mutated to an alanine, adding a single methyl group, this really reduces the block and changes gating. The idea was that the methylsulfonyl can stick in between the position of the pore helix and glycine 648 of S6. If this space is made smaller, then the methylsulfonyl can't fit.

Hancox: How do MK-499 and chloroquine compare in terms of size?

Sanguinetti: MK-499 is 20×7 Å. Chloroquine is smaller.

Hancox: But chloroquine is producing tail current cross-over, so it is not being trapped.

Sanguinetti: This may be because it is binding more to phenylalanine 656 lower in the cavity. Perhaps you don't have to have a foot in the door by literally having a drug blocking closure of the activation gate; you could have this effect by binding somewhere else without physically preventing closure. This is traditionally interpreted as 'foot in the door', though, for good reason.

Netzer: Your data are on channel block, but are these mechanisms also relevant for rescue of trafficking-deficient channels? Do we have any data on this?

January: Eckhard Ficker has mutagenesis data on the importance of the pore for the rescue process.

Ficker: The mutagenesis is rather complicated in position hERG F656. Drug binding to and rescue of hERG G601S by pharmacological chaperones can be abolished by mutating the phenylalanine in position 656 into a cysteine. However, when we introduced an alanine or valine into position 656, for whatever reason, trafficking of hERG G601S resumed without addition of a pharmacological chaperone. We have seen similar problems in position 652. We observed rescue when we made amino acid substitutions to destroy that part of the binding site.

January: There are other ways to affect rescue. Is it universally important for rescue? Probably not. This is a multifaceted issue. The data argue that this binding site or an overlapping binding site mediate some rescue. Drugs like thapsigargin, a SERCA inhibitor, also cause rescue. The mechanism underlying this isn't understood, and it is likely not to involve this binding domain.

Reference

Mitcheson JS, Chen J, Lin M, Culberson C, Sanguinetti MC 2000 A structural basis for drug-induced long QT syndrome. Proc Natl Acad Sci USA 97:12329–12333

In silico modelling — pharmacophores and hERG channel models

Maurizio Recanatini, Andrea Cavalli and Matteo Masetti

Department of Pharmaceutical Sciences, University of Bologna, Via Belmeloro 6, 40126 Bologna, Italy

Abstract. In computational drug design, modelling studies are undertaken following two main strategies that depend on which information is available. If experimental data exist only for the molecules displaying the biological property of interest, a so-called ligand-based approach is taken; if information is available on the macromolecular target(s) of the compounds (e.g. proteins' 3D structures), target-based studies can be carried out. Recently, in the field of hERG K^+-channel blocking drugs, pharmacophoric (ligand-based) studies started appearing aimed at determining the physicochemical features associated with the channel block, and also at predicting the hERG blocking potential of compounds. However, partial homology models (target-based) of the hERG channel have also been built and used as working tools to interpret electrophysiological and mutagenesis studies. Here, we review some of the ligand- and target-based *in silico* studies carried out on hERG, focusing on both their main characteristics and their meaning. In addition, we discuss some methodological aspects of the computational work that in our opinion should be considered, in view of the construction of reliable models possibly able to predict the functional behaviour of the channel system and the blocking potential of drugs.

2005 The hERG cardiac potassium channel: structure, function, and long QT syndrome. Wiley, Chichester (Novartis Foundation Symposium 266) p 171–185

In recent years, the critical role played by the QT-prolonging potential in the safety profile of drugs has prompted drug designers to investigate the physicochemical characteristics associated with the propensity of different molecules to elicit such an effect. In this context, the use of *in silico* methods has been proposed in an attempt both to rationalize the molecular basis of the QT prolonging effect, and to predict the QT prolonging potential for large numbers of compounds (De Ponti et al 2002, Ekins 2003, van de Waterbeemd & Gifford 2003). In this regard, different approaches can be taken, targeted either at the global pharmacological effect (e.g. the prolongation of the QT interval), or at the interaction with the macromolecular system considered as being responsible for this biological action (e.g. the hERG K^+ channel).

Traditionally, modelling studies in computational drug design are undertaken following two kind of strategies, depending on what information is available. If experimental data exist only for the molecules displaying the biological property of interest, a so-called ligand-based approach is taken, and QSAR (Quantitative Structure–Activity Relationship) models are derived; if information is available on the macromolecular target(s) of these compounds (protein 3D structures), target-based studies can be carried out, and docking models can be built. In the case of QT-prolonging drugs, there is clinical evidence on the action of many pharmacological agents, and there are potency data (IC_{50} values) regarding the inhibition of the hERG channel by such compounds, but the 3D atomic coordinates of the hERG potassium channel protein components are not available. On the other hand, some K^+ channel structures have been resolved (MacKinnon 2003), and can provide the basis for the homology modelling of hERG.

In this paper, we will review some recent reports on QT/hERG pharmacophore building and account for the efforts to model the channel. As regards the pharmacophores, here, we will consider published models, which are based on the chemical features of drugs inducing the QT prolongation through hERG channel inhibition, and which also pay much attention to the description of the stereoelectronic requirements for hERG blocking. Indeed, the predictive aspects of the models are also quite relevant, but they will not be discussed here. Different, more chemoinformatics-oriented approaches can be followed to build QSAR models more directly aimed at assessing the potential of compounds to inhibit the hERG channel.

Pharmacophores for hERG-blocking compounds

With regard to pharmacophores for hERG-blocking compounds, an early study was carried out by Morgan & Sullivan (1992), who took into consideration the so-called Class III electrophysiological agents (Singh & Vaughan Williams 1970), a class of drugs that owe their pharmacological action to the block of the hERG potassium channel (Spector et al 1996). Through a simple structure–activity relationship (SAR) analysis, these authors were able to depict a pharmacophoric scheme that illustrates the main structural characteristics associated with the action potential duration (APD)-prolonging activity of Class III antiarrhythmic drugs. The pharmacophore is shown in Fig. 1a and comprises a *para*-substituted phenyl ring (X-Ph) linked to a basic nitrogen (N) through a 1–4 atom linking chain. In addition, it includes substituents on the nitrogen atom, one (or both) of which can be aromatic ring(s) located close (1–3 atoms) to the basic centre. Despite being formulated without the intervention of sophisticated computational tools, this pharmacophoric scheme anticipated some of the main features detected in the subsequent studies regarding QT-prolonging drugs and hERG blockers.

In 2002, two *in silico* models aimed both at providing a pharmacophore and quantitatively predicting the hERG blocking activity appeared in the literature. First Ekins et al (2002) and then Cavalli et al (2002) reported 3D QSAR models for the inhibition of the hERG K^+-channel based on different sets of drugs known to induce QT prolongation through the blockade of that channel. The two groups worked independently but followed the same approach involving two main steps: (a) the building of a 3D pharmacophore derived from the alignment of selected conformations of the molecules of a training set, and (b) the development of a statistical model accounting for the 3D QSAR of the set of molecules. The computational tools used for the derivation of the models were different (Catalyst®, Accelrys Inc., San Diego, USA and CoMFA®, Tripos Inc., St. Louis, USA, respectively), as were the training sets on which the two models were built, although both were obtained from literature data. In addition, in both cases, the hERG inhibition data came from electrophysiological studies carried out on mammalian cells expressing the channel.

The similarities of the two pharmacophores are remarkable, and point out the relevance of the hERG blockade to some chemical functions that in part had already been identified by Morgan and Sullivan. The 'General hERG Pharmacophore' built by Ekins (Fig. 1B) consists of one positively ionisable feature (I, seemingly, the basic nitrogen common to all the drugs taken into consideration) and four hydrophobic centres (H), which are not necessarily simultaneously present in all of the molecules. In the Cavalli's pharmacophore (Fig. 1C), almost the same features appear, designated as a protonated nitrogen function (N) and three aromatic moieties (C0, C1, and C2) again not all simultaneously present in all of the molecules. Thus, in both cases, the structural scheme is centred around an ionisable function bearing some hydrophobic groups located at similar distances.

What is different between the two pharmacophoric frames is the position of the central function with respect to the surrounding hydrophobic or aromatic moieties. In fact, considering the pharmacophore as a pyramid with the ionisable moiety on top, the Ekins' structure shows a smaller base and a greater height with respect to Cavalli's one. The different shape of the two pyramidal structures representing the pharmacophoric schemes seemingly reflects the shape of the conformers that were selected to build the frames. It might be that in the Ekins' pharmacophore, folded conformations were mostly used, while in the Cavalli's model, the template molecule selected to align all the other molecules was in an extended conformation. The resulting pharmacophores thus probably represent the same molecular characteristics either in folded or in extended disposition.

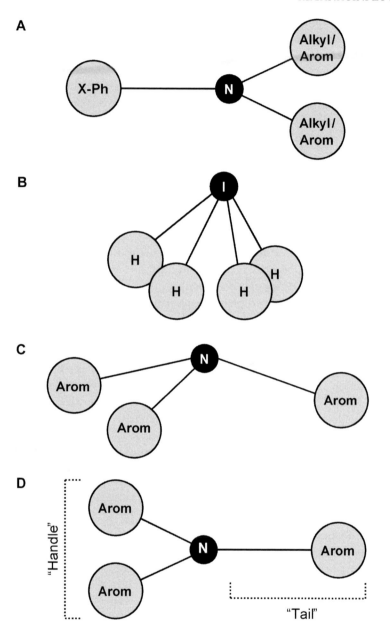

FIG. 1. Schematic representation of the pharmacophoric models for hERG blocking drugs reported in Morgan & Sullivan (1992) (A), Ekins et al (2002) (B), Cavalli et al (2002) (C), and Pearlstein et al (2003a,b) (D).

Early in 2003, Pearlstein et al (2003a) reported a CoMSIA (Klebe 1998) model aimed at interpreting the SAR of a series of sertindole derivatives able to block the hERG channel with varying potency. Twenty-eight molecules were used to build the model in a way similar to that described above for the 3D QSAR models of Ekins and Cavalli. The best CoMSIA model shows all the molecules in an extended conformation, except for fexofenadine, whose activity is best recalculated when the molecule is in a folded conformation. From the molecular alignment and the contour maps reported in the original article (Pearlstein et al 2003a), the pharmacophoric scheme of Fig. 1D can be derived, which appears quite similar to those previously derived by Morgan & Sullivan (1992), and Cavalli et al (2002), respectively.

Modelling the hERG K^+ channel

As mentioned above, no experimental determination of the structure of hERG has been disclosed up to date, so the only way to obtain a 3D description of this important pharmacological anti-target is to attempt to build it through a modelling procedure. While no report has yet appeared of a systematic and thorough modelling work on hERG, some groups have undertaken the building of working models, which have been used to complement and interpret other findings obtained by different approaches. Indeed, these models have contributed to a 3D framework in which hypotheses of drug binding have been advanced.

Most hERG modelling work has been carried out by the groups of Mitcheson and Sanguinetti, who provided the first computerized picture of the pore region of the hERG channel, and used the model to help rationalizing the role of some key amino acid residues on the binding of hERG blockers (Mitcheson et al 2000). The model was built through a classical homology modelling procedure by using the X-ray solved structure of the KcsA channel (Doyle et al 1998) as a template for the coordinate transfer. Actually, despite its prokaryotic origin and its two transmembrane helix (2TM) organization, this potassium channel from *Streptomyces lividans* shares some sequence homology with hERG, at least in some crucial regions of the channel like helices 5 and 6 (S5 and S6, respectively), and the selectivity filter region (Doyle et al 1998). A seemingly slightly refined version of the model was reported by Sanguinetti and co-workers who introduced a different template for the turret region (S5-P helix linker) and partially minimised the structure (Sanchez-Chapula et al 2002). Recently, Moreno et al (2003) described a model of the hERG channel built following the same procedure as that reported by the above-mentioned authors (Mitcheson et al 2000, Sanchez-Chapula et al 2002).

A second reported model of the hERG channel was due to Pearlstein and coworkers, who published it in two papers (Pearlstein et al 2003a,b). Also in this case, the hERG model was not the aim of the work, but rather a means through which the authors interpreted data obtained from another procedure. The substantial difference from the Mitcheson and Sanguinetti model was in the selection of the template that in the Pearlstein's case was the MthK potassium channel from the archaeon *Methanobacterium thermoautotrophicum* (Jiang et al 2002). This implied that the channel was modelled in an open state (contrarily to the one based on KcsA that represents a closed state), thus allowing them to take into account a different organization and conformation of the S6 gate region, admittedly crucial for the binding of drugs.

The two hERG K^+ channel models briefly described above cannot be used for a discussion of the 3D structural and functional properties of the channel, because in their development some computationally intensive refinement steps have been skipped to favour the rapid achievement of practical working tools. In this sense, these models have been employed to complement the interpretation of other data and indeed, they have accomplished this task, by allowing the researchers to advance and illustrate their hypotheses on the binding of drugs to hERG. The hERG binding modes of such compounds as MK-499, chloroquine and irbesartan together with their interactions with specific residues of helix 6 of the pore have been interpreted in physicochemical terms thanks to the discussion of mutagenesis and electrophysiological data in the light of hERG/drug docking models (Mitcheson et al 2000, Sanchez-Chapula et al 2002, Moreno et al 2003). On the other hand, the 3D QSAR model developed by Pearlstein was directly compared with the 3D structural model of the channel, allowing to tentatively extend the interaction hypotheses to a whole class of compounds, namely the sertindole analogues (Pearlstein et al 2003a,b).

Problems and perspectives in modelling hERG and hERG-binding compounds

In the above sections, a short account has been provided of the results of ligand- and target-based modelling strategies recently applied in the field of hERG and hERG-blocking drugs. Aims and applications of the models obtained can be manifold, but obvious prerequisites to the use of such models are their physicochemical reliability and agreement with experimental data. Taking these issues into consideration will allow us to point out some aspects of the modelling work that has been carried out so far and that is being presently performed in our laboratory.

A regards the pharmacophores of Fig. 1 (eventually integrated in the 3D QSAR models), they contain useful information for describing the general characteristics of molecules binding to the hERG channel, but they say little about the mode of binding and the ligand–protein interactions. Moreover, there might be some degree of arbitrariness in the alignment/disposition of the molecules from which the pharmacophore is derived. The experimental way to unravel these issues is to analyse the structure of hERG–inhibitor complexes by means of X-ray crystallography or NMR spectroscopy. However, this approach might be limited in that the ligands are taken into consideration individually, and consequently an overall view of the SAR of hERG blockers can be hard to obtain. A way to circumvent the lack of direct experimental evidence and the related drawbacks consists of the critical comparison of (statistical) ligand-based models and (graphical) 3D representations of the target (Hansch & Klein 1986, Cavalli et al 2000). In the case of hERG inhibitors, we started working in this way by discussing our CoMFA model (Cavalli et al 2002) in the light of the docking complex between the hERG channel and the inhibitor MK-499 reported by Mitcheson et al (2000). Actually, we detected a significant match between the features of our pharmacophore and those of the MK-499 molecule docked into its putative hERG binding pocket, and suggested the possible involvement of the protonated nitrogen atom featured by most inhibitors in π-cation interactions with nearby aromatic residues present in the channel wall (Cavalli et al 2002). A similar approach was taken by Pearlstein et al (2003a), which compared their CoMSIA model (here represented by the pharmacophoric scheme of Fig. 1D) with their model of hERG built on the basis of the open bacterial MthK potassium channel. Based on the comparison, the hypothesis was advanced that inhibitor molecules penetrating into the channel pore from the intracellular side might orient themselves with the long 'tail' pointing towards the selectivity filter, and the hydrophobic head ('handle') blocking the intracellular entrance of the pore, in a sort of 'drain–plug' interaction. Steric and electrostatic characteristics of the CoMSIA model and some critical residues of the channel like Tyr652 and Phe656 localised at the inner mouth of the pore seem to match ideally in this simple working model.

From the above considerations about the usefulness of comparing pharmacophoric with structural models of hERG-binding compounds, as well as from the ongoing studies on the molecular determinants of the drug binding to hERG (reviewed in Mitcheson & Perry 2003), the central role played by the availability of reliable 3D models of the protein complex constituting the channel unequivocally derives. Despite the rather extensive computational work recently carried out on potassium channels (see e.g. Sansom et al 2002, Giorgetti & Carloni 2003), no systematic modelling of hERG has been published up to now. Furthermore, most of the known simulations took into consideration K^+ channels belonging to the 2TM family, whereas in only one case a mammalian Kv6TM potassium channel has been studied computationally (Luzhkov et al 2003).

One of the central issues when undertaking a (homology) modelling study on K^+ channels and particularly on hERG regards the availability and subsequent

TABLE 1 Crystallographic structures of potassium channels available as templates for homology modelling

Channel class (nTM)	Template	PDB code	Gating state	Resolution ($Å$)	Chemical components	Structure deposited
K_{ir} (2TM)	KcsA	1BL8	closed	3.20	$3K^+$; $1H_2O$	tetramer
		1J95	closed	2.80	$4K^+$; TBA	tetramer
		1JVM	closed	2.80	$3 Rb^+$; TBA; $1H_2O$	tetramer
		1K4C	closed	2.00	$7K^+$; DGA; nonan-1-ol; Fab; H_2O	monomer
		1K4D	closed	2.30	$2K^+$; $1Na^+$; DGA; nonan-1-ol; Fab; H_2O	monomer
		1R3I	closed	2.40	$4Rb^+$; DGA; nonan-1-ol; Fab; H_2O	monomer
		1R3J	closed	1.90	$5Tl^+$; DGA; nonan-1-ol; Fab; H_2O	monomer
		1R3K	closed	2.80	$3Tl^+$; DGA; Fab; H_2O	monomer
		1R3L	closed	2.41	$5Cs^+$; DGA; nonan-1-ol; Fab; H_2O	monomer
	MthK	1LNQ	open	3.30	$8Ca^{2+}$	tetramer
	KirBac1.1	1P7B	closed	3.65	$4K^+$	dimer
K_v (6TM)	KvAP	1ORQ	open	3.20	$6K^+$; Cd^{2+}; Fab	full length monomer
		1ORS	open	1.90	H_2O; Fab	voltage sensor domain monomer

selection of a template structure, on which to fold the primary sequence of the protein subunits. Such templates are usually X-ray structures, which are supposed to retain the overall architecture of the channel and are requested to possess a sufficient level of amino-acid sequence identity with the target protein (hERG). In Table 1, currently available X-ray structures of K^+ channels are reported together with some of their characteristics: a point immediately evident is that all potential templates except for KvAP belong to the K_{ir} 2TM family of K^+

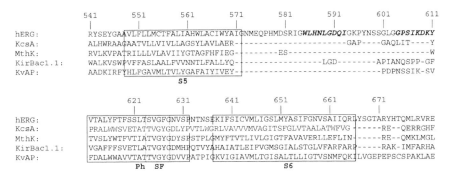

FIG. 2. Indicative multiple alignment of the pore regions of hERG channel and the available templates. Black boxes represent the hypothesized hERG topology according to Warmke & Ganetzky (1994), whereas the dotted box shows the conserved selectivity filter sequence. Residues forming the putative couple of α-helices in the S5-P helix linker as proposed by Torres et al (2003) are in bold italics. Numbering of the residues is from the hERG sequence.

channels. On the other hand, the structure of KvAP is still rather controversial (Cohen et al 2003), and this, at the moment prevents its use as a template for the modelling of hERG. Also, a rather serious problem in hERG modelling is the presence of a linker region between S5 and the P helix, which is definitely longer than the corresponding loop in other K^+ channels (Fig. 2), and has no direct structural reference. An interesting contribution to its modelling is that provided by the recent work of Torres et al (2003), who studied the topological and structural characteristics of the linker through NMR and CD spectroscopy.

Once the template(s) has been selected, a further crucial step in homology modelling is the alignment of the sequences of template and target protein, and the consequent building of the backbone structure of the channel system. In the case of hERG, besides the problem represented by its tetrameric organization, several other non trivial aspects emerge when considering even just the alignment of the sequence of its S5-linker-pore-S6 region with the corresponding M1-linker-pore-M2 region of KcsA and MthK, suitable for the open and closed states of the channel, respectively. Among others, the solution of the problem posed by the modelling of the S5-P helix linker is critical not only to address issues related to the voltage gating of the channel, but also to take into appropriate consideration the more basic problem of a correct description of the electrostatics of the pore region, as the linker sequence comprises four acid and four basic amino-acid residues.

The building of the backbone of the protein subunits and their assembly into the tetrameric system leads to a preliminary structure that needs to be subsequently refined by considering the amino acid side chains and the environment surrounding the pore region. As regards the latter, it is evident that the structural stability of the model needs a simulation of the effects exerted by the S1-S4 helices and by the membrane as well. This problem can be tackled at different levels, from the simple application of computational restraints to the simulation of the explicit membrane (Domene et al 2003). On the other hand, the detailed knowledge of the orientation of the side chains of the residues forming the pore region of the channel is the main goal of the hERG modelling efforts aimed at interpreting the hERG–drug interactions. Thus, the investigation of the conformational behaviour of the channel proteins is of utmost importance, and molecular dynamics simulations are the best suited computational tool to carry out this task. In this context, the available molecular mechanics force fields can be considered as adequately parameterized to properly account for all the interactions; and the advancements in computer technology afford the treatment of a molecular system of such a relevant size as that of tetrameric K^+ channels immersed in a hydrated phospholipid bilayer.

In conclusion, computational ligand- and target-based methods are emerging as useful tools to interpret at the physicochemical and atomic level the characteristics of the hERG K^+ channel and the hERG-blocking drugs. It has been shown that the integration of computational, electrophysiological and mutagenic studies can be truly synergistic in providing clues as to the behaviour of this complex system. It might be advanced that, through some more deepened methodological and computational efforts, modelling studies on hERG will lead to accurate predictions of both channel-functioning and drug-binding ability, which are the final goals of any investigation on this K^+ channel.

References

Cavalli A, Greco G, Novellino E, Recanatini M 2000 Linking CoMFA and protein homology models of enzyme-inhibitor interactions: an application to non-steroidal aromatase inhibitors. Bioorg Med Chem 8:2771–2780

Cavalli A, Poluzzi E, De Ponti F, Recanatini M 2002 Toward a pharmacophore for drugs inducing the long QT syndrome: insights from a CoMFA study of HERG K^+ channel blockers. J Med Chem 45:3844–3853

Cohen BE, Grabe M, Jan LY 2003 Answers and questions from the KvAP structures. Neuron 39:395–400

De Ponti F, Poluzzi E, Cavalli A, Recanatini M, Montanaro N 2002 Safety of non-antiarrhythmic drugs that prolong the QT interval or induce torsades de pointes: an overview. Drug Safety 25:263–286

Domene C, Bond PJ, Sansom MS 2003 Membrane protein simulations: ion channels and bacterial outer membrane proteins. Adv Protein Chem 66:159–193

Doyle DA, Morais Cabral J, Pfuetzner RA et al 1998 The structure of the potassium channel: molecular basis of K^+ conduction and selectivity. Science 280:69–77

Ekins S 2003 In silico approaches to predicting drug metabolism, toxicology and beyond. Biochem Soc Trans 31:611–614

Ekins S, Crumb WJ, Sarazan RD, Wikel JH, Wrighton SA 2002 Three-dimensional quantitative structure-activity relationship for inhibition of human ether-a-go-go-related gene potassium channel. J Pharmacol Exp Ther 301:427–434

Giorgetti A, Carloni P 2003 Molecular modeling of ion channels: structural predictions. Curr Opin Chem Biol 7:150–156

Hansch C, Klein TE 1986 Molecular graphics and QSAR in the study of enzyme-ligand interactions. On the definition of bioreceptors. Acc Chem Res 19:392–400

Jiang Y, Lee A, Chen J et al 2002 Crystal structure and mechanism of a calcium-gated potassium channel. Nature 417:515–522

Klebe G 1998 Comparative molecular similarity indices: CoMSIA. In: Kubinyi H, Folkers G, Martin YC (eds) 3D QSAR in drug design. Kluwer Academic Publishers, Great Britain, p 87–104

Luzhkov VB, Nilsson J, Arhem P, Aqvist J 2003 Computational modelling of the open-state Kv 1.5 ion channel block by bupivacaine. Biochim Biophys Acta 1652:35–51

MacKinnon R 2003 Potassium channels. FEBS Lett 555:62–65

Mitcheson JS, Perry MD 2003 Molecular determinants of high-affinity drug binding to HERG channels. Curr Opin Drug Discov Devel 6:667–674

Mitcheson JS, Chen J, Lin M, Culberson C, Sanguinetti MC 2000 A structural basis for drug-induced long QT syndrome. Proc Natl Acad Sci USA 97:12329–12333

Moreno I, Caballero R, Gonzalez T et al 2003 Effects of irbesartan on cloned potassium channels involved in human cardiac repolarization. J Pharmacol Exp Ther 304:862–873

Morgan TK Jr, Sullivan ME 1992 An overview of class III electrophysiological agents: a new generation of antiarrhythmic therapy. Prog Med Chem 29:65–108

Pearlstein RA, Vaz RJ, Kang J et al 2003a Characterization of HERG potassium channel inhibition using CoMSiA 3D QSAR and homology modeling approaches. Bioorg Med Chem Lett 13:1829–1835

Pearlstein R, Vaz R, Rampe D 2003b Understanding the structure-activity relationship of the human ether-a-go-go-related gene cardiac K^+ channel. A model for bad behavior. J Med Chem 46:2017–2022

Sanchez-Chapula JA, Navarro-Polanco RA, Culberson C, Chen J, Sanguinetti MC 2002 Molecular determinants of voltage-dependent human ether-a-go-go related gene (HERG) K+ channel block. J Biol Chem 277:23587–23595

Sansom MS, Shrivastava IH, Bright JN et al 2002 Potassium channels: structures, models, simulations. Biochim Biophys Acta 1565:294–307

Singh BN, Vaughan Williams EM 1970 A third class of anti-arrhythmic action. Effects on atrial and ventricular intracellular potentials, and other pharmacological actions on cardiac muscle, of MJ 1999 and AH 3474. Br J Pharmacol 39:675–687

Spector PS, Curran ME, Keating MT, Sanguinetti MC 1996 Class III antiarrhythmic drugs block HERG, a human cardiac delayed rectifier K^+ channel. Open-channel block by methanesulfonanilides. Circ Res 78:499–503

Torres AM, Bansal PS, Sunde M et al 2003 Structure of the HERG K^+ channel S5P extracellular linker: role of an amphipathic alpha-helix in C-type inactivation. J Biol Chem 278:42136–42148

van de Waterbeemd H, Gifford E 2003 ADMET in silico modelling: towards prediction paradise? Nat Rev Drug Discov 2:192–204

Warmke JW, Ganetzky B 1994 A family of potassium channel genes related to eag in Drosophila and mammals. Proc Natl Acad Sci USA 91:3438–3442

DISCUSSION

Sanguinetti: I'd like to ask the representatives from the pharmaceutical industry a question. How useful are pharmacophore models? Are you integrating what is known about the receptor? Are pharmacophore models within a clinical class sufficient?6

Nicklin: As we have seen, the models are currently quite general. If we set our entire chemical archive against some of the pharmacophore models we have seen there will be a large number of positives. What is important to us is pulling out the really potent ones and not pursuing those. When pharmacophore modelling is able to string things out in terms of potency in a reliable fashion, then it is likely to be more useful.

Recanatini: This is a basic starting point. With these models we have just explored a small amount of the molecular space. What we need to do is to increase our basis of biological data. Some problems may arise. We are experiencing this by trying to extend our model. The problem might be that when we try to extend the model by adding molecules to the training set, the model tends to change, i.e. the main characteristics reflected by the model change. This means that models based on small numbers of compounds are not stable. We need focused sets of compounds that allow us to study different portions of the molecular space covered by hERG binders step by step.

Nicklin: We are very interested in *in silico* prediction. We have invested time and effort into it. At this stage we invest more effort in measuring directly because these data are more useful to us.

Sanguinetti: Where you have a lot of chemistry already, do you make focused pharmacophore models and then rely on them?

Nicklin: We make the models but we don't solely rely on them.

Gosling: One of the problems with these models is the way in which we measure the hERG potency. The problem with hERG is that most of these molecules have to cross the membrane and get inside, so you have pharmacokinetic (PK) issues which are always confounding your data set. A medicinal chemist makes a molecule, it gets screened in a hERG patch clamp assay and say, for example, it is clean at 10 μM. Then the PK profile is determined and found to be terrible, so the medicinal chemist modifies the molecule, maybe adding a solubilizing group to improve the PK properties. When the new molecule is screened it may suddenly be found to be a 5 nM hERG inhibitor because it now has an enhanced ability to cross a cell membrane and gain access to the intracellular face of hERG and block it. This is the problem. If we can take this property out of the equation this would help. Currently with a lot of our measurements we are not looking at the affinity of a compound for hERG *per se*. It is the affinity plus a variety of pharmacokinetic parameters that contribute to whether or not the compound can gain access to the binding site(s) in the channel. This will make nailing a hERG pharmacophore very difficult unless we can segregate the two components.

Brown: What sort of throughput do you need for what you are talking about?

Gosling: It depends. If you have a better method of measuring, probably lower than we are currently doing now.

Brown: There is always this trade-off between throughput and sensitivity. What is your trade-off point?

Gosling: You still have a cell-based assay. For me that is not ideal because you aren't looking at purely the affinity of the molecule for hERG. The compound is going through transporters, is broken up and there are the issues of access and crossing membranes.

Brown: Those are real issues.

Gosling: Yes, they are indeed real issues with respect to predicting a molecule's ability to give you a problem in the clinic, but in my opinion they may be secondary to a molecule's ability to bind to hERG. We want our molecules not to have any hERG liability. Surely the most definitive way to do this is to make a molecule that has no affinity for hERG, rather than one that doesn't get access to hERG due to low membrane permeability or reduced stability.

Hoffmann: I can speak for Roche, and I have a certain overview from the ILSI IHESI initiative and from the safety pharmacology society. The current tendency seems to be that every company has an *in silico* tool. We have the model that you cited. The predictive value for new chemical entities is not good enough. But it has its value in optimizing within a chemical class, otherwise we look for biological data. Currently we don't invest very much in this SAR field because we feel during the next few years that there won't be much advance. The tendency is to go for high-throughput automatic patch clamp machines to try to get the biological data. I say this regretting that we can't predict better with *in silico* methods.

Recanatini: I agree. There are several issues that need to be addressed to refine the models, and the consistency of biological data is the first and most crucial of these. The high-throughput screening methods can be helpful. Some QSAR models are done just to obtain a classification. This might be useful for companies that have to obtain quick responses. Just classifying drugs into blockers and non-blockers might be a reliable goal.

Shah: Earlier on, Michael suggested that this channel might be highly plastic. Given the plasticity of this channel, do you think that the ligand-based approaches are more likely to be productive than a target-based approach? I have a general observation from a number of these drugs. Whatever model you go for, it would have to explain the following two observations: (1) Most of the hERG channel blockers (70–80%) are metabolized by cytochrome 3A4 and (2) I don't know of any hERG blocker that is metabolized by cytochrome 2C9.

Gosling: 3A4 has almost the same pharmacophore.

Shah: Then it is a ligand-based approach, as opposed to a target-based approach, that will probably succeed in identifying the drugs likely to bind to hERG.

Recanatini: I don't think the two approaches are exclusive. I distinguished them just for clarity. But work that is target-based often makes use of ligands. There are different philosophies of modelling. In some cases you might prefer the descriptive properties of the model. In other cases you may want just a predictive tool. With regards to the plasticity, we can address these characteristics for short time scales.

Netzer: Over the past few years we have seen an increasing number of companies starting virtual screening. If you already have information about active compounds or templates you could use this for generating a model, or you could generate information using structural biology, then you may add this as much as possible using *in silico* tests for ADMET (e.g. hERG) in such an investigation. Finally this will result in a number of compounds and templates that can be introduced to wet screening. It is a very elegant way to reduce the number of compounds for a sophisticated assay.

Recanatini: The idea is to take into consideration the potentially adverse effects as early as possible in the drug discovery process. Predictive tools can be incorporated into the filtering procedures for databases.

Noble: What sort of computational resources are needed for the kind of work you are describing?

Recanatini: We use a Linux cluster of five PCs—you don't need a supercomputer. Of course, if you have a supercomputer it makes things faster.

Noble: The reason I raise this is because the chances are in the next few years that grid or cluster computing will become very common. You may be able to jump up a whole level of complexity.

Recanatini: Typically, the greatest computational demand is on the molecular dynamics simulations and electrostatic calculations when they are carried out at a high level (*ab initio*).

Sanguinetti: Where are we with respect to high-capacity voltage clamp instrumentation? How close are we to every company having one and using it internally?

Traebert: Astra Zeneca claim that they have a fully validated automatic patch clamp technique. This seems to be the first one. It can do 5000 molecules a year.

Gosling: The problem with these machines is not the capital cost, it is the running cost, which is frightening. For Axon's Patch Xpress, their 16 channel array is US$170, and you can burn four of those in an hour.

Netzer: The success rate is low, too, at around 30%. It is variable.

Sanguinetti: Is that because of cell problems?

Brown: There are all kinds of problems. With one machine you have throughput but you don't have sensitivity. The others are reasonable, but if you look at the IC_{50} it comes out at 150. This is three or four times too high. One of them has problems in terms of exchange, because the hole sits at the bottom of a well. There is difficulty with exchange at the bottom.

Gosling: With the Molecular Devices IonWorks HT system, many companies are quite happy to use it for hERG. As a classical patch-clamp electrophysiologist, the system not having gigaohm seals is a problem for me, but Molecular Devices have gone to a lot of effort to say that you don't need a giga seal. It is however interesting to note that Molecular Devices have just bought Axon Instruments whose machine does have a giga seal capability.

Brown: How can you have the sensitivity if most of your current is going out of the side.

Gosling: Their blocker IC_{50}s looks fine.

Traebert: The problem these machines have is choosing the right cell culture. You can't use them for primary cardiac myocytes because the cells have to be in suspension.

The long QT syndrome: a clinical counterpart of *h*ERG mutations

Peter J. Schwartz

Department of Cardiology, University of Pavia and Policlinico S. Matteo IRCCS, V.le Golgi 19, 27100 Pavia, Italy

Abstract. The congenital long QT syndrome (LQTS) is a leading cause of sudden death in the young. While most patients die during conditions of sympathetic activation, such as physical exercise or emotions, other die suddenly while at rest or during sleep. Several genes responsible for the disease have been identified. The most important genes encode ion channels involved in the control of ventricular repolarization. The currents involved are I_{Ks}, I_{Kr}, and I_{Na}. Patients with mutations in the *h*ERG gene form the LQT2 subgroup. This chapter reviews several critical clinical aspects focusing on differences between LQT2 patients and those from the other main subgroups (LQT1 and LQT3). Presentation and discussion of the different phenotypes is followed by a number of still unanswered questions related to specific features of the LQT2 patients.

2005 The hERG cardiac potassium channel: structure, function, and long QT syndrome. Wiley, Chichester (Novartis Foundation Symposium 266) p 186–203

Seldom in medical history has a disease been kept outside of mainstream cardiology and regarded as an oddity for so long and then, over just a few years, with a sudden reversal has been embraced by many as a paradigm for sudden cardiac death (Members of the Sicilian Gambit 1998).

The identification at the end of March 1995 (Wang et al 1995, Curran et al 1995) of the first two long QT syndrome (LQTS) genes represented a major breakthrough not only for cardiac electrophysiology but also for cardiology as a whole, and paved the way for the understanding of how tight the relationship between molecular and clinical cardiology can be. Indeed, the impressive correlation between specific mutations and critical alterations in the ionic control of ventricular repolarization has made LQTS the best example to date for the specificity and value of the correlation between genotype and phenotype. Until the mid-1990s most cardiologists had been unimpressed by the clinical relevance of molecular biology, but they changed their minds largely on the basis of the very rapid developments that contributed the elucidation of this life-threatening disorder, which represents a sort of Rosetta stone for sudden cardiac death (Zipes

1991). Similarly, many basic science investigators who had not even ever heard of LQTS became involved in LQTS-related research because of its obvious potential in helping to elucidate the key mechanisms also underlying more common and complex clinical disorders.

Several LQTS-related genes have been identified; however, for only three of them (*KvLQT1*, *hERG*, *SCN5A*) disease-causing mutations have been identified in a sufficiently large number of LQTS patients to allow meaningful genotype–phenotype correlations. The three most frequently encountered genetic subgroups are LQT1 (due to mutations on the *KCNQ1*, previously *KvLQT1*, gene), LQT2 (due to mutations on the *hERG* gene), and LQT3 (due to mutations on the *SCN5A* gene). Very recently, the gene for the variant form of LQTS, associated with syndactyly, has been identified (Splawski et al 2004). Of note, the gene, $Ca_v1.2$ encodes the L-type calcium current, and all the 17 patients reported, albeit originating from different families, had the same mutation (G406R), which produces a gain of function.

LQT2 is due to mutations in *hERG*, the gene which encodes the α-subunit for the rapidly activating potassium current I_{Kr} (Curran et al 1995). Keeping in mind that currently 40–50% of the patients in whom LQTS has been diagnosed beyond doubt, on the basis of updated diagnostic criteria (Schwartz et al 1993) and typical clinical presentation, still remain without a molecular diagnosis.

The relative frequencies of the three main genetic subtypes, based on the two largest databases for genotyped LQTS patients (percentages shown in brackets from Splawski et al 2000 and Priori et al 2003 respectively), are as follows: LQT1 (44, 54%), LQT2 (53, 35%), LQT3 (6, 11%). These figures reflect the percentages of probands with mutations on *KvLQT1*, *hERG*, *SCN5A*, using as a denominator the sum of these individuals plus the few identified as LQT5 and LQT6 and those in whom no mutations were found, which equals approximately 40% of the total number of LQTS probands.

This chapter will primarily deal with the clinical features of LQT2 patients and with the main differences compared to LQT1 and to LQT3 patients. It will also focus on those clinical aspects that raise still unanswered questions on the specific links between *hERG* mutations and phenotype.

Genotype–phenotype correlation

ECG pattern

Following the first study by Moss et al (1995), Zhang et al (2000) extended the number of patients analysed and provided data on 284 gene carriers. They identified a series of typical patterns present in most of the patients with LQT1, LQT2, and LQT3 mutations.

The electrocardiogram (ECG) hallmark for LQT2 patients, as already suggested (Dausse et al 1996), is represented by notched T waves. Bifid is a term also often used to define this morphology, which is characterized by a second component ('notch') usually appearing before the peak of the T wave and which had been identified by Malfatto et al (1994) as frequently occurring among LQTS patients. Zhang et al (2000) identified four T wave patterns in LQT2 patients: 1) the obvious bifid T wave; 2) the subtle bifid T wave with the second component on top of the T wave or 3) on its downslope; and 4) the low amplitude bifid T wave. At variance with this view, Takenaka et al (2003) very recently reported that a third of their 31 LQT2 patients had a broad-based T wave morphology, typical of LQT1. Zhang et al (2000) concluded by saying that genotype identification by ECG can be useful for choosing which gene to screen first in molecular genetic studies.

The studies by Moss et al (1995) and by Zhang et al (2000) have provided unquestionable evidence for a significant relationship between genotype and morphology of the T wave. The finding is intriguing but the underlying mechanisms remain elusive, at least in part. The often quoted studies by Shimizu and Antzelevitch (Shimizu & Antzelevitch 1997, 1998) provide an elegant explanation which, however, fails to fully account for the existence of multiple 'typical' patterns per genotype and even within a single family. Figure 1 represents the case of an LQT2 family in which the proband has an ECG pattern quite different from those typical for LQT2 patients; the sister has a late second component in lead V4 and only the father has a clearly typical T wave morphology with clear notches. If only the proband's ECG had been available it would have not been possible to suspect LQT2 from this tracing.

ECG patterns should never be used to 'genotype' patients. They may indeed be useful, particularly in association with the conditions favouring the occurrence of cardiac events, to increase the probability for molecular screening to start on the culprit gene.

We still don't know why is it that patients with the same mutation on the *hERG* gene can present with very different T wave morphologies.

Triggers for cardiac events

The novel and unexpected observation, made in 1995 in a tiny group of genotyped patients (Schwartz et al 1995), that LQT3 patients had their cardiac events during sleep or at rest while LQT2 patients had their arrhythmic episodes during emotional stress was so richly endowed with significant implications that we concluded by stating our intention of revisiting the issue in a much larger population before drawing final conclusions.

Such a study was then performed, with worldwide cooperation, and did indeed provide final conclusions on the relationship between specific genotypes and the

**PROBAND
G.T. 7 years
QTc: 630 ms**

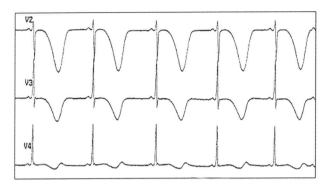

**SISTER
S.T. 10 years
QTc: 605 ms**

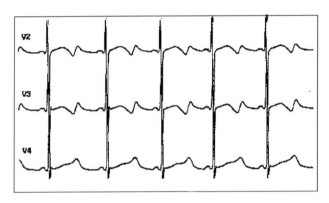

**FATHER
V.T. 37 years
QTc: 584 ms**

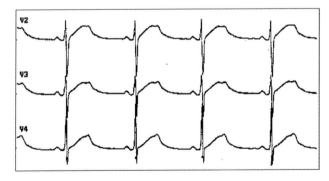

FIG. 1. Different T wave morphologies in affected members of the same family. The proband had a documented cardiac arrest as the first manifestation of LQTS. His sister is still asymptomatic, while his father has had two syncopal episodes. From Schwartz et al (2000) with permission.

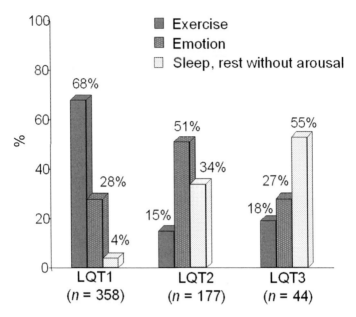

FIG. 2. Triggers for all lethal and non-lethal cardiac events in the three genotypes. Modified from Schwartz et al (2001).

conditions which cause (trigger) cardiac events in LQTS (Schwartz et al 2001). The study was based on 670 LQTS patients, all of known genotype (LQT1 = 371, 55%; LQT2 = 234, 35%; LQT3 = 65, 10%). Importantly for the data interpretation, all the patients had had cardiac symptoms (syncope, cardiac arrest, or sudden death); thus, asymptomatic patients were not included. Also, patients with the Jervell-Lange-Nielsen variant of LQTS were not included. Three distinct conditions, or triggers, associated with the onset of cardiac events were identified: exercise, emotion, and rest or sleep without arousal.

The analysis was performed by examining the relation between genotype and all three main cardiac symptoms but also restricting it, after elimination of syncope, to lethal cardiac events (resuscitated cardiac arrest and sudden death). The pattern clearly emerging from the first and broader analysis (Fig. 2) was further accentuated by the second, more selective one (Fig. 3). These data confirmed the initial observations (Schwartz et al 1995) and demonstrated that the probability of life-threatening cardiac events under specific circumstances varies in a gene-specific manner.

The most prominent findings are the following. LQT1 patients have a major propensity to develop arrhythmias under conditions of increased sympathetic activity, with physical exercise playing the biggest role, while their probability of

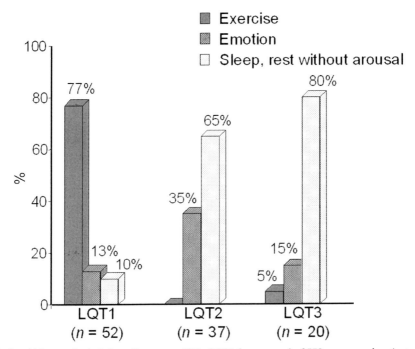

FIG. 3. Triggers for lethal cardiac events (CA, SCD) from a total of 579 genotyped patients. Modified from Schwartz et al (2001).

having cardiac events at rest is minimal. By contrast, both LQT2 and LQT3 patients have only a modest probability of suffering cardiac events during exercise. LQT2 patients experience most of their arrhythmic episodes during emotional periods and also at rest. LQT3 patients are at greatest risk at rest or while they are asleep. When the analysis is limited to lethal episodes (Fig. 3), this pattern is greatly accentuated with not a single LQT2 patient dying during exercise and with over 80% of the lethal episodes for LQT3 patients occurring at rest or during sleep.

Finally, it became evident that for LQT1 and LQT2 patients there are also highly specific triggers; namely, swimming for LQT1 and auditory stimuli for LQT2 (Fig. 4). Among patients having cardiac events while swimming, 99% are LQT1 whereas among patients having cardiac events in association with loud noises, 80% are LQT2. Both swimming and loud noises had been identified as important triggers for LQTS patients in the pre-molecular era (Schwartz et al 1991); while the relation between swimming and LQT1 was largely expected, given the unique importance of physical exercise as a trigger for LQT1 patients, the relation between auditory stimuli was largely unexpected and was first described by Wilde et al (1999).

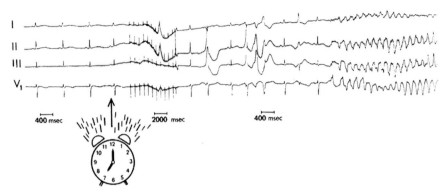

FIG. 4. Tracings of a 14-year old patient with LQTS. This patient had a history of losing consciousness being awakened by thunder or the noise of an alarm clock. Initiation of ventricular fibrillation following an auditory stimulus (alarm clock) is shown. The middle tracing is obtained at a slower speed than the beginning and end tracings. This patient was treated with β-blockers and remained free of syncopal episodes for several years until her boyfriend convinced her that she did not need medications. She interrupted her therapy and died suddenly within a few weeks. From Wellens et al (1972) with permission.

These gene-specific differences in the propensity toward life-threatening arrhythmias under specific conditions can be explained, at least in part, by our current understanding of the underlying molecular mechanisms. LQT1 patients are at much higher risk when sympathetic activity increases and especially when this increase is accompanied by a marked and sustained increase in heart rate, as it occurs more with exercise than with a sudden startle. This is a consequence of the impairment in the ability of the mutant channels to increase the I_{Ks} current and to respond appropriately to β-adrenergic receptor signalling (Marx et al 2002). Conversely, the relative low risk for both LQT2 and LQT3 during exercise is likely due to the presence of an intact I_{Ks} and is matched by the greater than normal ability of LQT3 patients to shorten their QT interval when heart rate increases (Schwartz et al 1995). The fact that LQT3 patients are at an excessively high risk while asleep or at rest is in agreement with the observations that during the night their QT interval is prolonged (Stramba-Badiale et al 2000).

The information originating from the largest data set ever used for genotype–phenotype correlation studies (Schwartz et al 2001) provides definitive evidence for gene-specific differences in the propensity to suffer life-threatening arrhythmias under specific conditions while also offering insights into some of the underlying mechanisms.

As to patients with mutations on *hERG*, a number of questions are still unanswered. Why is it that they are at higher risk when exposed to conditions, such as startle reactions induced by emotions and sudden noises, in which the main physiological event is an abrupt release of norepinephrine not preceded, at variance with exercise, by a progressive increase in heart rate? And what is the link between triggers as apparently diverse as startle and sleep? Perhaps periods of rapid eye movement (REM) sleep, during which sudden increases in sympathetic and in vagal activity are known to occur?

Whatever the answers to these questions, the fact remains that while the triggering events for LQT1 have provided valid guidance for the choice of effective therapies, in the case of LQT2 such a straightforward indication is missing.

Onset of cardiac events

In theory, one would expect LQTS patients to become symptomatic according to some individual characteristic and not based on genetic subgroup. The actual findings show the opposite. Fig. 5 demonstrates that LQT1 patients become symptomatic significantly earlier than LQT2 and LQT3 patients. Within a large population of individuals all destined to become symptomatic, 86% of the LQT1 patients had their first event before the age of 20.

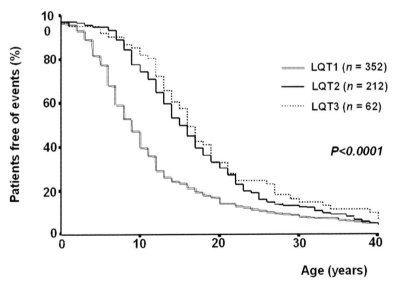

FIG. 5. Kaplan–Meier cumulative survival curves showing time interval between birth and first cardiac event (syncope, resuscitated cardiac arrest, sudden death). Numbers in parentheses indicate patients. It is important to remember that the figure and study do not include asymptomatic patients. LQT1 vs. LQT2, $P<0.0001$. LQT1 vs. LQT3, $P=0.0001$. LQT3 vs. LQT2, $P=$ NS. From Schwartz et al (2001) with permission.

Why is it that LQT2 patients become symptomatic significantly later than LQT1 patients? And why, for this aspect, their pattern is superimposable on that of the LQT3 patients? Is the presence of a normal I_{Ks} sufficient to explain this difference?

Natural history and risk stratification

The most complete set of data that provide useful information on gene-specific natural history was based on 193 families (104 LQT1, 68 LQT2, and 21 LQT3) consecutively genotyped at a single centre and included a total of 647 LQTS patients (Priori et al 2003).

This study focused on the cumulative probability of a first cardiac event, also differentiating between all cardiac events (syncope, cardiac arrest and sudden death) and lethal cardiac events (cardiac arrest and sudden death occurring below age 40 and prior to initiation of therapy). A lower cumulative survival was observed among LQT2 vs. LQT1 patients ($P < 0.001$) and a similar trend was present among LQT3 vs. LQT1 ($P = 0.07$); in other words, mortality was lower among LQT1 patients. Compared to LQT1 the relative risk for cardiac events was significantly higher for both LQT2 (relative risk [RR] 1.6, 95% confidence interval [CI] 1.16–2.25) and LQT3 (RR1.8, 95% CI 1.07–3.04).

Gender had a different influence across different genotypes. It had no influence among LQT1 patients, whereas a higher risk was present for LQT2 females and LQT3 males. Among LQT2 patients the female gender carries a very high risk as even females with a QTc < 500 ms have a fourfold greater risk than males with similar QTc.

Why is it that LQT2 patients are at higher risk of events than LQT1 patients? Is it solely because the percentage of 'silent' mutation carriers is higher (36% vs. 19%) in LQT1 than in LQT2 patients? And why are LQT2 females at much higher risk than LQT2 males with a similar QTc?

Site of mutation

Risk stratification for specific genetic subgroups now extends to the actual site of the mutation. Donger et al (1997) had the merit of being the first to call attention to the fact that the clinical severity of LQTS might have been correlated to the specific site of the mutations. In a relatively small study they suggested that mutations occurring in the C-terminal domain might be less malignant than mutations occurring in the pore region. This observation preceded the molecular evidence that pore mutations are more likely to be dominant-negative, thus producing > 50% reduction in channel function, while non-pore mutations are more likely to result in co-assembly or trafficking abnormalities which produce haplotype insufficiency and a 50% reduction in channel function (January et al 2000, Huang et al 2001). The most convincing demonstration of the clinical importance of these observations came with the study by Moss et al (2002) who studied 201 LQT2 patients, of whom 35 had mutations in the pore region of *hERG* and 166 had mutations in the non-pore region. The major finding was that LQT2 patients with pore mutations were at considerably greater risk for cardiac events than patients with non-pore mutations; they had the first event at younger age and more frequent cardiac events (74% vs. 35%, $P<0.001$), even though the difference in the incidence of aborted cardiac arrest and sudden death (15% vs. 6%) was not statistically significant. These data provide important evidence on the different clinical consequences that the site of different mutations may have for the affected patients.

Response to therapy

Since the first large studies on LQTS patients including data on therapy were reported (Schwartz et al 1975, Schwartz 1985), β-blockers represented the first-choice therapy. Information on the effect of β-blockers in genotyped patients has only recently been presented in adequate numbers (Schwartz et al 2001, Priori et al 2004).

In the study by Schwartz et al (2001) data on the efficacy of β-blockers were available for 271 genotyped patients (LQT1 = 161, LQT2 = 91, and LQT3 = 18). No asymptomatic patients were included. The analysis considered two aspects important for clinical management: the recurrence of cardiac events and the occurrence of cardiac arrest or sudden death (Table 1). The LQT1 subgroup did better in terms of prevention of recurrences during treatment as 81% of the patients remained free from symptoms, compared to significantly lower percentages among LQT2 (59%) and LQT3 (50%). When the cumulative incidence of cardiac arrest and sudden death was examined, there was no difference between LQT1 and LQT2 (4% for both) while the incidence of lethal events despite therapy was much higher among LQT3 patients (17%). The number of LQT3 patients was small.

TABLE 1 Genotype and β blocker therapy

Genotype	Prevention of recurrences n (%)	Recurrences n (%)	Cardiac arrest/sudden cardiac death n (% of all patients; % of patients with recurrences)
LQT1 ($n=162$)	131 (81%)	31 (19%)	7 (4%; 23%)
LQT2 ($n=91$)	54 (59%)	37 (41%)	4 (4%; 11%)
LQT3 ($n=18$)	9 (50%)	9 (50%)	3 (17%; 33%)

Very recently, Priori et al (2004) provided data on 335 genotyped patients (LQT1 = 187; LQT2 = 120; LQT3 = 28) treated with β-blockers. During a 5 year follow-up period there were 4 (1.2%) sudden deaths (1 LQT1, 3 LQT3) and 14 (4.2%) cardiac arrests. The latter occurred more frequently in LQT3 (14%) and in LQT2 (7%) compared to LQT1 (1%).

Another important modality of therapy is represented by left cardiac sympathetic denervation (LCSD), used in LQTS patients who continue to have syncope despite β-blockers or who cannot tolerate β-blockers (Schwartz et al 2004). LCSD is highly effective but the scanty data available in genotyped patients, if confirmed in larger numbers, may suggest a higher efficacy in LQT1 and in LQT3 patients compared to LQT2 patients. The current differences are not statistically significant.

The issue of the severity of LQT2 surfaces also when examining data on patients implanted with an ICD. Preliminary data from the European Registry of ICD in LQTS (Crotti et al 2004) indicate a disproportionate number of LQT2 patients compared to LQT1 and even LQT3 patients. Despite the not always appropriate use of the ICDs, such therapeutic choice usually reflects the difficulty in managing patients with traditional therapy.

Why are LQT2 patients less well protected by β-blockers? Is this just a consequence of the greater role of sympathetic activation as a trigger for cardiac events among LQT1 patients?

Conclusion

Patients with mutations affecting the $hERG$ gene and the I_{Kr} current represent a well-defined subgroup of the LQTS. While some of their gene-specific clinical manifestations can be explained on the basis of our current understanding of cardiac electrophysiology and of molecular biology, many others still represent unanswered questions. Correct answers will move forward not only our basic knowledge but probably also our ability to find the most appropriate therapeutic strategy for the individual patients affected by LQTS.

References

Crotti L, Spazzolini C, De Ferrari GM et al 2004 Is the implantable defibrillator appropriately used in the long QT syndrome? Data from the European Registry. Heart Rhythm 1(suppl):582

Curran ME, Splawski I, Timothy KW, Vincent GM, Green ED, Keating MT 1995 A molecular basis for cardiac arrhythmia: HERG mutations cause long QT syndrome. Cell 80:795–803

Dausse E, Berthet M, Denjoy I et al 1996 A mutation in HERG associated with notched T waves in long-QT syndrome. J Mol Cell Cardiol 28:1609–1615

Donger C, Denjoy I, Berthet M et al 1997 KVLQT1 C-terminal missense mutation causes a forme fruste long-QT syndrome. Circulation 96:2778–2781

Huang FD, Chen J, Lin M, Keating MT, Sanguinetti MC 2001 Long-QT syndrome-associated missense mutations in the pore helix of the HERG potassium channel. Circulation 104:1071–1075

January CT, Gong Q, Zhou Z 2000 Long QT syndrome: cellular basis and arrhythmia mechanism in LQT2. J Cardiovasc Electrophysiol 11:1413–1418

Malfatto G, Beria G, Sala S, Bonazzi O, Schwartz PJ 1994 Quantitative analysis of T wave abnormalities and their prognostic implications in the idiopathic long QT syndrome. J Am Coll Cardiol 23:296–301

Marx SO, Kurokawa J, Reiken S et al 2002 Requirement of a macromolecular signaling complex for β adrenergic receptor modulation of the KCNQ1-KCNE1 potassium channel. Science 295:496–499

Members of the Sicilian Gambit 1998 The search for novel antiarrhythmic strategies. Eur Heart J 19:1178–1196 and Jpn Circ J 62:633–648

Moss AJ, Zareba W, Benhorin J et al 1995 ECG T-wave patterns in genetically distinct forms of the hereditary long QT syndrome. Circulation 92:2929–2934

Moss AJ, Zareba W, Kaufman ES et al 2002 Increased risk of arrhythmic events in long QT syndrome with mutations in the pore region of the human ether-a-go-go-related gene potassium channel. Circulation 105:794–799

Priori SG, Schwartz PJ, Napolitano C et al 2003 Risk stratification in the long-QT syndrome. N Engl J Med 348:1866–1874

Priori SG, Napolitano C, Schwartz PJ et al 2004 Association of long QT syndrome loci and cardiac events among patients treated with β-blockers. J Am Med Assoc 292:1341–1344

Schwartz PJ 1985 Idiopathic long QT syndrome: progress and questions. Am Heart J 109:399–411

Schwartz PJ, Periti M, Malliani A 1975 The long QT syndrome. Am Heart J 89:378–390

Schwartz PJ, Zaza A, Locati E, Moss AJ 1991 Stress and sudden death. The case of the long QT syndrome. Circulation 83 (suppl 4):II71–II80

Schwartz PJ, Moss AJ, Vincent GM, Crampton RS 1993 Diagnostic criteria for the long QT syndrome: an update. Circulation 88:782–784

Schwartz PJ, Priori SG, Locati EH et al 1995 Long QT syndrome patients with mutations on the SCN5A and HERG genes have differential responses to Na^+ channel blockade and to increases in heart rate. Implications for gene-specific therapy. Circulation 92:3381–3386

Schwartz PJ, Priori SG, Napolitano C 2000 The long QT syndrome. In: Zipes DP, Jalife J (eds) Cardiac electrophysiology. From cell to bedside. III edn. WB Saunders Co., Philadelphia, p 597–615

Schwartz PJ, Priori SG, Spazzolini C et al 2001 Genotype-phenotype correlation in the long QT syndrome. Gene-specific triggers for life-threatening arrhythmias. Circulation 103:89–95

Schwartz PJ, Priori SG, Cerrone M et al 2004 Left cardiac sympathetic denervation in the management of high-risk patients affected by the long QT syndrome. Circulation 109:1826–1833

Shimizu W, Antzelevitch C 1997 Sodium channel block with mexiletine is effective in reducing dispersion of repolarization and preventing torsade de pointes in LQT2 and LQT3 models of the long-QT syndrome. Circulation 96:2038–2047

Shimizu W, Antzelevitch C 1998 Cellular basis for the ECG features of the LQT1 form of the long-QT syndrome: effects of β-adrenergic agonists and antagonists and sodium channel blockers on transmural dispersion of repolarization and torsade de pointes. Circulation 98:2314–2322

Splawski I, Shen J, Timothy KW et al 2000 Spectrum of mutations in long-QT syndrome genes. Circulation 102:1178–1185

Splawski I, Timothy KW, Decher N et al 2004 $Ca_v1.2$ calcium channel dysfunction causes a multisystem disorder including arrhythmia and autism. Cell 119:19–31

Stramba-Badiale M, Priori SG, Napolitano C et al 2000 Gene-specific differences in the circadian variation of ventricular repolarization in the long QT syndrome: a key to sudden death during sleep-? Ital Heart J 1:323–328

Takenaka K, Ai T, Shimizu W et al 2003 Exercise stress test amplifies genotype-phenotype correlation in the LQT1 and LQT2 forms of the long-QT syndrome. Circulation 107:838–844

Wang Q, Shen J, Splawski I et al 1995 SCN5A mutations associated with an inherited cardiac arrhythmia, long QT syndrome. Cell 80:805–811

Wellens HJJ, Vermeulen A, Durrer D 1972 Ventricular fibrillation occurring on arousal from sleep by auditory stimuli. Circulation 46:661–665

Wilde AAM, Jongbloed RJE, Doevendans PA et al 1999 Auditory stimuli as a trigger for arrhythmic events differentiate HERG-related (LQTS2) patients from KvLQT1-related patients (LQTS1). J Am Coll Cardiol 33:327–332

Zhang L, Timothy KW, Vincent GM et al 2000 Spectrum of ST-T-wave patterns and repolarization parameters in congenital long QT syndrome. ECG findings identify genotypes. Circulation 102:2849–2855

Zipes DP 1991 The long QT interval syndrome. A Rosetta stone for sympathetic related ventricular tachyarrhythmias. Circulation 84:1414–1419

DISCUSSION

Hondeghem: In the patients with LQTS, when the alarm clock goes off and 15–20 s later they go into TdP, what happens in the interim? Does the heart slow because of a vagal reaction? Does the RR interval become unstable? Does the QT interval start oscillating?

Schwartz: I don't know how long the interval is between the sudden noise and the onset of TdP. I suspect it is just a few seconds. The typical clinical history is that the phone rings and the patient falls on the ground. It seems that is a very short period of time. What happens in between? Release of norepinephrine is the first response to startle. I don't know why it takes a few seconds to initiate TdP. I am not sure that heart rate will oscillate that much. The vagal tone probably is still whatever it was. On the background of tonic vagal activity there is a sudden release of norepinephrine. The QT interval seems to further prolong and its initial value may be an important contributor. The little we know is based on anecdotes. Unfortunately, most patients do not have a Holter recording at the time of the episode. These are random events, not planned as in a laboratory experiment.

Hondeghem: Even a discharge of norepinephrine could immediately slow the heart rate.

Schwartz: The first response to norepinephrine actually is an increase in heart rate. It is only when, a little bit later, blood pressure increases that there will be a vagal response through a baroreceptor reflex. Moreover, if norepinephrine initiates a ventricular tachyarrhythmia there will be no decrease in heart rate. I don't expect significant changes in vagal activity in the immediate proximity to a startle, besides a likely immediate and partial withdrawal.

Shah: We have had some high profile cases of infant deaths in the UK recently. Do you think there should be a requirement for forensic examination of the arrhythmia-related genotypes of these victims and their immediate relatives?

Schwartz: The issue of a requirement for forensic examination should probably be limited to cases of 'multiple SIDS' or those in which an apparently healthy parent is accused of smothering. We are completing a large case-control study, with blind molecular screening, in almost 200 cases of SIDS and having opened the code in 150 cases we are finding that almost 10% of SIDS victims actually have LQTS mutations. With approximately 40% of typical LQTS patients without identifiable mutations, this suggests that between 10% and 15% of all SIDS victims are actually affected by LQTS and that their deaths could have been prevented by early diagnosis which supports our previous findings (Schwartz et al 2001, Schwartz & Samba-Badiale 2004).

Shah: I haven't had an ECG yet: should I cancel my wakeup call tomorrow morning?!

Schwartz: As you have already survived a goodly number of wake-up calls so far, I see no need to add costs to the NHS. But you may still wish to cancel your wake-up call tomorrow morning and take it easy for once.

Hoffmann: TdP is typically spontaneously reversible to sinus rhythm. How often does this evolve into fatal ventricular fibrillation?

Schwartz: The natural history of this disorder clearly points to the fact that TdP is transient and terminates spontaneously most of the time. Long QT syndrome patients have a number of these episodes and most of the time they revert to sinus rhythm. Yet at one point TdP evolves into ventricular fibrillation. We don't know why one specific episode leads to sudden death. As the denominator (number of TdP episodes) is unknown and as this is only the tip of the iceberg, no quantification is currently possible.

Rosenbaum: One of the differences seen in the lab is the contrast between TdP and the polymorphic VT that occurs in the setting of ischaemic heart disease (IHD). In IHD there is rapid and immediate shortening of the action potential with ongoing ischaemia with the generation of fibrillatory wave fronts. However, in TdP the wave fronts tend to be much broader. There are collisions of the wavefronts with the boundaries of the tissue and they have an opportunity to extinguish themselves which isn't seen in fibrillation. This is thought to be one of the reasons why TdP episodes tend to be self limiting.

Hoffmann: That still doesn't answer the question of why people die sometimes and not others.

Rosenbaum: This is because sometimes the rhythm becomes more disorganized. Talking about denominators, when you screen tissues from people who have died suddenly, in what percentage do you find a mutation or a polymorphism?

Schwartz: The only data I have is from the first group of 20 individuals who died suddenly below the age of 30. Four of them had an *LQT1* mutation.

Robertson: Regarding the specificity of the auditory stimuli, could it just be that our startle response is tuned to auditory stimuli? This is how our system is set up. Those things that surprise us tend to be auditory stimuli.

Schwartz: That is possible. It would make sense from an ontogenic and teleologic point of view, in terms of survival.

Sanguinetti: The important thing here is that the background is sleep with an auditory stimulus.

Schwartz: Not just sleep. Startle is the important aspect here, and it may well happen during wakefulness at rest. So, what probably matters is that the startling event manifests itself on a background of low sympathetic activity, relatively high vagal activity, in other words when heart rate is relatively low.

Brown: Do we know anything about the distribution of hERG in the auditory cortex?

Schwartz: I don't.

Arcangeli: This leads to an interesting question. hERG is indeed expressed in the nervous system, so there could be a link between the auditory effect on TdP and the expression and role of hERG in specific areas of the CNS. Another question is: in cardiac hypertrophy, there is the possibility for sudden death and arrhythmia. Can this type of arrhythmia be linked with dysfunction of I_{Kr} or I_{Ks}?

Schwartz: Probably not.

Hondeghem: The failing heart has a slow repolarization (triangulation).

Schwartz: It is also associated with an increase in QT and heart failure patients with longer QT have a greater risk of sudden death.

Gosling: Are there any animal models of this? Have people tried to generate knockouts or transgenics?

Schwartz: There have been some transgenics but they have not led to very close representations of the clinical picture. It is not ideal to use an animal with a heart rate of 500–600 as a model for human disease.

January: In mice I_{Kr} and I_{Ks} don't play much of a role in repolarization. In the next year or two we will see rabbit models emerging. These may be better.

Schwartz: Rabbits are weird animals. They may die suddenly just holding them by their ears. The British physiologist Sidney Hilton once defined rabbits — 'they are not animals, they are flowers'!

Hoffmann: You mentioned that there may be a sex difference for QT intervals?

Schwartz: Yes, as a rule women tend to have longer QTc. This could explain why drug-induced TdP is much more frequent among women, especially those light in weight. This is a dose effect. Among LQT2 patients there is a clear difference in risk between males and females, with the latter remaining at much higher risk throughout life. There is also a difference in LQT1. For an LQT1 male, to have a first event after age 20 is exceptional, whereas women may have events throughout their lives.

Hancox: You mentioned three T wave morphologies in a single family with the same mutation. What were the differences in age and gender? Are there any data on patients similar in age and of the same gender, such as twins? Can you still get the same differences in the T wave? Or are these different T waves due to age or gender differences superimposed onto the same defect?

Schwartz: I have no data on twins. You can find these patterns independently of age. They are not strictly age-related. Overall, there are some patterns that are more frequent, according to a specific genotype. My point is that despite this, we may find within a certain family, in people with the same mutation, a very different morphology of the T wave. This is difficult to explain.

Shah: I want to respond to Peter Hoffmann's earlier point about the frequency of TdP in the regulatory evaluation of a drug. When we evaluate the torsadogenic potential of a QT-prolonging drug, we don't just restrict ourselves to cases of TdP or other nasty arrhythmias in the dossier. Any symptoms such as blackouts, syncope or convulsions go into the evaluation. Someone earlier made a point about I_{Kr} increasing serum K^+. There have been some reports from Japan (Hatta et al 1999, 2000) that patients with psychosis have normal K^+ but during an acute psychotic episode, their K^+ level drops markedly. This is why some of these patients might be at risk of tachyarrhythmias and sudden death when they are given intravenous antipsychotics to sedate them.

January: There have been varying drugs that have been proposed to be potentially useful in LQTS. These include gene-specific drugs such as flecainide for LQT3, drugs that may not be gene-specific such as spironolactone to raise serum potassium, and I_{Ks} activators. What is the view on drug development in rare diseases?

Nicklin: Novartis does this. The Novartis model for its research portfolio is quite simple. The vertical axis is unmet medical need, the horizontal axis is knowledge of mechanism. Market size doesn't come into it. We are considering diseases such as cystic fibrosis and primary pulmonary hypertension.

January: This is a disease where the mechanisms are being steadily understood, and are therefore addressable. Then we get into the issue of drug development for rare diseases.

Nicklin: Ideally, what we would start with is the genetic linkage and then through functional genomics we would functionalize the genetic linkage so you have a compelling hypothesis for how you treat the disease. Then we hope that the market follows, but at this point we are not going to be bound by market size. Marketeers have been wrong too many times. Mark Fishman, our new head of research, is driving this. It will be a number of years before these potential products are hitting development.

January: Does Europe have the equivalent of an orphan drug law?

Shah: Our orphan drug legislation was introduced in April 2000, and we are now seeing the development and approval of a lot more products for orphan diseases — drugs such as arsenic trioxide. Perhaps if we had a more potent, slightly safer class IA antiarrhythmic drug, it could be marketed for LQT3. A β-blocker could be developed and marketed for LQT1 or LQT2. Nobody has considered these LQT syndromes in the context of orphan drug development, but there is scope for this.

Abbott: Of the people who had episodes while swimming, you mentioned that 106 out of 107 had *KCNQ1* mutations. Was this a group consisting only of LQT1, 2 and 3, or was it everyone?

Schwartz: This was from the study of 670 patients who had already been genotyped as 1, 2 or 3. We had 371 LQT1, 234 LQT2 and 65 LQT3 patients. Among these groups, of the patients who had events while swimming, 99% were LQT1.

Abbott: I was wondering whether patients with MinK mutations also have a higher predisposition for cardiac episodes while swimming.

Schwartz: Yes, we had a very small number of MinK patients. I don't recall if they also had events while swimming.

Abbott: Earlier we were talking about whether MinK mutations might disrupt I_{Kr}, I_{Ks} or both. Looking at the phenotype and seeing how many MinK patients suffered from specific types of arrhythmia — particularly those typically associated with *KCNQ1* versus *hERG* mutations — might give a clue.

Shah: Someone had suggested that if people with LQT1 or LQT2 (because they have diminished K$^+$ channel function) were to marry LQT3 patients (who have augmented repolarizing Na$^+$ current), they might have siblings with a normal ECG phenotype!

Hoffmann: What is puzzling to me is that as well as I_{Kr}-mediated congenital TdP, we also have I_{Ks} and I_{Na} involvement. But drug-induced TdP only involves I_{Kr}. Are we expecting drugs with other mechanisms to show up with TdP in humans soon?

Schwartz: There are still very few I_{Ks} blockers available and widely used. We'll have to wait and see.

References

Hatta K, Takahashi T, Nakamura H, Yamashiro H, Asukai N, Yonezawa Y 1999 Hypokalemia and agitation in acute psychotic patients. Psychiatry Res 86:85–88

Hatta K, Takahashi T, Nakamura H, Yamashiro H, Yonezawa Y 2000 Prolonged QT interval in acute psychotic patients. Psychiatry Res 94:279–295

Swartz PJ, Stramba-Badiale M 2004 Prolonged repolarization and sudden infant death syndrome. In: Zipes DP, Jalife J (eds) Cardiac electrophysiology. From cell to bedside. 4th edn. WB Saunders Co, Philadelphia, p 711–719

Schwartz PJ, Priori SG, Bloise R et al 2004 Molecular diagnosis in a child with sudden infant death syndrome. Lancet 358:1342–1343

Cellular mechanisms of Torsade de Pointes

Steven Poelzing and David S. Rosenbaum[1]

The Heart and Vascular Research Center, and The Department of Biomedical Engineering, MetroHealth Campus, Case Western Reserve University, 2500 MetroHealth Drive, Hamman 330 Cleveland, Ohio 44109-1998, USA

Abstract. Torsade de Pointes (TdP) is a life-threatening arrhythmia closely linked to abnormal cardiac repolarization. It has been demonstrated that cardiac ion channel alterations underlying cellular repolarization results in the phenotypic expression of long QT syndrome, which is closely associated with TdP. However, the mechanisms by which prolonged repolarization leads to TdP remain controversial. Prolonged repolarization is associated with triggered activity, and multiple foci of triggered activity can underlie a TdP phenotype. Action potential shortening associated with rapid ventricular rhythms, in theory, removes conditions for triggered activity. Therefore, while triggered activity may initiate TdP, another mechanism may be responsible for the maintenance of TdP. Re-entrant arrhythmias can also give rise to a TdP phenotype. In intact myocardium significant inhomogeneities of repolarization are manifest in the presence of I_{Kr} blockade. Large repolarization gradients between subepicardial and midmyocardial cells formed zones of conduction block responsible for sustained reentrant TdP. Gap junction proteins responsible for intercellular coupling between subepicardial and midmyocardial cells are reduced in normal myocardium which may maintain arrhythmogenic gradients of repolarization. Therefore, the mechanism of TdP is multi-factorial and related to triggered activity and spatial inhomogeneities of ion channel expression combined with regional expression patterns of gap junctions.

2005 The hERG cardiac potassium channel: structure, function, and long QT syndrome. Wiley, Chichester (Novartis Foundation Symposium 266) p 204–224

Torsade de Pointes (TdP) is a life-threatening arrhythmia closely linked to QT prolongation of the electrocardiogram (ECG) and abnormal cardiac repolarization (Dessertenne 1966, Roden 2000, Ben-David & Zipes 1993). It is well recognized that mutations of ion channels responsible for cellular repolarization underlie the various forms of congenital long QT syndrome

---

[1]This paper was presented at the symposium by David S. Rosenbaum to whom correspondence should be addressed.

(LQTS) (Keating 1993, Curran et al 1995, Wei et al 1999). A variety of antiarrhythmic drugs that target the same ion channels implicated in the congenital form of LQTS cause an acquired form of LQTS, suggesting that both forms of the disease share common electrophysiological mechanisms (Sanguinetti et al 1995). Despite the identification of the molecular basis for LQTS, mechanisms by which channel protein dysfunction promotes the formation of an arrhythmogenic substrate leading to TdP remain unknown.

Currently, there are two hypotheses concerning the mechanisms underlying TdP, based on clinical (Schwartz et al 1975) and experimental (Antzelevitch & Sicouri 1994, El-Sherif et al 1996, Verduyn et al 1997) observations. On one hand, prolongation of action potential duration has been associated with increased risk of triggered activity caused by early and delayed after depolarizations (EAD and DAD respectively) (Luo & Rudy 1994). It has been demonstrated that competing foci of triggered activity can give rise to the electrocardiographic manifestation of a ventricular tachyarrhythmia similar to TdP (Dessertenne 1966). The clinical presentation of TdP is associated with prolonged ventricular repolarization, which supports the hypothesis that multiple foci of triggered activity produces TdP.

However, TdP is a ventricular tachyarrhythmia, and is consequently associated with rapid ventricular activation rates. High rates of ventricular activation significantly shorten repolarization of cardiac myocytes, which in theory eradicates the prerequisite conditions of TdP. Therefore, while triggered activity may cause the initiating beats of TdP, an alternate mechanism may be responsible for the maintenance of the tachyarrhythmia.

The second proposed mechanism of TdP is based on the association between dispersion of repolarization (DOR) and TdP. DOR relates to the time difference between repolarization of one group of cells and another. Cells capable of repolarizing quickly can support a second action potential which may be incapable of propagating into cells with markedly prolonged repolarization times creating a situation of conduction block leading to reentrant excitation. Recent observations from surrogate models of LQTS suggest a role for reentrant activity involving relatively large circuits around the cardiac chambers (El-Sherif et al 1996, Verduyn et al 1997). However, focal (i.e. non-re-entrant) patterns of activation were also observed in these models, which raises the intriguing possibility that EADs and DADs may serve as triggers for reentrant arrhythmias. Additionally, one would expect that normal cell-to-cell coupling through gap junctions would attenuate transmural DOR. In this article we review a hypothesis that ion channel and gap junction distributions in a model of LQT2 underlie the formation of transmural DOR and thereby create a substrate for conduction block and reentrant arrhythmias similar to the clinical manifestation of TdP.

Transmural optical mapping in an experimental model of LQT2

Transmural repolarization and its role in TdP were investigated using high-resolution optical mapping of the arterially perfused canine wedge preparation. This technique allowed for detailed and simultaneous measurement of cellular repolarization from all cell types across the ventricular wall of a multicellular, three-dimensional preparation. Additionally, we were able to precisely align optical maps with the immunohistochemical measurement of gap junction proteins. This approach provided a quantitative assessment of transmural DOR and underlying gap junction distribution under conditions of bradycardia and blockade of the rapid component of the delayed rectifier potassium current (I_{Kr}) with d-sotalol (LQT2).

This model of LQT2 exhibited many features similar to the clinical manifestation of LQTS. For example, QT interval was significantly prolonged (by 133 ± 26 ms) compared to control. Arrhythmias induced in this model of LQT2 exhibited characteristics typical of TdP (Fig. 1). Specifically, ventricular tachycardias were rapid (~ 300 beats/min), self-terminating, readily induced by long-short coupling sequences, only initiated in the setting of a prolonged QT-interval, and displayed a polymorphic undulating ECG morphology. Moreover, TdP induction was dependent on bradycardia (BCL 2000 ms) and I_{Kr} blockade. TdP was never induced in the absence of d-sotalol, and was induced in 10 of 14 (71%) animals during bradycardia and perfusion with d-sotalol.

Unlike arrhythmias in the canine wedge preparation, clinical TdP can persist for long time periods, often degenerating into ventricular fibrillation (VF), possibly due to the superimposed influences of myocardial ischaemia during haemodynamically significant tachycardia. Alternatively, the relatively confined muscle mass of the wedge preparation may result in early termination of TdP because of wavefront collision and annihilation along tissue boundaries. In the surrogate model of LQT2, TdP did not initiate spontaneously. However, spontaneous TdP was reported previously in this model at warmer temperatures (Shimizu & Antzelevitch 1997). In preliminary experiments, we observed the initiation of spontaneous episodes of TdP when perfusion temperature was raised to 37 °C. By perfusing at 35 °C, we intentionally suppressed spontaneous activity in order to measure the electrophysiological substrate and propagation patterns underlying TdP in a controlled fashion.

Propagation was not significantly affected by this model of LQT2 as evidenced by the nearly identical spread of depolarizing wavefronts under control and LQT2 conditions (Fig. 2). Unlike depolarization, gradients of repolarization were highly sensitive to experimental conditions. In control, repolarization gradients were relatively small (Fig. 2, control). The maximum spatial gradient of repolarization (∇R_{max}) measured as the maximum local gradient of

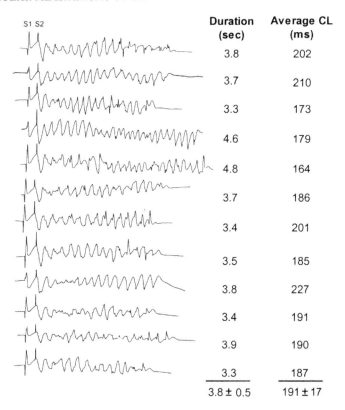

FIG. 1. The electrocardiogram during twelve inductions of TdP in the canine wedge model of LQTS. The duration and average cycle length of each episode are given. The canine wedge model exhibits many of the features of clinical TdP. From Akar et al (2002) with permission.

repolarization times (Pastore & Rosenbaum 2000) was 2.7±0.9 ms/mm. In contrast, bradycardia or perfusion with d-sotalol enhanced transmural gradients of repolarization significantly (Fig. 2, bradycardia, d-sotalol). During both bradycardia and perfusion with d-sotalol, transmural gradients of repolarization increased substantially (∇R_{max} 12.2±2.1 ms/mm), as zones of delayed repolarization extending from the mid-wall towards subepicardial or subendocardial borders emerged (Fig. 2, d-sotalol+bradycardia) thereby enhancing transmural DOR.

Transmural dispersion of repolarization

The underlying mechanism of transmural DOR is well established and attributed to the findings that epicardial cells exhibit the shortest action potential duration

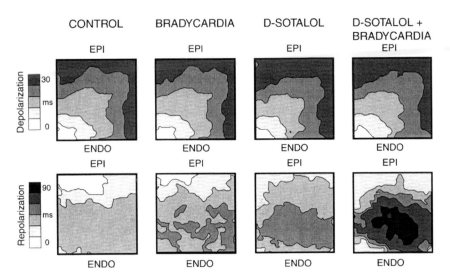

FIG. 2. Transmural depolarization and repolarization contour maps for a wedge endocardially paced at (i) 500 ms BCL (control), (ii) 2000 ms (bradycardia), (iii) 500 ms BCL and perfusion with 100 μmol/l d-sotalol (d-sotalol), and (iv) 2000 ms BCL and perfusion with 100 μmol/l d-sotalol (d-sotalol+bradycardia). While depolarization wavefronts were relatively similar in all four conditions tested, transmural repolarization was exquisitely sensitive to experimental conditions. In d-sotalol+bradycardia, a subpopulation of cells underwent a disproportionate prolongation of their APD forming a broad refractory barrier. The mapping field was ∼15×15 mm. BCL, basic cycle length. From Akar et al (2002) with permission.

(APD) and midmyocardial cells exhibit the longest APD in normal intact left ventricular (LV) myocardium (Akar et al 2002, Yan et al 1998, Poelzing et al 2004). The close juxtaposition of relatively short APD epicardial cells and their midmyocardial neighbours is presumably responsible for the largest APD gradients within the epicardial–midmyocardial interface (Yan et al 1998, Poelzing et al 2004).

Previously, the underlying electrical heterogeneities that cause APD gradients between transmural muscle layers have been reported in a wide variety of animals (Sicouri & Antzelevitch 1991, McIntosh et al 2000, Yao et al 1999) and humans (Drouin et al 1995). One such electrophysiological heterogeneity was proposed with the discovery of a subpopulation of cells within the mid-myocardium of several species (M-cells) (Sicouri et al 1996, Drouin et al 1995, Yan et al 1998). When studied in *isolation*, M-cells have a longer APD, a steeper rate dependence of APD, and a stronger sensitivity to class III anti-arrhythmic agents compared to other myocardial cell types (Liu et al 1993). In *intact* myocardium, M-cells, thought to define the end of the T-wave on the ECG (Fish et al 2004), were characterized by relatively longer APD at baseline and a disproportionate prolongation of their

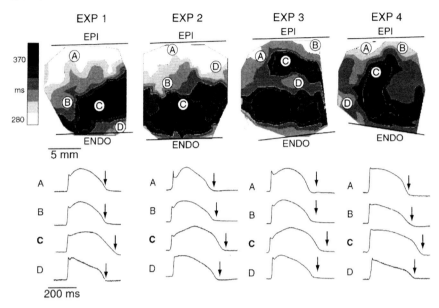

FIG. 3. The topography of M-cells across the left ventricular wall in the canine wedge model of LQTS in four experiments. The distribution of APD across the transmural wall is shown. M-cells exhibiting the longest APDs (black) were located in midmyocardial layers and extended towards endocardial and/or epicardial layers. Typically, epicardial cells exhibited the shortest action potentials (white). In LQTS, a 90 ms gradient of APD was measured across the transmural wall and was associated with a relatively large spatial gradient of repolarization (∼50 mm/ms). The mapping field was ∼15×15 mm. M, midmyocardial; LQTS, long QT syndrome; APD, action potential duration. From Akar et al (2002) with permission.

action potential in response to bradycardia. Unlike their epicardial neighbours, M-cells have a smaller current density of the slow component of the delayed rectifier current (I_{Ks}) (Liu & Antzelevitch 1995) and larger late sodium conductance (I_{Na}) (Zygmunt et al 2001). Both factors result in prolonged M-cell APD compared to epicardial myocytes (Wettwer et al 1993).

Although the location and spatial extent of transmural regions exhibiting the greatest degree of APD prolongation (black) varied between experiments (Fig. 3), the apparent 'M-cell' zones consistently localized within the mid-wall region. Clearly M-cells were not distributed along uniform bands at a given depth of myocardium. Instead, they formed myocardial clusters that typically included the mid-wall and extended towards the endocardial or epicardial surfaces. The M-cell potentials (black) are juxtaposed near other cells located in mid-myocardial zones (white) that did not exhibit comparable APD prolongation. Importantly, the distribution of functionally distinct M-cells is not restricted anatomically to midmyocardial layers.

Previously, M-cells were encountered predominantly in sub-endocardial layers of the anterior left ventricle and sub-epicardial layers of the posterior left ventricle (Yan et al 1998). In our studies (Akar et al 2002), wedges were isolated from the anterior, antero-lateral, and posterior surfaces, which may account for the seemingly random variability in the location and distribution of M-cells. Importantly, these findings may explain the discrepancy in earlier studies regarding the functional existence of M-cells in intact myocardium. Anyukhovsky et al (1996) reported that M-cells were present in all myocardial layers except for the most superficial endocardial and epicardial layers, but were not functionally present *in vivo*. Since M-cell clusters do not exist uniformly at a given depth of myocardium and often extend to the epicardial and endocardial surfaces, it is not surprising that a large variability in midmyocardial APD was measured in earlier studies. (Anyukhovsky et al 1996, Sicouri & Antzelevitch 1991). Finally, M-cells often formed relatively small ($\sim 25\,\text{mm}^2$) zones of increased refractoriness within the mid-myocardium. Therefore a relatively high (<5 mm) spatial resolution is required for their detection.

When intact tissue was subjected to agents that preferentially target one ion channel, the transmural ionic heterogeneities gained greater importance in terms of susceptibility to arrhythmias. Pharmacological block of I_{Kr} and bradycardia (LQT2) selectively prolonged M-cell APD relative to epicardial and endocardial cell types, thereby prolonging the QT interval (Akar et al 2002). By decreasing a repolarizing current in the M-cells which already had reduced repolarization reserve due to low I_{Ks} expression, M-cells displayed a stronger sensitivity to interventions that prolonged APD and created an intrinsic driving force for increased DOR between epicardial and mid-myocardial cell types. Despite the finding that the spatial extent of M-cells may be heterogeneous within a layer of tissue, the sharpest transition in repolarization consistently occurred within the epicardial-midmyocardial interface (within 3 mm of the epicardium) (Poelzing et al 2004), consistent with previous reports in canine left ventricular myocardium (Yan et al 1998).

Cell-to-cell coupling underlies transmural heterogeneities of repolarization

There is considerable controversy regarding whether or not functional expression of ion channel heterogeneities in the intact heart can be maintained in the presence of normal coupling (Lesh et al 1989). Strong cell-to-cell coupling in the intact heart is expected to attenuate the transmural electrophysiological heterogeneities and thereby reduce DOR within the epicardial-midmyocardial interface where the intrinsic driving force is greatest under control and particularly LQT2 conditions. Thus, the extent to which M-cells may functionally influence transmural DOR or arrhythmogenesis in the intact heart was unclear (Anyukhovsky et al 1996) given that some have suggested that significant heterogeneities may not be able to form in the normally coupled intact heart (Anyukhovsky et al 1996). Therefore, we hypothesized that gap junction expression patterns may play an important role in the maintenance of DOR which forms the substrate for the TdP.

Cardiac myocytes are coupled to neighbouring cells through low resistance channels called gap junctions. Gap junctions are found at the border of adjacent cell membranes and permit the flow of ions and small molecules. These channels are composed of proteins called connexins whose isomers are classified by their molecular mass in kDa. Six connexins form a connexon which docks with a connexon from an adjacent cell to form a gap-junction channel, and thereby join the cytoplasmic space of neighbouring cells. Cell–cell coupling, as measured by total gap-junction conductance is presumably due to two main factors: the unitary conductance of the individual channels, and the number of individual gap-junction channels that are functionally expressed. The unitary conductance of a channel is, in part, dependent on the types of connexins composing a channel. Although the principal ventricular gap-junction protein connexin 43 (Cx43) has a main conductance state of 120 pS and connexin 45 (Cx45), another connexin found in ventricle, has a main conductance state of 38 pS, the coexpression of these two proteins results in channels with multiple conductance states between 35 through 105 pS (Martinez et al 2002). Therefore the relative stoichiometry of Cx43 and Cx45 co-expression is an important factor regulating one of the two parameters governing total gap-junctional conductance. However, Cx43 is expressed to a much greater extent than Cx45 in the ventricle (Saffitz et al 1995). Therefore, total gap-junctional conductance is likely to be dependent predominantly on the quantity of Cx43 expression in working channels. Previously, the expression patterns of Cx43 and the effect on transmural DOR across the transmural wall were unknown.

In order to determine the underlying mechanism of enhanced APD gradients between the subepicardial and midmyocardial layers, Cx43 protein expression was measured within each transmural muscle layer. Subepicardial, mid-myocardial, and sub-endocardial Cx43 expressions were compared in representative immunofluorescence (IF) images shown in Fig. 4 (top). Within each transmural layer Cx43 localized to the ends of individual myocytes, consistent with patterns reported previously (Li et al 2002). This pattern was observed across all experiments. It is also demonstrated in Fig. 4 (top) that subepicardial Cx43 related signal is reduced compared to mid-myocardial and sub-endocardial Cx43 signal. Although Cx43 quantity was similar in mid-myocardial and endocardial layers, subepicardial Cx43 expression was significantly decreased by $24 \pm 17\%$ (Fig. 4, bottom) (0–20% of the transmural distance) compared to deeper layers of tissue.

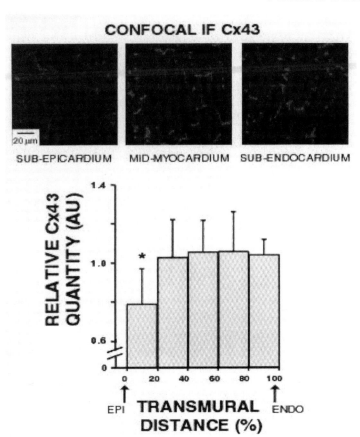

FIG. 4. Transmural Cx43 protein expression. Top: Longitudinally sectioned myocardium showing Cx43 (grey signal) distributed predominantly at the ends of myocytes. Subepicardial Cx43 images have a lower density of Cx43 signal compared to deeper myocardial layers. Bottom: Average Cx43 quantity measured from transmural muscle layers spanning from epicardium to endocardium of the canine wedge preparation. Subepicardial Cx43 expression is reduced relative to mid-myocardial and sub-endocardial layers. From Poelzing et al (2004) with permission.

Reduced subepicardial Cx43 protein expression closely followed other indices of intercellular coupling. For example, transmural conduction velocity was significantly reduced in the subepicardium compared to deeper myocardial layers (33 ± 5 cm/s versus 41 ± 6 cm/s in deeper layers). Importantly, we found that conduction slowing, which was localized exclusively to subepicardial layers was associated with significant increases in dV/dt_{max} (by 18%) suggesting that subepicardial conduction slowing could not be attributed to altered sodium channel availability.

To further establish the functional significance of heterogeneous Cx43 expression across the transmural wall, and specifically reduced Cx43 expression in sub-epicardial layers, we used the effective transmural space constant (λ_{TM}) as an empirical index of transmural coupling between muscle layers (Poelzing et al 2004). Subepicardial λ_{TM} was significantly reduced compared to deeper layers, which nicely corresponds with our findings on subepicardial dV/dt_{max} and θ_{TM}, thus reaffirming the presence of regional intercellular uncoupling within sub-epicardial compared to deeper transmural layers. Importantly, the largest APD gradients within the epicardial-midmyocardial interface occurred precisely where Cx43 and cell–cell coupling was lowest in normal intact myocardium.

The functional consequences of reduced subepicardial Cx43 are not necessarily straightforward, because there is a complex interplay between spatial gradients of ion channels across the transmural wall and cell–cell coupling. Nonetheless, these data suggest that patterns of expression of Cx43 across the transmural ventricular wall can be a mechanism for modulating DOR in syncytial preparations. The question remains though whether the large repolarization gradients within the epicardial mid-myocardial interface are of sufficient magnitude to cause TdP under conditions of LQT2.

Re-entrant mechanism of TdP

The mechanism by which increased DOR, particularly within the epicardial-midmyocardial interface underlies TdP in this model of LQT2 is demonstrated in Fig. 5. As mentioned previously, M-cells did not constitute uniform bands of myocardium as stylized previously. Instead, they formed clusters that typically included the mid-wall and extended towards the endocardial or epicardial surfaces (Fig. 4). As a result of their complex distribution, and their exquisite sensitivity to rate and I_{Kr} blockade, M-cells formed significant ($P < 0.01$) gradients of repolarization within the subepicardial-midmyocardial interface (~ 12 ms/mm) in LQT2, sufficient to cause functional conduction block in response to a premature stimulus (Fig. 5). An epicardial premature stimulus (S2) introduced during the effective refractory period of the baseline beat (240 ms) blocked in the region of most delayed repolarization (Fig. 5, M1 and M2), propagated decrementally, and subsequently blocked in the antidromic direction (Fig. 5A, sites a–c). The S2 wavefront, however, successfully propagated in the orthodromic direction (Fig. 5A, sites a'–e'), circumventing the region of delayed repolarization (Fig. 5A, hatched area). The former sites of block along the epicardium (sites c and d) were undergoing their final phase of repolarization, so the impulse conducted from site e back to site a (Fig. 5B); thereby, completing the first beat of re-entry. An area of functional conduction block was still present in the M-cell region exhibiting the longest APDs (Fig. 5B, hatched area). However, due

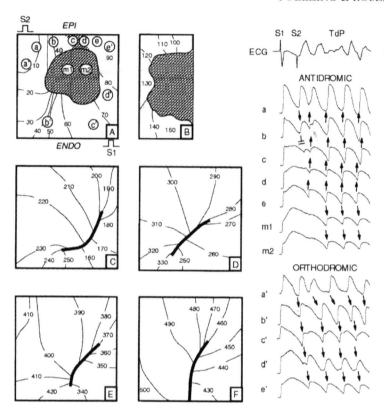

FIG. 5. Activation maps (10 ms isochrones) during TdP. The S2 stimulus was applied on the epicardial surface during the effective refractory period of the S1 beat. In the wedge model of LQT2 a properly timed premature beat blocked transmurally into the M-cell region and propagated around it in a reentrant fashion. The mapping field was ~18×18 mm. TdP, Torsade de Pointes; S2, premature stimulus; M, midmyocardial; LQTS, long QT syndrome. From Akar et al (2002) with permission.

to pronounced rate-adaptation of M-cells, the broad areas of block during the S1 beat rapidly collapsed and were replaced by functional lines of block on subsequent beats (panels C–F). Although the initial beats of reentry occurred within the mapped transmural surface where the premature impulse originated, the reentrant circuit typically meandered into deeper layers of myocardium on subsequent beats.

Conduction slowing was not a requirement for reentry since the path length as dictated by the M-cell zone, was sufficiently long to allow partial recovery of excitability of the former sites of block (Fig. 5). Despite relative normalization of the M-cell APD on subsequent beats, re-entry persisted as the leading edge of the wavefront propagated into the recovering tail of the circuit. Such dynamic M-cell APD adaptation undoubtedly accounted for the rapidly changing trajectory of the reentrant circuit producing the characteristic TdP ECG morphology. The self-limiting nature of TdP may have been due to the wavefront's dynamic head–tail interactions resulting in the instability of the reentrant circuit as the M-cell refractory zone collapsed. Epicardial activation, on the other hand, was uniform, showing no evidence of reentry. The existence of uniform propagation on the epicardium may explain the appearance of a monomorphic waveform configuration in certain ECG leads but not others. Taken together, these findings suggest the existence of a single rotor during TdP that initially forms in the transmural wall and subsequently meanders into deeper layers of myocardium. Consistent with our findings, a single meandering rotor was previously reported to account for a polymorphic ECG morphology resembling TdP in another model of LQTS (Gray et al 1995).

These investigations highlight the importance of transmural heterogeneities of repolarization in the heart. Even in the absence of structural abnormalities (such as myocardial infarction, fibrosis, etc.), electrical heterogeneities can provide an arrhythmogenic substrate, which in the presence of an appropriate trigger can lead to life-threatening arrhythmias. Under normal conditions, transmural DOR is insufficient to cause unidirectional block or reentry. Since connexin expression patterns may also underlie transmural DOR, gap junctions may be an important target for future therapies aimed at reducing the risk of arrhythmias (Eloff et al 2003).

Acknowledgements

These studies were supported by a National Institutes of Health grant RO1-HL54807 (to Dr Rosenbaum) and an American Heart Association Pre-Doctoral Fellowship (to Steven Poelzing).

References

Akar FG, Yan GX, Antzelevitch C, Rosenbaum DS 2002 Unique topographical distribution of M cells underlies reentrant mechanism of Torsade de Pointes in the long-QT syndrome. Circulation 105:1247–1253

Antzelevitch C, Sicouri S 1994 Clinical relevance of cardiac arrhythmias generated by afterdepolarizations. Role of M cells in the generation of U waves, triggered activity and Torsade de Pointes. J Am Coll Cardiol 23:259–277

Anyukhovsky EP, Sosunov EA, Rosen MR 1996 Regional differences in electrophysiological properties of epicardium, midmyocardium, and endocardium—*in vitro* and *in vivo* correlations. Circulation 94:1981–1988

Ben-David J, Zipes DP 1993 Torsades de pointes and proarrhythmia. Lancet 341:1578–1582

Curran ME, Splawski I, Timothy KW, Vincent GM, Green ED, Keating MT 1995 A molecular basis for cardiac arrhythmia: *HERG* mutations cause long QT syndrome. Cell 80: 795–803

Dessertenne F 1966 La tachycardie ventriculaire a duex foyers opposes variables (Ventricular tachycardia with 2 variable opposing foci). Arch Mal Coeur Vaiss 59:263–272

Drouin E, Charpentier F, Gauthier C, Laurent K, Le Marec H 1995 Electrophysiologic characteristics of cells spanning the left ventricular wall of human heart: evidence for presence of M cells. J Am Coll Cardiol 26:185–192

El-Sherif N, Caref EB, Yin H, Restivo M 1996 The electrophysiological mechanism of ventricular arrhythmias in the long QT syndrome — tridimensional mapping of activation and recovery patterns. Circ Res 79:474–492

Eloff BC, Gilat E, Wan X, Rosenbaum DS 2003 Pharmacological modulation of cardiac gap junctions to enhance cardiac conduction: evidence supporting a novel target for antiarrhythmic therapy. Circulation 108:3157–3163

Fish JM, Di Diego JM, Nesterenko V, Antzelevitch C 2004 Epicardial activation of left ventricular wall prolongs QT interval and transmural dispersion of repolarization. Implications for biventricular pacing. Circulation 109:2136–2142

Gray RA, Jalife J, Panfilov A et al 1995 Nonstationary vortexlike reentrant activity as a mechanism of polymorphic ventricular tachycardia in the isolated rabbit heart. Circulation 91:2454–2469

Keating M 1993 Evidence of genetic heterogeneity in the long QT syndrome (response). Science 260:1962

Lesh MD, Pring M, Spear JF 1989 Cellular uncoupling can unmask dispersion of action potential duration in ventricular myocardium: a computer modeling study. Circ Res 65:1426–1440

Li WEI, Waldo K, Linask KL et al 2002 An essential role for connexin43 gap junctions in mouse coronary artery development. Development 129:2031–2042

Liu D-W, Antzelevitch C 1995 Characteristics of the delayed rectifier current (I_{Kr} and I_{Ks}) in canine ventricular epicardial, midmyocardial, and endocardial myocytes: a weaker I_{Ks} contributes to the longer action potential of the M cell. Circ Res 76:351–365

Liu D-W, Gintant GA, Antzelevitch C 1993 Ionic bases for electrophysiological distinctions among epicardial, midmyocardial, and endocardial myocytes from the free wall of the canine left ventricle. Circ Res 72:671–687

Luo C, Rudy Y 1994 A dynamic model of the cardiac ventricular action potential: II. Afterdepolarizations, triggered activity, and potentiation. Circ Res 74:1097–1113

Martinez AD, Hayrapetyan V, Moreno AP, Beyer EC 2002 Connexin43 and connexin45 form heteromeric gap junction channels in which individual components determine permeability and regulation. Circ Res 90:1100–1007

McIntosh MA, Cobbe SM, Smith GL 2000 Heterogeneous changes in action potential and intracellular Ca^{2+} in left ventricular myocyte sub-types from rabbits with heart failure. Cardiovasc Res 45:397–409

Pastore JM, Rosenbaum DS 2000 Role of structural barriers in the mechanism of alternans-induced reentry. Circ Res 87:1157–1163

Poelzing S, Akar FG, Baron E, Rosenbaum DS 2004 Heterogeneous connexin43 expression produces electrophysiological heterogeneities across ventricular wall. Am J Physiol Heart Circ Physiol 286:H2001–H2009

Roden DM 2000 Acquired long QT syndromes and the risk of proarrhythmia. J Cardiovasc Electrophysiol 11:938–940

Saffitz JE, Davis LM, Darrow BJ, Kanter HL, Laing JG, Beyer EC 1995 The molecular basis of anisotropy: Role of gap junctions. J Cardiovasc Electrophysiol 6:498–510

Sanguinetti MC, Jiang C, Curran ME, Keating MT 1995 A mechanistic link between an inherited and an acquired cardiac arrhythmia: *HERG* encodes the I_{Kr} potassium channel. Cell 81:299–307

Schwartz PJ, Periti M, Malliani A 1975 Fundamentals of clinical cardiology: the long Q-T syndrome. Am Heart J 89:378–390

Shimizu W, Antzelevitch C 1997 Sodium channel block with mexiletine is effective in reducing dispersion of repolarization and preventing torsade de pointes in LQT2 and LQT3 models of the long-QT syndrome. Circulation 96:2038–2047

Sicouri S, Antzelevitch C 1991 A subpopulation of cells with unique electrophysiological properties in the deep subepicardium of the canine ventricle: the M cell. Circ Res 68:1729–1741

Sicouri S, Quist M, Antzelevitch C 1996 Evidence for the presence of M cells in the guinea pig ventricle. J Cardiovasc Electrophysiol 7:503–511

Verduyn SC, Vos MA, Van der Zande J, Kulcsàr A, Wellens HJJ 1997 Further observations to elucidate the role of interventricular dispersion of repolarization and early after-depolarizations in the genesis of acquired Torsade de Pointes arrhythmias — a comparison between almokalant and d-sotalol using the dog as its own control. J Am Coll Cardiol 30:1575–1584

Wei J, Wang DW, Alings M et al 1999 Congenital long-QT syndrome caused by a novel mutation in a conserved acidic domain of the cardiac Na^+ channel. Circulation 99:3165–3171

Wettwer E, Amos G, Gath J, Zerkowski H-R, Reidemeister J-C, Ravens U 1993 Transient outward current in human and rat ventricular myocytes. Cardiovasc Res 27:1662–1669

Yan GX, Shimizu W, Antzelevitch C 1998 Characteristics and distribution of M cells in arterially perfused canine left ventricular wedge preparations. Circulation 98:1921–1927

Yao JA, Jiang M, Fan JS, Zhou YY, Tseng GN 1999 Heterogeneous changes in K currents in rat ventricles three days after myocardial infarction. Cardiovasc Res 44:132–145

Zygmunt AC, Eddlestone GT, Thomas GP, Nesterenko VV, Antzelevitch C 2001 Larger late sodium conductance in M cells contributes to electrical heterogeneity in canine ventricle. Am J Physiol Heart Circ Physiol 281:H689–H697

DISCUSSION

Tseng: You mentioned that heart failure reduces I_{Kr} expression. What is the basis for this? Did you measure the current, protein, or mRNA?

Rosenbaum: We didn't. Others have reported this: in a canine model of heart failure there is a reduction in I_{Kr} as well as I_{Ks} and I_{to}. It seems to be fairly uniform across the wall of the heart. There doesn't seem to be any selective reduction in one particular layer or another.

Tseng: Is this induced by fast pacing in dog heart failure?

Rosenbaum: Yes. But in virtually every heart failure model and in most hypertrophy models we can see a very similar electrophysiological footprint. There is down-regulation of K^+ channels, SERCA and ryanodine Ca^{2+} release channels, with up-regulation of the Na^+/Ca^{2+} exchanger. This is common.

Tseng: I raised this question because we recently published a paper in *Circulation* (Jiang et al 2004) in which we studied dog heart hypertrophy induced by myocardial infarction through intracoronary microembolizations. We saw an increase of I_{Kr} which we attributed to down-regulation of KCEN2, the auxiliary subunit that can suppress I_{Kr} current amplitude. With Western blot we didn't see changes in dog ERG protein, but there was a decrease in the KCNE2 protein.

Functionally, I_{Kr} channel density was increased in ventricular myocytes from the microembolized hearts.

Rosenbaum: In heart failure and hypertrophy models it is common to see prolongation of the action potential.

Tseng: That can be due to a decrease of I_{to}, I_{Ks} or I_{K1}.

Robertson: When you showed levels of connexin, was this protein or RNA?

Rosenbaum: Protein.

Robertson: Are there other things that can decouple these gap junctions during high levels of activity? Mike Bennett's group published a paper some years ago showing that Ca^{2+} and pH levels reduced the conductance of gap junctions (Spray et al 1982).

Rosenbaum: There is increasing awareness of a number of factors that regulate the conductance of gap junctions. It is clear that intracellular acidosis has a potent effect on intercellular coupling, particularly through its action on the carboxyl group and its relationship to scaffolding proteins in the myocyte junctions. Gap junctions also exhibit voltage gating. Depolarization of neighbouring myocytes causes gap junctions to close. This occurs with very long time constants that are probably not physiological. There is also an important story regarding the relationship of gap-junction activity to the state of phosphorylation of gap junctions. This is evolving. There is another factor which has to do with cell size. In species with smaller myocytes, such as mice, resistive coupling via gap junctions becomes less important. For example, in mice impulse propagation is supported by capacitative coupling. There is a problem with knockout mice: you can knockout connexin 43 and it will only reduce conduction velocity by 50%. In larger mammals with bigger myocytes gap junctions play a larger role.

Robertson: Do you think a physiological regulation of gap junctions will contribute to the arrhythmia?

Rosenbaum: It is possible, but not in any way that we completely understand as yet. The extent to which intracellular Ca^{2+} in the physiological range influences coupling. Under normal metabolic conditions, with no acidosis, we don't see a significant effect on coupling.

Robertson: Does acidosis occur in the early stages of TdP?

Rosenbaum: Not in the early stages. Development of the arrhythmia is not dependent on gap junctions closing. We have taken these preparations and have further reduced coupling by introducing acidosis using a pH of 6.5. There are now some compounds that prevent closing of gap junctions, so we could prevent this phenotype from developing.

Tristani-Firouzi: You mentioned that TdP didn't occur without marked QT prolongation. Under what circumstances can you get TdP without QT prolongation?

CELLULAR MECHANISMS OF TdP 219

Rosenbaum: In drug-induced TdP it is not all that uncommon to have only a minimal prolongation of the QT interval. The relationship between the extent of QT prolongation and the development of TdP is not strong. If QT prolongs a great deal we worry more, but this probably causes more concern in the physicians than it does in the TdP patients. It is a rather weak relationship. In the model, one striking finding is that we can prolong QT interval in many different ways, but we only get the arrhythmia when we prolong QT interval in association with a heterogeneity of repolarization across the transmural wall. This is what seems to be specific for the substrate for the development of TdP.

Brown: If I understand you, then the most vulnerable spot for developing a line of block will be between the mid-myocardium and the epicardium. So if you do frequency-induced arrhythmias in dog, does the line of block prediction hold?

Rosenbaum: Let me describe an example of what heart failure looks like. There are 100 ms differences over a span of a couple of millimetres. In those conditions, when you introduce a premature stimulus, you get conduction block exactly at that interface between the epicardium and mid-myocardium. Then you get re-entrant excitation.

Brown: How does the torsadogenic event get extinguished?

Rosenbaum: These are isolated perfused preparations so they don't become ischaemic. Excitability and action-potential duration is relatively normal. The stability of a re-entrant circuit is often determined by the relationship between the path length of the circuit and the wave length (action potential duration×conduction velocity) of the excitatory impulse. With longer action potential duration, the excitatory head of a re-entrant wave can interact and collide with its repolarizing tail (i.e. wavelength=pathlength) leading to termination of reentry. Since we believe TdP is caused by meandering reentrant rotors, they eventually collide with the lateral boundaries and extinguish themselves. This is why, we believe, the arrhythmia self-terminates. By contrast, *in vivo*, the onset of tachycardia can be associated with haemodynamic collapse, ischaemia, altered excitability, and marked shortening of action potential duration. Under these circumstances, wavelength is more likely to be much shorter than pathlength leading to the sustenance of TdP, or even degeneration to VF.

Brown: So it is not that the line of block disappears; it is that you get collision extinction. The line of block is not moving.

Rosenbaum: That's right. Bigger wedges tend to have longer sustained arrhythmias.

Noble: You made the valid point that whatever the conditions were for getting an early after depolarization initially, they wouldn't persist during the re-entrant arrhythmia. So we have a trigger situation, if that was the trigger for getting it going. One of the conclusions that can be drawn from the mathematical simulations in two- and three-dimensional blocks of tissue is that even in a homogeneous medium, once you have initiated a rotor, it is difficult to stop it. Buzz Brown's question is the right one to ask: not 'how does a rotor keep on going when the conditions for getting it going have already disappeared?', but instead, 'how do they stop?' I also have a question: how big a block do you need for the rotor to break down into fibrillation?

Rosenbaum: I don't have the answer to that interesting question, because there is a technical limitation to these experiments. We have to take a great amount of care to select the wedges and actually trim them so that they can be fully perfused. In the early phase of doing these experiments we did quite a bit of validation with fluorescently labelled microspheres to assure that the flow was uniform. The last thing we wanted to do was introduce heterogeneity simply because perfusion wasn't uniform. As a consequence we take care to keep these preparations from becoming too large. This prevents us from answering that question. They are typically 3–4 cm long by 2 cm deep.

Noble: What enables a re-entrant arrhythmia to break down into fibrillation is clearly going to be one of the most important questions. In simulations we find to some extent that it is conduction velocity that matters. This is why I would suggest that Na^+ channel blockers have a tendency to produce fatal arrhythmia. This leads me to a question for Peter Schwartz: has anyone looked at this issue? It could be seen in the width of the QRS. If that is wide, is there a bigger tendency for TdP to break down into fibrillation?

Schwartz: For all I remember, there was never any hint of differences between those developing arrhythmias and not developing them in terms of QRS. With regards to Na^+ channel blockers, mexiletine was clearly associated in at least two clinical trials with increased mortality. The only reason physicians have continued to prescribe it is because the studies were terminated before reaching a P level below 0.05. This was because the sponsor did not allow the investigators to continue to enrol patients to reach the n necessary to show significance. The issue was then settled by CAST. The subset of LQT3 patients that I have been mentioning is at variance because we are dealing with a very different mechanism of action.

Hondeghem: I don't know whether we can extrapolate from atrial tissue, but Allessie has shown that in atria, if you have fewer than three re-entry wavelets, then fibrillation will usually terminate, but if you have more than six fibrillation will usually be sustained. Between 3 and 6 re-entry wavelets normally yields paroxysmal or intermittent fibrillation.

Rosenbaum: There are many people who argue about this question in a much more passionate way. There are two schools of thought: one is the multiple wavelet idea, where a critical number of wavelets is needed and this is related to conduction velocity and curvature, and then there is the mother–daughter concept where if there are restitution slopes greater than one then wavelets will break off from the mother wave and tend to maintain the fibrillation.

Hondeghem: I don't think those two are mutually exclusive. I think both mechanisms may occur.

Hoffmann: Is there a physiological role for transmural dispersion?

Rosenbaum: That is a great question. We see two different types of transmural dispersion. With this system we also measure intracellular Ca^{2+} transients from the same cells, and compare these to expression of Ca^{2+}-cycling proteins. What we see consistently is that the endocardium tends to be a region of relatively slow Ca^{2+} cycling. There is reduced SERCA and weaker ryanodine release in the endocardium, and so we and other people have postulated that there may be some advantage from the standpoint of mechanical contraction. There is a more sustained contraction of the endocardium and the epicardium might give a more advantageous pump function. The other major heterogeneity is in ion channels, which tends to result in a shorter action potential in the epicardium and endocardium. The endocardium is the first to depolarize. If there is synchronized contraction you would like all the cells to be repolarizing around the same time. Having the epicardial cells have shorter action potentials is, therefore, advantageous in this regard, because they are the last to depolarize. Why mid-myocardial cells are present, what they are doing, and how they got there are interesting questions. There are a number of theories, but we don't really know.

Sanguinetti: What is the evidence for M cells in the mid-myocardial layer having longer action potentials? Can you summarize this in other species, particularly human?

Rosenbaum: This has been fairly well worked out in the dog. In guinea-pig, one of the things we notice in the lab when we do a procedure in which we endocardially cryoablate the guinea-pig heart to make a model of ventricular tachycardia, the moment we remove the endocardial and midmyocardial layers the action potentials in the epicardium get shorter. This is because they are no longer under the electrotonic influence of the M cells in the deeper layers. In rabbit there are also midmyocardial cells. I don't know whether this has been looked at systematically in humans.

Tseng: There is a study of human heart M cells (Li et al 1998). This shows that human heart also has M cells.

Rosenbaum: It is a difficult issue. In the literature some people find M cells and others don't. If you look at the topography and map the entire extent of the wall, if you are putting a microelectrode in it is hit or miss. Unless you have optical imaging to see where you are going you could easily miss the cells. There are some regions where there were no apparent M cells all the way through the transmural wall. You can be isolating myocytes or impaling cells and miss them because their topography is very complex.

Terrar: You mentioned the difference in connexin expression maintaining the inhomogeneity between epicardial and midmyocardial cells. What maintains it between mid-myocardial and endocardial cells? Are there APD differences there that are maintained?

Rosenbaum: Yes, but they are not nearly as marked. Connexin expression is higher in the midmyocardial region, and the action potential duration differences are lower. The action potential profiles show the steepest gradients in the epicardial to mid-myocardial interface. It is a complicated process: action potential gradients represent the sum total of the degree of coupling present between those interfaces and the differences in I_{Ks} expression between the two layers. We don't have a good handle on what the patterns of I_{Ks} expression are. We know they are lower in the mid-myocardium, but we don't know exactly where they become lower.

Terrar: In the slide you showed of the LQT model when you put I_{Kr} blocker on, there was quite a big difference between the mid-myocardium and endocardium action potentials, so presumably something is maintaining this.

Rosenbaum: There still is reduced I_{Ks} in the midmyocardial cell compared with the endocardial cell, so there is a greater sensitivity of midmyocardial cells to I_{Kr} blockade. Uncoupling still exaggerates the action-potential difference.

Schwartz: Some of the major differences of M cells become even more evident at very long pacing intervals. This may help to understand their potential arrhythmogenic role at very low heart rates. How does this fit with arrhythmias occurring at increased heart rates?

Rosenbaum: In the TdP and LQT models, this is correct: a component of the model, like the clinical syndrome, is brachycardia. At physiological heart rates of 120 for a dog, the gradients are relatively mild. This is probably a good thing. But in heart failure there are significant gradients even at faster heart rates. These are further magnified at slower heart rates. We were happy to see this because there is a nice correspondence there. In the LQT model they are reduced at faster heart rates.

Sanguinetti: Why is drug block a great model for LQT2? You would think that there would be other genetic and epigenetic factors that would come into play during development. Yet the drug model looks similar to the LQT2 phenotype.

Rosenbaum: I think it is problematic. We don't know at all how well this represents the hereditary LQT syndrome: it is a better model for acquired LQT. Until we have better animal models such as transgenic rabbits, this is probably the best we can do.

Schwartz: One interesting feature that makes a difference between drug-induced TdP (particularly due to I_{Kr} blockade) and LQT2, is that in the former no one has ever identified the trigger mechanism. These episodes may occur at any time and usually they are unrelated to a particular event. I don't recall a single drug-induced TdP triggered by a stimulus such as an alarm clock. It is a relevant model, because it has to do with an impairment in I_{Kr} current, but there must be something else. In LQT2 patients one is able to identify for most events a specific stimulus.

Abbott: What about cases of intravenous erythromycin or clarithromycin therapy, in which there is a high concentration rapidly administered into the bloodstream? Is the TdP precipitated quickly?

Schwartz: I don't know.

Brown: This is what Eckhard Ficker was trying to say earlier. A lot of these drugs are doing things other than just directly blocking the channels. Some are specifically affecting trafficking, such as arsenic trioxide, pentamidine and the geldamycin derivatives. There are other drugs that seem to have both properties. I think historically in the literature we find that people take the drugs for two or three days and then have the event.

January: There is a list of risk factors for drug-induced LQT that has evolved. Anything from IV infusion to achieve high drug doses to concomitant drugs that interfere with metabolism. Patients can get them at first dose and they can appear months later. It is a complicated area to understand.

Schwartz: Mike Sanguinetti has raised the issue of high temperature as a trigger. A number of patients have their first events when they had high fever. This brings in a number of novel considerations. It seems that the fever-related episodes are much more frequent in LQT2.

Hoffmann: This model of arterially-perfused ventricular wedge-preparation looks quite sophisticated. How realistic is it to think of introducing this model in a drug development process?

Rosenbaum: I think the model is very powerful because it gives us a bird's eye view of the whole action-potential profile across the wall, especially in conjunction with optical mapping. It is fairly labour intensive: there is certainly an art to this. We have a technician who does two or three of these a week. One dog might give us three or four wedges, and once you overcome the art of recognizing where to select wedges from, and how to perfuse them, it is not that difficult. But it is certainly not a rapid throughput process. I view this as a secondary screen tool that would give more specificity.

Schwartz: Do you use male or female dogs?

Rosenbaum: We have not tracked sex. It is an interesting question. It would also be interesting to do these experiments after chronic exposure to the drugs versus acute exposure alone to see whether there is a difference.

Abbott: How old are the dogs? Cardiac ion channel expression changes markedly with age.

Rosenbaum: They are fully-grown mongrels. Going back to Mike Sanguinetti's earlier comment, the other important difference between LQT and the drug models is that we induce the arrhythmia with programmed stimulation. In humans with congenital LQT there is no known role for testing susceptibility to arrhythmia with programmed stimulation of the heart. We don't know about drug-induced LQT. No one has looked at this.

Sanguinetti: The inducibility of arrhythmia by programmed stimulation must have been studied in combination with quinidine, right?

Rosenbaum: In the era of serial drug testing there were a lot of drugs tested during programmed stimulation. But it certainly wasn't well recognized that inducibility of arrhythmia with drugs could have been a proarrhythmic effect.

References

Jiang M, Zhang M, Tang DG et al 2004 KCNE2 protein is expressed in ventricles of different species and changes in its expression contribute to electrical remodeling in diseased hearts. Circulation 109:1783–1788

Li G-R, Feng J, Yue L, Carrier M 1998 Transmural heterogeneity of action potentials and Ito1 in myocytes isolated from the human right ventricle. Am J Physiol 275:H369–H377

Spray DC, Stern JH, Harris AL, Bennett MV 1982 Gap junctional conductance: comparison of sensitivities to H and Ca ions. Proc Natl Acad Sci USA 79:441–445

Expression and role of hERG channels in cancer cells

Annarosa Arcangeli

Department of Experimental Pathology and Oncology, University of Firenze, Viale Morgagni 50, 50134 Firenze, Italy

Abstract. Increasing evidence indicates that ion channels are involved in the pathophysiology of cancer. The *human ether-á-go-go-related* gene (*hERG*) can be considered one of the most critical ion-channel encoding genes involved in the establishment and maintenance of neoplastic growth. In this review, evidence is presented to demonstrate that hERG channels are frequently over- and/or mis-expressed in many tumour cell lines as well as in primary human cancers. Moreover, many tumour cells, especially leukaemia cells, express a truncated isoform (hERG1B) along with the full length hERG1 protein, to form heterotetrameric channels. Three main functions relevant to tumour cell biology can be ascribed to hERG channel activity: (i) the control of cell proliferation, especially in leukaemias; (ii) the regulation of tumour cell invasiveness, possible through a physical and functional interaction with adhesion receptors of the integrin family; and (iii) the control of tumour cell neoangiogensis, through the modulation of angiogenic factor secretion. hERG channels are thus considered novel diagnostic and prognostic factors in human cancers, as well as targets for anti-neoplastic therapies.

2005 The hERG cardiac potassium channel: structure, function, and long QT syndrome. Wiley, Chichester (Novartis Foundation Symposium 266) p 225–234

Although we are still a long way from cataloguing cancer as a channelopathy (as we do the long QT syndrome which is the main topic of the present symposium), there is mounting evidence to indicate that cancer can be ascribed, at least in part, to ion channel malfunction. Emerging data demonstrate that the expression of certain oncogenes directly affects Na^+, K^+ and Ca^{2+} channel function (cited in Bubien et al 1999), and that genes encoding K^+ channel proteins have an oncogenic potential themselves (Pardo et al 1999, Mu et al 2003). On the other hand, some ion channel encoding genes have been appointed as tumour suppressor genes in human cancers (Ivanov et al 2003, Li et al 2004). Moreover, the expression pattern of ion channels are often altered in tumour cells as compared to normal counterparts: many types of K^+ and Na^+ channel are mis- or overexpressed in cancer cells, and a mislocalization of inward K^+ channels has been demonstrated in astrocytic tumours (reviewed in

Arcangeli & Becchetti 2004). The contribution of ion channels to the neoplastic phenotype is as diverse as the ion-channel families themselves: the majority of studies are concerned with the involvement of ion channels, especially K^+ channels, in cell-cycle regulation (Wonderlin & Strobl 1996, Wang 2004, Arcangeli & Becchetti 2004). Not surprisingly, the same ion-channel mechanisms that regulate cell proliferation are also implicated in the control of apoptosis (Yu 2003). For instance, cell shrinkage, due to K^+ efflux, is one of the early events marking the onset of apoptotic cell death. The potential role of ion channels in tumour cell invasion and angiogenesis has not been properly addressed until now. There are, however, increasing data concerning ion-channel involvement in cytoskeleton reshaping and cell–cell interactions (Davis et al 2002); these phenomena represent the initial steps towards increased cell motility, a prerequisite for tumour invasion. Data have also been gathered concerning the interaction of ion channels with integrin receptors (Davis et al 2002, Arcangeli & Becchetti 2004); indeed integrins are pivotal adhesion receptors dictating tumour cell invasion (Juliano 2002). Finally, it is becoming increasingly clear that tumour cell invasion can be halted by the use of channel blockers (Fraser et al 2003, Soroceanu et al 1999, Lastraioli et al 2004).

Our and other groups have demonstrated that the hERG channels can be included in the list of ion channels mis/overexpressed in cancer cells and whose activity is somehow involved in the regulation of neoplastic growth (Arcangeli et al 1995, Bianchi et al 1998, Cherubini et al 2000, Smith et al 2002, Pillozzi et al 2002, Wang et al 2003, Lastraioli et al 2004). hERG channels belong to the EAG family of voltage-activated, outward-rectifying K^+ channels (Warmke & Ganetzki 1994), although, in contrast to EAG channels, they are characterized by a quick inactivation, resulting in functional inward rectification. These biophysical characteristics are important during the repolarization of spiking cells, such as cardiac myocytes, as extensively reported and discussed in the present book. On the other hand, hERG steady-state properties also appear to be important in regulating the V_m of non excitable cells, especially cycling cells like tumours. In fact, the crossover of the steady-state activation and inactivation curves produces an appreciable current at a V_m of around $-40\,mV$, which is a crucial range for cycling cells, that are often more depolarized than resting, differentiated cells (Binggeli & Weinstein 1986).

After our first demonstration that human and murine neuroblastoma cells express hERG currents and that these currents are regulated by the cell cycle clock (Arcangeli et al 1995), we showed that in many other tumour cell lines of different histogenesis (Bianchi et al 1998), the resting V_m depends on the activation state of hERG channels. Moreover, in tumour cells, the midpoint of channel activation depends on the cycle phase, shifting during the G_1/S phase, to produce a membrane depolarization during the S phase when compared to G_1

(Arcangeli et al 1995). When we further investigated the molecular features of hERG currents in cancer cells, it emerged that the *hERG1* gene was constantly overexpressed in various types of tumour cell line, whereas the *hERG2* gene was expressed in a human retinoblastoma cell line, and *hERG3* in SkBr3 mammary adenocarcinoma cells. Other Kv channel encoding genes, such as *EAG* or *Kv1.3*, which have been linked to cell proliferation in different models, are not so widely overexpressed in tumour cell lines as the *hERG1* gene (Crociani et al 2003).

We subsequently demonstrated that tumour cells also express the N-truncated *hERG1b* isoform along with the full length *hERG1* RNA. The entire transcript of this gene was cloned for the first time in our laboratory from human neuroblastoma and leukaemia cells (Crociani et al 2003). Moreover, the encoded protein hERG1b is expressed on the plasma membrane of tumour cells and forms heterotetramers with the hERG1 protein. The expression of the two isoforms on the plasma membrane oscillates during the cell cycle of tumour cells: while hERG1 protein is up-regulated in G_1 and down-regulated in the S phase, the N-truncated hERG1b isoform is up-regulated in S. Interestingly, the biophysical properties of hERG1b facilitate cell depolarization as compared to hERG1. These results add a molecular dimension to the above cited modulation of the hERG channels biophysical features during cell cycle progression of neuroblastoma cells (Arcangeli et al 1995).

Since all the above results were obtained from established tumour cells lines, it was possible that these cells might have possessed altered gene expression profiles under *in vitro* culture conditions. We therefore asked whether both the *hERG1* and *hERG1b* genes and their corresponding proteins were indeed expressed in primary human cancers as well. We first studied this expression in peripheral blood mononuclear cells (PBMNC), and PBMNC enriched in $CD34^+$ cells ($PBCD34^+$), taken as representative of non-proliferating haemopoietic progenitors, and compared with expression in primary leukaemias (Pillozzi et al 2002). The *hERG1* RNA was never detected in PBMNC or in $PBCD34^+$; however, when $PBCD34^+$ cells treated *in vitro* with cytokines had entered into the S phase of the mitotic cycle, *hERG1* expression was switched on. *hERG1* was constitutively expressed in blasts obtained from primary human acute myeloid leukaemias (AML), that are continuously cycling cells. The highest incidence of *hERG1* expression was found in the M1, M2, M3 and M4 FAB group (Bennet et al 1976). Interestingly, the vast majority of the primary AML tested showed the simultaneous expression of the *hERG1b* transcript, stressing the involvement of this splice variant in haemopoietic tumours. Also the two proteins, hERG1 and hERG1b, could be detected in AML blasts, with a peculiar expression pattern, limited to definite blast cells (Pillozzi et al 2002). These data were confirmed by results from Dr Schlichter's group (Smith et al 2002) and by our demonstration that *hERG1* and *hERG1b* genes are also overexpressed in peripheral blasts from paediatric acute lymphoblastic leukaemias (ALLs) (S. Pillozzi and A. Arcangeli, unpublished data). In summary, a selective up-regulation of hERG channels occurs in proliferating and neoplastic haemopoietic cells.

It is emerging that hERG channels are found also in various types of primary human solid tumours: a high percentage of primary endometrial adenocarcinomas expressed both the *hERG1* gene and the corresponding protein *in situ*. Interestingly both the gene and the protein were absent in normal endometrium as well as in endometrial hyperplasias (Cherubini et al 2000). A similar expression pattern was observed in primary human colorectal cancers: around 60% of primary tumours expressed the hERG1 protein, while no expression could be detected either in normal colonic mucosa or in hyperproliferative lesions, such as adenomas (Lastraioli et al 2004). Interestingly, the highest incidence of hERG1 expression occurred in metastatic cancers. The same association was observed in primary gastric cancers (E. Lastraibli, F. Gasperi-Campani, A. Arcangeli, unpublished data), as well as in primary human sarcomas, with the highest incidence in Ewing Sarcomas (G. Hofmann and A. Arcangeli, unpublished data).

The expression and role of hERG currents was recently studied on primary gliomas (astrocytomas, oligodendroglyomas and ependimomas) (A. Masi and A. Arcangeli, unpublished data). It emerged that while 'inward rectifier' K^+ (K_{IR}) channels were expressed both in normal astrocytic cells and in primary gliomas (irrespective of the histotype and WHO grade), hERG channels are never expressed in normal human glial cells, while they are barely detectable in low grade gliomas, and are greatly present (around 80%) in high grade, very aggressive, astrocytomas (WHO grade IV or 'glioblastoma multiformis').

Therefore, although hERG channels are overexpressed in solid human cancers, their expression does not correlate with mere hyperproliferative lesions, but is restricted to the more aggressive types of cancers. Interestingly, the hERG1b isoform was never detected in all the above described primary human solid cancers.

The question now arises: what is the putative role of hERG channels in tumour establishment and progression? Some answers to the above question are beginning to emerge and are briefly summarized below.

hERG channels and the control of cell proliferation

Data gathered on haemopoietic precursors (see above) suggested to us that *hERG*-encoded currents could be somehow involved in regulating proliferation of haemopoietic cells. To address this question, we evaluated the clonogenic activity of granulocyte–macrophage- colony forming units (CFU-GM) present in the PBCD34$^+$ population, after cytokine stimulation, in the absence or presence of specific blockers of hERG currents (Way 123,398 [Way] and E4031). CFU-GM expansion *in vitro* was drastically inhibited by hERG channel blockers. The same effect on colony formation *in vitro* was also exerted by hERG inhibitors on blasts from AML, that constitutively express hERG channels (see above). The mechanism of hERG-dependent control of cell proliferation in leukaemia cells was studied: the inhibitory effect of hERG blockers could be ascribed to a retardation of cell cycle progression, through a partial block of the cells in the G_1 phase of the cell cycle. On the whole these data and those obtained from Dr Schlichter's laboratory (Smith et al 2002, Pillozzi et al 2002) lend weight to configuring hERG channels as proliferation related proteins. We can further speculate that the hERG1b isoform may be primarily involved in the regulation of cell proliferation. This notion is supported by the fact that this particular isoform is overexpressed in the S phase of the cell cycle, and is able to clamp the V_m to more depolarized values compatible with cell proliferation (Binggeli & Weinstein 1986). Finally, the antiproliferative effect of hERG inhibitors has been observed only in those human tumour cell lines expressing hERG1b along with hERG1, like in leukaemia (Pillozzi et al 2002) and neuroblastoma cells (Crociani et al 2003).

hERG channels and the control of tumour cell invasion

We recently obtained data indicating that hERG K^+ ion channels are also important determinants for the acquisition of an invasive phenotype in solid cancers like colorectal cancers (Lastraioli et al 2004). The latter was determined by studying cell migration through Matrigel in various colon cancer cell lines expressing hERG channels. When the hERG inhibitor Way was added to the cells, it significantly reduced cell migration in cell lines with high hERG expression, whilst having a lesser effect on cell lines displaying lower expression of hERG, and no effect in non neoplastic epithelial cells devoid of hERG channels. Since the cell lines used in this set of experiments derived from divergent sources, and thus might have many genetic differences other than hERG expression, we developed a set of colon cancer cell clones with various levels of hERG expression from a homogeneous background. The clones with high hERG expression (H-clones) displayed an invasion capacity through Matrigel almost three times higher than the clones with low hERG expression. The invasive phenotype of H-clones reverted to the less invasive phenotype when the activity of the hERG channel was specifically blocked by Way addition. The relationship between the amount of hERG1 expression and invasion capacity was confirmed and further stressed showing that the migration through Matrigel of HEK 293 cells transfected with a *hERG1*-containing plasmid (HEK-hERG1) was higher than that of HEK 293 cells transfected with an empty vector (HEK-MOCK). These results correspond with *in vivo* observations that both the *hERG1* gene and hERG1 protein are expressed in a high percentage of primary human colorectal cancers, with the highest incidence occurring in highly invasive, metastatic cancers.

The regulatory role of hERG channels in cell invasion may be traced back to the well documented functional association between hERG channels and β_1 integrins in neoplastic cells (Arcangeli et al 1993, 1996, Hofmann et al 2001). Our recent data on this topic also showed that a physical link between the above two molecules exists in neuroblastoma cells (Cherubini et al 2002), as well as in HEK–hERG1 cells (see below).

hERG channels and the control of tumour angiogenesis

Data obtained from primary astrocytomas (see above) prompted us to study whether hERG channels could somehow regulate some aspect of gliomagenesis, in particular neoangiogenesis, which is a feature of high grade astrocytomas (Kitange et al 2003). For this, various glioblastoma cell lines were tested for their expression of hERG channels. We focused our attention on two glioblastoma cell lines, U138 and A172, the former expressing and the latter lacking functional hERG currents on their plasma membranes. The amount of the angiogenic factor VEGF secreted into the medium by the two cell lines was evaluated in the absence or presence of various hERG inhibitors (see above). It emerged that a block of hERG currents significantly impaired VEGF secretion, only in hERG expressing glioblastoma cells.

A few intriguing questions have since arisen: how can the hERG protein(s) and hERG channel activity dictate such diverse cellular functions? Has the hERG protein(s) a signalling role? Our current hypothesis is that hERG channels are certainly endowed with signalling properties. This hypothesis is supported by recent data obtained from HEK-hERG1 cells (see above). In these cells, a macromolecular signalling complex was shown to occur between the β_1 integrin subunit and the hERG1 protein, after cell adhesion to proteins of the extracellular matrix. This complex is localized in focal adhesion, in particular, into lipid rafts/caveolae, that represent plasma membrane microdomains acting as signalling platforms (Lai 2003). Moreover the β_1/hERG1 complex can recruit the tyrosine kinase p125^{FAK} and the small GTPase Rac1. Interestingly, both p125^{FAK} recruitment and phosphorylation and Rac1 activation were strictly dependent on hERG channel expression and activity (A. Arcangeli, unpublished data).

Taken together, we can conclude that hERG channels are indeed overexpressed in various types of human cancers and that their activity can regulate different aspects of cancer cell behaviour. Furthermore, we propose that hERG protein expression should be considered a novel independent prognostic factor of tumours, as well as a molecular target for innovative antineoplastic therapies.

Acknowledgements

I wish to thank Professors M. Olivotto and E. Wanke for advice and encouragement, and all the collaborators who participated to experiments reported in this paper. The kind revision of the English by L. M. Costa, University of Oxford is warmly acknowledged. Experiments here described were supported by Ministero dell'Università e Ricerca Scientifica e Tecnologica, Associazione Italiana per la Ricerca sul Cancro (AIRC) and Associazione Italiana contro le Leucemie (AIL, Firenze).

References

Arcangeli A, Becchetti A 2004 Ion channels and cell proliferation. In: Janigro D (ed) Cell cycle in the central nervous system. Humana Press, New Jersey, in press

Arcangeli A, Becchetti A, Mannini A et al 1993 Integrin-mediated neurite outgrowth in neuroblastoma cells depends on the activation of potassium channels. J Cell Biol 122:1131–1143

Arcangeli A, Bianchi L, Becchetti A et al 1995 A novel inward-rectifying K^+ current with a cell-cycle dependence governs the resting potential of mammalian neuroblastoma cells. J Physiol 489:455–471

Bennet JM, Catocsky D, Daniel MT et al 1976 Proposal for the classification of the acute leukaemias. French-American-British (FAB) co-operative group. Br J Haematol 33:451–458

Bianchi L, Wible B, Arcangeli A et al 1998 HERG encodes a K^+ current highly conserved in tumors of different histogenesis: a selective advantage for cancer cells? Cancer Res 58:815–822

Binggeli R, Weinstein RC 1986 Membrane potentials and sodium channels: hypotheses for growth regulation and cancer formation based on changes in sodium channels and gap junctions. J Theor Biol 123:377–401

Bubien JK, Keeton DA, Fuller CM et al 1999 Malignant human gliomas express an amiloride-sensitive Na+ conductance. Am J Physiol 276:C1405–C1410

Cherubini A, Taddei GL, Crociani O et al 2000 HERG potassium channels are more frequently expressed in human endometrial cancer as compared to non-cancerous endometrium. Br J Cancer 83:1722–1729

Cherubini A, Pillozzi S, Hofmann G et al 2002 HERG K^+ channels and beta1 integrins interact through the assembly of a macromolecular complex. Ann NY Acad Sci 973:559–561

Crociani O, Guasti L, Balzi E et al 2003 Cell cycle-dependent expression of HERG1 and HERG1B isoforms in tumor cells. J Biol Chem 278:2947–2955

Davis MJ, Wu X, Nurkiewicz TR et al 2002 Regulation of ion channels by integrins. Cell Biochem Biophys 36:41–66

Fraser SP, Salvador V, Manning EA et al 2003 Contribution of functional voltage-gated Na^+ channel expression to cell behaviours involved in the metastatic cascade in rat prostate cancer: I. Leteral motility. J Cell Physiol 195:479–487

Hofmann G, Bernabei PA, Crociani O et al 2001 HERG K^+ channels activation during beta(1) integrin-mediated adhesion to fibronectin induces an up-regulation of alpha(v)beta(3) integrin in the preosteoclastic leukemia cell line FLG 29.1. J Biol Chem 276:4923–4931

Kitange GJ, Templeton KL, Jenkins RB 2003 Recent advances in the molecular genetics of primary gliomas. Curr Opin Oncol 15:197–203

Ivanov DV, Tyazhelova TV, Lemonnier L et al 2003 A new human gene KCNRG encoding potassium channel regulating protein is a cancer suppressor gene candidate located in 13q14.3. FEBS Lett 539:156–160

Juliano RL 2002 Signal transduction by cell adhesion receptors and the cytoskeleton: functions of integrins, cadherins, selectins, and immunoglobulin-superfamily members. Annu Rev Pharmacol Toxicol 42:283–323

Lai EC 2003 Lipid rafts make for slippery platforms. J Cell Biol 162:365–370

Lastraioli E, Guasti L, Crociani O et al 2004 herg1 gene and HERG1 protein are overexpressed in colorectal cancers and regulate cell invasion of tumor cells. Cancer Res 64:606–611

Li X, Cowell JK, Sossey-Alaoui K 2004 CLCA2 tumor suppressor gene in 1p31 is epigenetically regulated in breast cancer. Oncogene 23:1474–1480

Mu D, Chen L, Zhang X et al 2003 Genomic amplification and oncogenic properties of the KCNK9 potassium channel gene. Cancer Cell 3:297–302

Pardo LA, del Camino D, Sanchez A et al 1999 Oncogenic potential of EAG K^+ channels. EMBO J 18:5540–5547

Pillozzi S, Brizzi MF, Balzi M et al 2002 HERG potassium channels are constitutively expressed in primary human acute myeloid leukemias and regulate cell proliferation of normal and leukemic hemopoietic progenitors. Leukemia 16:1791–1798

Smith GA, Tsui HW, Newell EW et al 2002 Functional up-regulation of HERG K+ channels in neoplastic hemopoietic cells. J Biol Chem 277:18528–18534

Soroceanu L, Manning TJ Jr, Sontheimer H 1999 Modulation of glioma cell migration and invasion using Cl^- and K^+ ion channel blockers. J Neurosci 19:5942–5954

Wang H, Zhang Y, Cao L et al 2003 Herg K^+ channel, a regulator of tumor cell apoptosis and proliferation. Cancer Res 62:4843–4848.

Wang Z 2004 Roles of K^+ channels in regulating tumour cell proliferation and apoptosis. Pflügers Arch 448:274–286

Warmke JW, Ganetzky B 1994 A family of potassium channel genes related to eag in Drosophila and mammals. Proc Natl Acad Sci USA 91:3438–3442

Wonderlin WF, Strobl JS 1996 Potassium channels and G1 progression. J Memb Biol 154:91–107

Yu SP 2003 Regulation and critical role of potassium homeostasis in apoptosis. Prog Neurobiol 70:363–386

DISCUSSION

Gosling: There has been some interesting work by Mustafa Djamgoz' group at Imperial College which suggests that in prostate cancer there is an up-regulation of an Na^+ channel that is not normally expressed (Grimes et al 1995). Have you looked for any other ion channels that you wouldn't expect to be there, or have you just looked at hERG?

Arcangeli: We have just looked for hERG. We didn't test the EAG. We have studied Inward Rectifier channels in astrocytomas. We have focused our attention on hERG. I know the paper you mentioned, which is very interesting. There could be a relationship between various types of ion channels to maintain a proper electrostatic profile of the membrane. We think that it is not the channel *per se* that is important but the electrostatic profile of the membrane. You can get a proper electrostatic profile by regulating the conductance of different ion channels, either K^+ or Na^+ channels. This profile is important for regulating signalling proteins like FAK, Src, small GTPases, etc. Stuart McLaughlin in New York has demonstrated that many signalling proteins, like PKC or heterotrimeric G proteins can be considered as 'voltage-dependent' proteins (Murray et al 2001).

Robertson: How?

Arcangeli: Through biophysical experiments I am unable to explain to you properly.

Robertson: It is easier for me to see how the integrin signalling system could modify hERG currents. It is more difficult to imagine how hERG currents could modify the integrin signalling system. You are suggesting that integrins are sensitive to transmembrane voltage changes.

Netzer: As you know, Walter Stuehmer and colleagues have found EAG expressed in a lot of tumour tissues. They have very interesting results from blocking these channels. Have you done a direct comparison of hERG to EAG?

Arcangeli: No. I think that, if we want to treat patients with channel blockers we must define which type of channel is expressed in a specific tumour and what its function might be.

Netzer: It might not always be the best policy to eradicate the main current. It is sometimes better to have slight modulation and take a current that is not expressed in different tissue to immediately avoid side effects. If I had a choice to give an EAG blocker or a hERG blocker, I would prefer the EAG blocker.

Sanguinetti: What is the difference with Pardo and Stuehmer's tumour experiments? What kind of tumour cells did they examine? Are you saying that there is no change in EAG, or even a down-regulation?

Wanke: We can only find EAG in 5% of the cells, and only before differentiation. There is a paper from Stephan Heinemann (Meyer & Heinemann 1998) showing this change in expression of EAG and hERG currents. After 1 d you cannot record EAG currents.

Ficker: I think the crucial experiment in Stuehmer's lab was that they put EAG in some cell line and got a transformed phenotype. If we assume that you collect bone marrow cells and put a retrovirus expressing hERG in there stably, what do you get?

Arcangeli: I don't know. I know that a transgenic mouse model could be a good tool to discover whether hERG expression versus that of EAG or other channels is important in tumorigenesis. We had got mice where the hERG expression could be switched on in specific tissues. These mice were transgenic for the *hERG1* gene, but in the founder mice *hERG1* expression was silenced by a reporter gene (GFP) floxed by two LoxP sites. The *hERG1* expression could be switched on after mating with a Cre mouse, expressing the recombinase, for example in the intestinal mucosa. I had prepared the mice, but animal-rights activists came into our animal house and took all the mice. The genetic approach could be crucial to answering some of these questions, but must be postponed.

Shah: These are fantastic data. You showed that hERG blockade decreases the secretion of VEGF. A lot of tumours rely on angiogenesis for their survival. A number of cytotoxic agents have an effect on hERG. Are those cytotoxic drugs that have an activity at hERG more effective than the ones that don't?

Arcangeli: There have been no reports on this.

Shah: I also have a word of caution about concluding at present anything from hERG overexpression in tumours. We have seen this with a COX-2 isoform of cyclooxygenase, which is overexpressed in brain areas related to memory in Alzheimer's disease. This was thought to offer a pharmacological target for the treatment of Alzheimer's. Yet all the studies with COX-1 and COX-2 selective inhibitors have proved to be negative in this disease. These enzymes could just be innocent bystanders which are overexpressed, and they don't necessarily have a role in pathogenesis of the disease.

Schwartz: What is the experimental evidence that hERG inhibitors are efficacious in *in vivo* models?

Arcangeli: We are thinking of including hERG inhibitors in cocktails of adjuvant therapy. Moreover, we are also trying to encourage the use of hERG expression as a prognostic factor. We hope that it will help to identify groups of patients with more aggressive tumours. After this we will start testing the effects of hERG blockers *in vivo*.

Sanguinetti: In your Boyden chamber experiments in which you investigate the effects on invasiveness and show that you can block it with a hERG blocker, do you transfect those cells with IRK or some other channel that will change the membrane potential and rescue their ability to invade?

Arcangeli: We haven't done this experiment.

Sanguinetti: Specifically which IRK is expressed in these cells?

Aracangeli: Kir2.1.

References

Grimes JA, Fraser SP, Stephens GJ 1995 Differential expression of voltage-activated Na^+ currents in two prostatic tumour cell lines: contribution to invasiveness *in vitro*. FEBS Lett 369:290–294

Meyer R, Heinemann SH 1998 Characterization of an eag-like potassium channel in human neuroblastoma cells. J Physiol 508:49–56

Murray D, McLaughlin S, Honig B 2001 The role of electrostatic interactions in the regulation of the membrane association of G protein beta gamma heterodimers. J Biol Chem 276:45153–45159

TRIad: foundation for proarrhythmia (triangulation, reverse use dependence and instability)

Luc M. Hondeghem

HPC n.v., Westlaan 85, B-8400 Oostende, Belgium

Abstract. There exist both safe and dangerous prolongations of the QT interval. Proarrhythmia can be induced by triangulation of the cardiac action potential, reverse use dependence and instability, a set of three features termed TRIad. TRIad leads to dispersion (spatial, transmural and temporal), stalling of fast repolarization or even early after-depolarization (EAD). EAD can progress to Torsade de Pointes (TdP) especially in the presence of prolonged APD; as the QT interval shortens (e.g. by reverse use-dependence), the cardiac wavelength shortens and TdP can progress to ventricular fibrillation (VF). In the absence of TRIad and QT prolongation, chemicals exhibit on average neither pro- nor anti-arrhythmia. However, QT prolongation in the absence of TRIad becomes antiarrhythmic. Furthermore, this desirable effect increases as TRIad components are replaced by squaring of the action potential, use-dependent prolongation of APD and stabilization of the action potential. It is concluded that proarrhythmic characteristics of drugs can readily be recognized and that hope exists for an effective and safe class III antiarrhythmic agent after all.

2005 The hERG cardiac potassium channel: structure, function, and long QT syndrome. Wiley, Chichester (Novartis Foundation Symposium 266) p 235–250

QT prolongation

Physiological and pathological QT prolongation

Eating, sleeping, getting up, exercise and other pleasurable activities can prolong the QT interval (Molnar et al 1996, Nagy et al 1997). Although they occur billions of times each day, no reports of induction of Torsade de Pointes (TdP) by these activities could be found in healthy individuals. Thus, physiological QT prolongation appears not to be torsadogenic.

In contrast, numerous pathological conditions exist where QT prolongation is frequently associated with TdP (inherited gene mutations in cardiac ion channels, cardiomyopathies; Meyer et al 2003).

Drug-induced prolongation of QT is not always torsadogenic

There exist long lists of drugs that lengthen the QT interval and are associated with TdP (Shah 2002). However, there also exists a list of 37 drugs that prolong the QT interval without reports of TdP (Redfern et al 2003) and 29 drugs, for which only isolated instances of TdP have been reported. Thus, drug-induced QT prolongation can be associated with TdP, but does not necessarily cause TdP.

Causal/casual association of QT prolongation and TdP

The link between QT prolongation and TdP appears to be poorly established: some torsadogenic drugs prolong QT by only a few milliseconds when TdP develops, while others can prolong it substantially before TdP occurs (Haverkamp et al 2001). As a matter of fact, some drugs have been reported to precipitate TdP without QT prolongation (Zehender et al 1989).

Similarly, in families with 'long QT syndrome' (LQTS) some patients have no identifiable electrocardiographic abnormalities but sudden death is their first symptom (Meyer et al 2003). Could it be that the cause for TdP is not QT prolongation, but some other proarrhythmic force(s)?

Can QT prolongation be antiarrhythmic?

During TdP, the QT prolongation may decline and this shortening may facilitate the development of ventricular fibrillation (VF). Actually, VF is the direct cause of death among 10–20% of patients; in the remainder the TdP is not sustainable: could QT prolongation actually impede the onset of fibrillation so that the true proarrhythmic force(s) can elicit 'only' TdP? Since TdP is frequently self-terminating (VF is not), could QT prolongation then actually protect against sudden death?

In the present article, I will propose that the action potential duration (APD) can be prolonged in both good and bad ways, that at least three repolarization-associated proarrhythmic forces have been identified and that these become more proarrhythmic as the APD is shortened.

TRIad

If QT prolongation is not the cause of TdP, then the question of interest becomes: what are the proarrhythmic forces and when do they lead to TdP? In a large study on blockers of the hERG channel, a strong proarrhythmic association was found with **T**riangulation, **R**everse use-dependence and **I**nstability (TRIad; Hondeghem et al 2001).

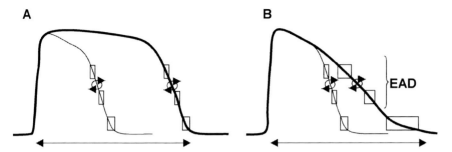

FIG. 1. Prolongation of APD without (A) and with (B) triangulation. A reference action potential is represented by a fine line, while the prolonged action potentials are shown by a thick line. The top rectangle represents the calcium window current; the arrowed circle indicates the Na^+/Ca^{2+} exchange current; the middle rectangle refers to the Na^+ window current; the bottom rectangle represents the voltage-time window over which the sodium channels reactivate and conduction returns towards normal. Although both action potentials have similar total length, as indicated by the double arrows, in (A) the plateau is prolonged, while in (B) it is fast repolarization that is lengthened.

Triangulation

The APD is primarily the sum of the plateau and delayed fast repolarization (phase 3). So APD prolongation primarily results from prolongation of the plateau (Fig. 1A), prolongation of phase 3 (Fig. 1B) or both (not shown). Because slowing of repolarization during phase 3 triangulates the action potential, $APD_{90}-APD_{30}$ was defined as a measure of triangulation. Triangulation most commonly results from reduced outward current (primarily I_{Kr}, which flows through channels encoded by hERG). This lengthens the period for reactivation of calcium and sodium currents, as well as the time during which the Na^+/Ca^{2+} exchange current can exert a depolarizing action upon repolarization (Patterson et al 1997). These three changes may contribute to inward currents overcoming outward currents, and result in excessive slowing of repolarization or even early after-depolarization (EAD).

Triangulation also markedly lengthens the reactivation of the sodium current, i.e. prolonging the time during which conduction is slowed. Reduced outward current at the end of repolarization and early diastole, increases the likelihood of re-excitation in the presence of dispersion of repolarization (see below). Triangulation thus markedly prolongs the vulnerable period for VF. It is then not surprising that reduction of triangulation (squaring) was recently confirmed by Milberg et al (2002) to be a major antiarrhythmic force.

Reverse use-dependence

To be effective, class III antiarrhythmic action (prolongation of APD) should be most marked during tachycardia and minimal at normal heart rate. However, the prolongation of the QT interval in response to many drugs instead declines as the cardiac cycle length shortens; this is reverse frequency-dependence (Hondeghem & Snyders 1990). Thus, following long cycle lengths the prolongation may become excessive, lead to extreme slowing of repolarization, EADs and proarrhythmia (including TdP, VT [ventricular tachycardia] and VF). Reverse use/frequency-dependence is widely described (>1300 publications) for hundreds of drugs.

Instability

Reverse use-dependence destabilizes the APD: at a given cycle length a short APD is followed by a long diastole, which in turn is followed by a long APD. A longer APD results in a shorter diastole and thus the APD starts oscillating (see bottom of Fig. 2). Beat-by-beat instability of the APD may well be a consequence of reverse use-dependence, but instability can frequently be detected long before reverse

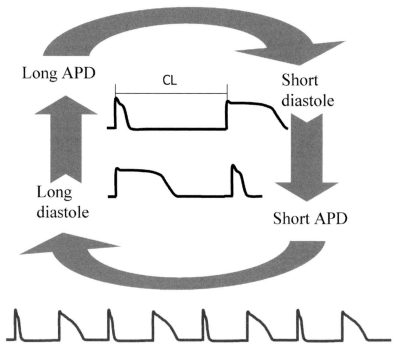

FIG. 2. Reverse use-dependence results in instability of the APD. At a given cycle length (CL), a short APD is followed by a long diastole and a long APD by a short diastole. Reverse use-dependence dictates that longer diastolic intervals are followed by longer APDs and that shorter diastolic intervals are followed by shorter APDs. Such positive feedback leads to a vicious circle that results in oscillation of APD or instability.

Dofetilide

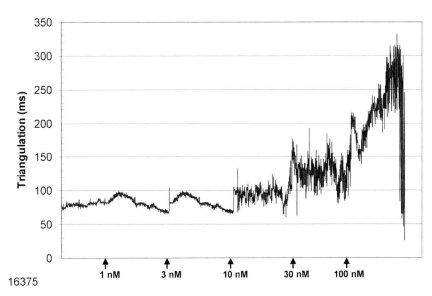

FIG. 3. Dofetilide induced triangulation and instability. Triangulation, measured as APD_{90}–APD_{30} augments with increasing concentration of dofetilide (concentrations indicated started at arrows). In addition to prolongation, the duration of fast repolarization also becomes unstable, i.e. exhibiting long-short oscillation of the type described in Fig. 2. Note that the instability of phase 3 may increase, before phase 3 lengthens.

use-dependence, i.e. at 10 to 100-fold lower concentrations and hence becomes a potent predictor of proarrhythmia.

Instability is highly frequency-dependent, so it is important that it be evaluated at various frequencies and during irregular stimulation. For example, the instability of cisapride only occurs at relatively short cycle lengths, while that of terfenadine only at longer cycle lengths (Fossa et al 2002).

Instability is not limited to APD, it can also be the duration of fast repolarization that exhibits instability. This instability may occur before the APD or fast repolarization is lengthened (see Fig. 3).

Dispersion

When two regions of the ventricle are at different potentials (ΔE), then Ohms law mandates that there will be a current flow (I) proportional to ΔE and inversely proportional to the resistance (R). This resistance will be proportional to the specific tissue resistance (r_s) times the length (l) of the path between the two regions:

$$I = \frac{\Delta E}{l * r_s}$$

Thus, for a given potential difference and specific tissue resistance, the current increases as the distance declines. Such currents are known to lead to re-excitation, membrane oscillations and even pacemaker activity, especially under conditions where the outward repolarizing potassium current is reduced (Katzung et al 1975).

Spatial dispersion

Drug-induced APD prolongation does not develop uniformly throughout the heart. Tissues may exhibit variable sensitivity to prolongation of APD (e.g. Purkinje fibres may be more sensitive than ventricular myocardium), while the effects may also differ between regions (e.g. right versus left ventricle, apex versus base). These regional differences can lead to QT dispersion in the ECG. While such long-distance potential differences may not set up currents that are strong enough to trigger re-excitation, they nevertheless can contribute to re-entry paths.

Transmural dispersion

When the voltage gradients develop over shorter distances, the current flow increases. For example, mid-myocardium frequently appears to be much more sensitive to prolongation of APD than epicardium (Antzelevitch & Shimizu 2002). Thus, substantial differences in APD may create voltage differences exceeding 100 mV over just a few millimetres. Such gradients can become strong enough for the longer APD to depolarize the cells with the shorter APD, leading to EAD and even ectopics or TdP.

Temporal dispersion

Beat-to-beat variability in APD is not synchronized, so that long and short APDs can 'collide'. This can lead to large voltage gradients over very short distances, i.e. result in strong depolarizing currents. It is therefore not unexpected that, in patients, temporal QT dispersion is also a strong predictor of susceptibility to proarrhythmia (Figueredo et al 2001).

Proarrhythmia

EAD and ectopics

It has already been described above how TRIad may lead to dispersion, slowed repolarization, stalling of repolarization or even EAD. When these EADs reach

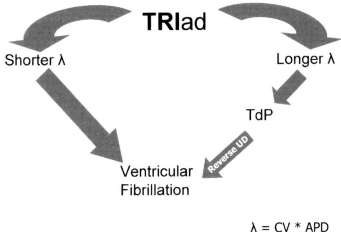

$$\lambda = CV * APD$$

FIG. 4. TRIad induced TdP/VF. TRIad leads to EADs and ectopics, which in turn can lead to re-entry. If the cardiac wavelength (λ), determined by conduction velocity (CV) and APD, is short so that multiple wavelets can fit into the ventricle, then it promptly may deteriorate into VF. If λ is long (long APD/QT), then the wave front encounters too much refractory tissue for VF to be established. Instead, propagation twists around the refractory obstacles: Torsade de Pointes. Commonly, the wave front runs out of non-refractory tissue so that the TdP becomes self-terminating. However, reverse use-dependence may shorten APD and therefore λ. As a result, the critical number of wavelengths for VF fit more easily into the ventricular mass and the TdP can progress into VF.

the threshold potential they can lead to ectopic beats. The latter can trigger re-entry, which can in turn lead to VT or VF, especially in non-uniform substrates (regions of longer refractoriness may serve as islands for the wave front to encircle).

TdP

As the QT interval is lengthened the cardiac wavelength increases (Fig. 4). The larger λ becomes, the fewer re-entry wavelets can fit into the ventricles and the less likely VF becomes. Instead, the wave front attempts to meander around the beat-by-beat varying refractory obstacles resulting in a twisting of the ECG vectors: TdP. However, as long as λ is too long, the arrhythmia may not always find sufficient excitable tissue to maintain itself. For this reason, TdP is usually self-terminating.

VF

When triangulation is more pronounced, the window during which conduction is slowed lengthens (Fig. 1). In addition, due to reverse frequency-dependence, TdP may shorten the APD. Thus, TdP predominates with prolonged QT. However, as the reverse use-dependence shortens the QT interval and the triangulation slows conduction of re-entry waves, the cardiac wavelength shortens. Hence, TdP increasingly can shift to VF and in 10–20% of cases, TdP indeed deteriorates into VF.

QT prolongation can be antiarrhythmic

It can be seen from the above that QT prolongation may actually inhibit VF (frequently a lethal arrhythmia) and instead 'only' lead to TdP (usually self-terminating, only becoming lethal after it deteriorates into VF). Thus, QT prolongation can be protective against VF and thus be antiarrhythmic.

Unfortunately, the QT prolongation that leads to TdP is contaminated by TRIad. Thus, in order to evaluate the antiarrhythmic effect of QT prolongation it is important that this be done in the absence of TRIad. Clearly, if TRIad is indeed proarrhythmic, then reversal of TRIad might be antiarrhythmic (especially when combined with QT prolongation)?

In order to address these questions, we revisited the data from Hondeghem et al (2001) and plotted them in bins of ±20 ms APD changes (Fig. 5). For each bin, the average proarrhythmic rate of all observations is connected by the thick black line. The average proarrhythmic rate for agents that have any triangulation, reverse use-dependence or instability is increased ($\triangle$), while for those where these proarrhythmic forces are all significant (triangulation >29 ms, reverse use-dependence >6 ms and instability >14 ms; Hondeghem et al 2003) the increase is more marked (▲). Conversely, for those agents without any TRIad ($\square$) there is less pro-arrhythmic activity. Actually, when there is neither TRIad nor APD prolongation, then there is zero effect upon ectopic activity, suggesting that no other major proarrhythmic forces exist. But in the absence of TRIad, as the APD lengthens an antiarrhythmic action develops.

Drugs with significant reduction of all TRIad components could not be found in this data. However, for drugs with significant reduction of reverse use-dependence (>10 ms) and instability (>11 ms), there was marked reduction of ectopic activity ($\diamond$) and this reduction was significant. Whether significant reduction of triangulation or prolongation of APD would further enhance the antiarrhythmic property could not be established from the present series of chemicals, as there were no chemicals with all four actions combined.

That TRIad must include the primary proarrhythmic forces is further supported by the fact that TRIad suffices to separate proarrhythmic agents from safe agents in blinded trials (Hondeghem & Hoffman 2003, Hondeghem et al 2003). Furthermore, in the latter studies, prolongation of the APD without TRIad never resulted in EAD or TdP.

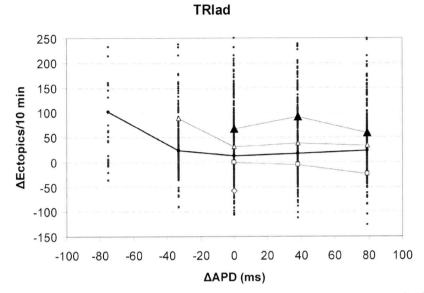

FIG. 5. Proarrhythmia as a function of TRIad and APD. The small dots represent individual experimental points grouped into ±20 ms bins around their average. The solid line represents the average proarrhythmic levels for this series of hERG blockers. Agents endowed with TRIad exhibit more proarrhythmia (△), while agents devoid of TRIad exhibit less proarrhythmia (□). When all three components of TRIad are significant, the proarrhythmia becomes more marked (▲). No agents with significant reduction of all three components could be found, but for significant reduction of reverse use-dependence and instability, the antiarrhythmic action increased markedly (◇).

Conclusion

Drugs can lengthen the QT without becoming proarrhythmic. Actually if they do not induce TRIad, then QT prolongation becomes antiarrhythmic. Thus, QT prolongation without TRIad should not be sufficient ground for impeding a drug's development, neither should lack of QT prolongation be assumed to indicate cardiac safety.

Acknowledgement

The author wishes to thank Dr B. Katzung for his helpful suggestions, Dr G. Duker (AstraZeneca) for providing the compounds and Dr F. De Clerck (Johnson & Johnson) for his assistance with establishing confidence intervals in the absence of drugs.

References

Antzelevitch C, Shimizu W 2002 Cellular mechanisms underlying the long QT syndrome. Curr Opin Cardiol 17:43–51

Figueredo EJ, Ohnishi Y, Yoshida A, Yokoyama M 2001 Usefulness of beat-to-beat QT dispersion fluctuation for identifying patients with coronary heart disease at risk for ventricular arrhythmias. Am J Cardiol 88:1235–1239

Fossa AA, DePasquale MJ, Raunig DL, Avery MJ, Leishman DJ 2002 The relationship of clinical QT prolongation to outcome in the conscious dog using a beat-to-beat QT-RR interval assessment. J Pharmacol Exp Ther 302:828–833

Haverkamp W, Eckardt L, Mömmig G et al 2001 Clinical aspects of ventricular arrhythmias associated with QT prolongation. Eur Heart J Suppl 3(suppl K):K81–K88

Hondeghem LM, Snyders DJ 1990 Class III antiarrhythmic agents have a lot of potential but a long way to go: Reduced effectiveness and dangers of reverse use-dependence. Circulation 81:686–690

Hondeghem LM, Hoffman P 2003 Blinded test in isolated female rabbit heart reliably identifies action potential duration prolongation and proarrhythmic drugs: importance of triangulation, reverse use dependence, and instability. J Cardiovasc Pharmacol 41:14–24

Hondeghem L, Carlsson L, Duker G 2001 Instability and triangulation of the action potential predict serious proarrhythmia, but APD prolongation is antiarrhythmic. Circulation 103:2004–2013

Hondeghem LM, Lu HR, van Rossem K, De Clerck F 2003 Detection of proarrhythmia in the female rabbit heart: blinded validation. J Cardiovasc Electrophysiol 14:287–294

Katzung BG, Hondeghem LM, Grant AO 1975 Cardiac ventricular automaticity induced by current of injury. Pflügers Arch 360:193–197

Milberg P, Eckardt L, Bruns HJ et al 2002 Divergent proarrhythmic potential of macrolide antibiotics despite similar QT prolongation: fast phase 3 repolarization prevents early afterdepolarizations and torsade de pointes. J Pharmacol Exp Ther 303:218–225

Molnar J, Zhang F, Weiss J, Ehlert FA, Rosenthal JE 1996 Diurnal pattern of QTc interval: how long is prolonged? Possible relation to circadian triggers of cardiovascular events. J Am Coll Cardiol 28:799–801

Meyer JS, Mehdirad A, Salem BI, Kulikowska A, Kulikowski P 2003 Sudden arrhythmia death syndrome: importance of the long QT syndrome. Am Fam Physician 68:483–488 (Erratum in Am Fam Physician 69:2324)

Nagy D, DeMeersman R, Gallagher D et al 1997 QTc interval (cardiac repolarization): lengthening after meals. Obes Res 5:531–537

Patterson E, Scherlag BJ, Lazzara R 1997 Early afterdepolarizations produced by d,l-sotalol and clofilium. J Cardiovasc Electrophysiol 8:667–678

Redfern WS, Carlsson L, Davis AS et al 2003 Relationships between preclinical cardiac electrophysiology, clinical QT interval prolongation and torsade de pointes for a broad range of drugs: evidence for a provisional safety margin in drug development. Cardiovasc Res 58:32–45

Shah RR 2002 The significance of QT interval in drug development. Br J Clin Pharmacol 54:188–202

Zehender M, Meinertz T, Hohnloser S et al 1989 The Multicenter Ketanserin Research Group. Incidence and clinical relevance of QT prolongation caused by the selective serotonin antagonist ketanserin. Am J Cardiol 63:826–832

DISCUSSION

Rosenbaum: When you described the provocation of sudden death, it reminded me of a patient who I saw recently. This woman was recovering from myocardial infarction. She asked me whether it was safe for her to have sexual relations when she was recovering from her infarct. I told her only with her husband because I didn't want her to get too excited! I have a question. The rabbit is known for having unique action-potential dynamics. It does some very funny things at slow cycle lengths. For example, the action potential can shorten rather than lengthen. Have you validated this approach in other mammals?

Hondeghem: No. I selected the rabbit for a good reason: I am interested in proarrhythmia. If you are interested in showing APD lengthening, then there exist better preparations. If you want a QT surrogate, you are much better off taking the guinea-pig, because this has more repolarizing currents and therefore you have more chance of recovering QT lengthening because there are more currents you can block and therefore more targets. However, QT prolongation is not my interest; proarrhythmia is. I have observed that I_{Kr} block can be a primary problem. So I select a preparation where I_{Kr} is the only major current present. In the guinea-pig there is a good chance that the other repolarizing currents will cover up a proarrhythmia problem; the dog is similar. If you mess up I_{Kr} in rabbit, there is no other important repolarizing current to cover it up! I have done terfenadine in rabbits and compared the results with what is known in the literature about this drug. Data in rabbit show that terfenadine proarrhythmic signals are five times more likely to be induced than in the dog heart. Fivefold may not look like much. However, if you put this in a mathematical formula to calculate the number of experiments needed to have just a 1 in 10 000 chance of missing such a chemical, the answer is: either you do 700 dog hearts or get the same confidence with seven rabbits.

Rosenbaum: If the results are unique to the rabbit, how applicable are they to humans?

Hondeghem: That is a good question, and we could be concerned that the rabbit is too sensitive. But the facts are there: the industry has run 374 blinded tests in SCREENIT. So far, not a single no-effect drug (such as salts, sugars, penicillin, aspirin) have ever caused a spontaneous TRIad signal. In over 17 000 experiments, rabbit hearts also do not spontaneously exhibit TRIad signals during baseline observations.

Rosenbaum: There is now a well established literature on the role of alternans in the APD, with a mechanistic explanation that relates the progression to fibrillation through the type of alternating activity that you see. There are fairly compelling data that the alternans arises from oscillations in Ca^{2+} cycling in the myocytes, and the voltage change is secondary. This doesn't mean that hERG blockade couldn't play a role in there. You prolong the action potential and increase Ca^{2+} load in the myocyte and you can start an alteration of Ca^{2+} cycling. What may be happening is that we are seeing concentrations of Ca^{2+} that are leading to voltage changes. For instance, it is possible to perform these kinds of experiment under voltage-clamped conditions where you can eliminate any possibility of voltage oscillation and get very nice Ca^{2+} oscillation.

Hondeghem: I agree. It is rare to find only one mechanism operating. I am sure I have seen enough evidence that Ca^{2+} oscillations contribute to this. The evidence is overwhelming. However, I would be willing to take on the argument that this is not the only way we can make the system oscillate. This can also be done on the basis of repolarizing channels. There are many ways of making a complex system such as this go into oscillation. The rabbit hearts analysed using SCREENIT only give the answer to the question whether there is triangulation, reverse use-dependence or instability. The mechanism(s) I leave for other people, but it is clearly more complex than just calcium oscillations.

Brown: I want to congratulate you. I have been looking at circuits and re-entrant pathways and I think this scheme is one of the few times that someone has connected the cellular level to the intact organismal level in a way that is understandable at least to me. I have a couple of questions about the mechanism. We have done literally thousands of drugs on hERG to look for reverse use-dependence. We have seen this perhaps in two cases. With standard drugs such as terfenadine and cisapride reverse use-dependence is not apparent, at least in our hands. Mechanistically, might you want to refine things a little?

Hondeghem: One of the reasons is that the system is much more complex than just a hERG channel. This doesn't mean that we can't make the system oscillate by interfering with hERG, and as a result make the system go into oscillation. For example, you are probably preventing the system from going into oscillation just by imposing the voltage. Indeed, the voltage-dependent binding to and unbinding from the hERG channel is quite steep for some of these drugs. This could be one of the mechanisms that you by definition eliminate oscillation because you don't allow the voltage changes. Another mechanism is the Ca^{2+} oscillations. If that plateau is a little bit higher or lower, this will have dramatic consequences. There are many ways that the system can go into oscillation without it necessarily having to be the hERG binding *per se* (hERG block may just facilitate oscillations).

Brown: So perhaps in your diagram you have hERG and should allow for some other mechanism in terms of the reverse use-dependence. I don't believe that it is hERG itself that is showing reverse use-dependence. It may be something a little more complicated.

Hondeghem: There are missing pieces to the puzzle.

Brown: How do you explain verapamil? In our hands verapamil is a triangulator par excellence. This is what you would expect to see when you have a drug that blocks hERG and also blocks Ca^{2+} current. In many instances this kind of property is a result of mixed ion effects, such as with verapamil or tolteridine where the drugs can prolong APD_{90} but shorten APD_{30}. The effect is that you get triangulation but they are not proarrhythmic. Also, if we can't use the QT interval as a surrogate marker, we can't go around doing these TRIad experiments, so what should the surrogate marker be?

Hondeghem: This can be measured in patients. Triangulation will lead to changes in their T wave, so you can monitor for this. People have seen reverse use-dependence many times in patients. I recently typed in 'reverse use-dependence' and the National Library of Medicine returned more than 1300 citations. With regards to instability, there is an increasing number of papers coming out on unstable T waves being a prologue to serious arrhythmias. These three equivalents to the system are also seen in the clinic.

Brown: And verapamil?

Hondeghem: It's complex. First, it has use-dependence. Reverse use-dependence, in my opinion, is a requirement for instability. If the drug doesn't show reverse use-dependence you don't get instability and don't get in trouble. One of the major precursors for instability is the presence of inward Ca^{2+}/Na^+ currents, and verapamil blocks both. This gives a protective action on top of that. It is not surprising that the triangulation alone will not suffice.

Brown: So the TRIad is actually a little more complicated than it appears. You weight the different terms T, R and I.

Hondeghem: That is true. The more you have of either one of them, the more TdP is seen. If all three of them are present, the incidence of TdP increases with their presence.

Brown: For example, R and I are more important than T in the verapamil case.

Hondeghem: I wouldn't want to say for sure that if you just have triangulation you could get TdP. The reason you don't get it with verapamil is that on top of that there is Ca^{2+} channel and Na^+ channel block. If you block the inward currents it becomes very difficult to get these oscillations. Actually, when the system starts oscillating, addition of lidocaine will usually suppress the instability. The Na^+ current is also involved in that oscillation. I don't know how, but it is involved, because Na^+ channel block is one of the most effective ways to stop the oscillation.

Schwartz: The big issue there is determining the starting point of the QT interval. Never mind that the QT prolongs for 20–30 ms when you start from 400, it is a different situation if you start from 450 or 460. The other thing I wanted to comment on is that you said some of these drugs lengthen QT and some shorten it. It depends on your definition of 'shorten': what is your threshold for this definition?

Hondeghem: I simply took my spreadsheet, and anything that was negative was shortening.

Schwartz: So if it was 2 or 5 ms it was shortened?

Hondeghem: Yes.

Schwartz: I would call this no change.

Hondeghem: Lots of hERG blockers shorten the APD and the proarrhythmic effect is progressive and increases rather steeply with shortening.

Schwartz: This is a measurement that varies a lot and there is the issue of subjectivity in defining the end of the T wave. What matters are changes that are biologically relevant. Unless there is a relatively gross change of the QT interval in either direction, it seems unlikely that they will make a difference in terms of survival. Unfortunately, too often people discuss lengthening or shortening when the actual data show minimal changes that may not be statistically significant and that are biologically irrelevant.

Hondeghem: Even if the APD doesn't change, you still get TdP. It is not a requirement. Even in patients who suffer from inherited LQT, there are reports of siblings who dropped dead without ever showing QT prolongation.

Schwartz: You are right, in part. There is a small fraction, about 5%, of relatives of LQTS patients who have apparently normal QTc and who can die suddenly. However, we don't know if they had sudden QT prolongation in response to certain stimuli. One observation you made which I found interesting was the relationship between short action potential and long diastole, and vice versa. This has to do with the constancy or lack of constancy of the RR interval. The diagrams you showed had a fixed RR interval; then, of course, if you shorten the action potential you will lengthen diastole. In a whole animal, physiology is different.

Hondeghem: I have two comments on that. You have a condition first where the heart is going regularly when the sinus node is pacing steadily. When you get one ectopic, this is where you can see this short diastole, short APD resulting in longer diastole, long APD. But even when we look at the TdPs, it is not a random phenomenon like fibrillation. If you put a sheet of paper in fibrillation and you make a guess, you can't even draw one cycle, because it is totally random noise. TdP is not. If you have a short diastolic interval, I will draw the next cycle of this TdP as being short. If there is a long diastolic interval, I will draw the next cycle as being long. TdP is a *deterministic* arrhythmia: it is oscillating by the rule short–long–short–long.

Schwartz: I am not questioning that. I thought you were referring to general changes in action potential.

Hondeghem: It is true even when the heart becomes totally irregular. Short ones are followed by short ones and long ones are followed by long ones. It is a result from the reverse use dependence. To the extent that your drug does this, it is more dangerous. If a drug exhibits little reverse frequency dependence then it will be better tolerated: the TdP will most likely terminate spontaneously. Drugs that exhibit much reverse use-dependence will be the ones that move on into fibrillation.

Rosenbaum: You have provided a very useful way of looking at the problem, but in parallel it is important that we try to make the connections to mechanisms. The instability of the action potential that you are seeing on the surface could be eliminated with verapamil because this blocks inward Ca^{2+} flux into the cytoplasm. You can do exactly the same thing using BAPTA. This reduces the cytosolic load of Ca^{2+} and eliminates the oscillations. I think there are fairly compelling data that this is emanating from the sarcomere. Anything you do to increase the Ca^{2+} load will increase these types of oscillations. This is a possible explanation that may bring all the observations together.

Hondeghem: I am sure there is some truth in this, but it is not the full picture: if that were the only thing that was important, Na^+ channel blockers wouldn't be expected to stop the oscillation although they usually do.

Brown: The exchanger generates current too.

January: QT intervals have been used because they are easy to measure. Action potentials are complicated because they have multiple time- and voltage-dependent properties. This goes back to the Ca^{2+} window current days. You can't measure the action-potential plateau voltage on a surface ECG, but you can measure the time to repolarization. QT is an interval that is giving us only limited amounts of information, but it has been easy to measure. The question, I guess, is instability. You have shown us examples of microinstability (APD_{60} oscillating) and macroinstability (the entire action potential). Is one worse than the other? Is there a certain area where there is instability that you think is key?

Hondeghem: No, it is just a progressive thing. It is not always on APD_{60} that oscillations occur. Some drugs elicit oscillations high in the plateau (probably primarily Ca^{2+} oscillation) and will rarely lead to TdP. If you see the oscillations around APD_{60} or $_{70}$, these drugs will be the most problematic.

January: Some of those would not be expected to show up in an electrocardiogram. They are different mechanisms.

Hondeghem: You can't see them by eye on the traces, even if you look carefully. You need a computer to unravel these things so that they become visible.

January: Presumably it reflects a multiplicity of mechanisms.

Hondeghem: It is a progression. It starts at one level and expands.

Shah: In this context, I have a point to make about the drugs that have so far proved to be the most potent torsadogens, such as prenylamine, terodiline and cisapride. These drugs are potent hERG blockers and associated with their clinical use, we see reports of tachyarrhythmias, bradyarrhythmias or heart blocks, with or without QT interval prolongation. This has raised some question marks on the reliability of the QT interval alone in the assessment of the proarrhythmic potential of drugs. The fact that you can measure something is not necessarily a good guide to its use. We have seen this in other cardiovascular areas. In cardiac failure, for example, the exercise test is easy to measure but it is a poor predictor of the clinical outcome.

Wanke: Using β blockers is a normal therapy for LQT2. This means that for an attack to occur in LQT2 patients, an increase in the Ca^{2+} currents capable of producing the prolongation is needed. Otherwise, the patient can live normally. We have investigated whether in LQT patients there could be an increase of the Ca^{2+} currents induced by circulating adrenaline. We have shown that in rat chromaffin cells, the adrenalin-releasing cells use hERG channels (Gullo et al 2003). If we block these channels we can see an enormous increase in the firing of those cells and in catecholamine release. This effect could be connected with LQT2 because the cardiac action potential is already long (hERG is missing) and the adrenaline-induced potentiation of Ca^{2+} currents originating from the hyperexcitable (hERG is missing) adrenal medulla cells, create a sort of positive feedback generating the heart action potential block. A classical example is that of a patient sleeping with a remarkably long cardiac action potential. Anything that produces an enormous increase in adrenaline release will produce block of the heart. A normal firing chromaffin cell is using hERG channels to make spontaneous firing. When the cell has no hERG channel (LQT2 patient) it is hyperexcitable and this can produce an increase of the catecholamine release which reflects an enormous increase in Ca^{2+} current in the heart cells, helping the prolongation of the action potential.

Hondeghem: In LQT2 patients, if there is no challenge, you don't see it. There is just a substrate that is ready to be destabilized if you give it a trigger. In our case it is possibly an ectopic that leads to the oscillation and then TdP; in this case perhaps it is a catecholamine burst that leads to overt oscillations in an unstable system.

January: Have you studied any of the drugs that are used as surrogates for the congenital syndromes, apart from I_{Kr} blockers?

Hondeghem: I do everything blind, so I don't know.

Reference

Gullo F, Ales E, Rosati B et al 2003 ERG K^+ channel blockade enhances firing and epinephrine secretion in rat chromaffin cells: the missing link to LQT2-related sudden death? FASEB J 17:330–332

Drug-induced QT interval prolongation — regulatory guidance and perspectives on hERG channel studies

Dr Rashmi R Shah[1]

Senior Clinical Assessor, Medicines and Healthcare products Regulatory Agency (MHRA), Market Towers, 1 Nine Elms Lane, Vauxhall, London SW8 5NQ, UK

Abstract. Drug-induced QT interval prolongation with or without Torsade de Pointes has led to the withdrawal, or severe prescribing restrictions being placed on the use, of many drugs. Other drugs have been denied regulatory approval because of their 'QT liability'. This mechanism-based toxicity results principally from inhibition of I_{Kr}, the rapid component of delayed rectifier potassium current. The *KCNH2* (hERG) gene encodes the physiologically germane α-subunits of the channels that conduct this current. Among the battery of non-clinical tests used to characterize a drug for its 'QT liability' are the hERG channel studies. Studies with the hERG channel have been used for early screening of lead compounds and making 'go–no-go' decisions. However, the predictive value of these studies is limited by inter-laboratory variations, a high false-positive rate and lack of a consensus on the definition of a negative study ('safety margins'). From a regulatory perspective, it is reassuring that clinical torsadogens have always been hERG positive with no false negatives. hERG channel studies are useful in guiding electrocardiographic safety monitoring in early human studies, evaluating the relative risks of metabolites and enantiomers of a drug and clarifying uncertain mechanisms of action. One emerging issue of concern is the effect of drugs on hERG trafficking. Classical hERG channel studies will not identify these drugs. For adequate risk assessment, hERG channel data should be integrated with all other non-clinical and clinical data; otherwise there is a risk of novel and valuable drugs being rejected from development and/or denied regulatory approval.

2005 The hERG cardiac potassium channel: structure, function, and long QT syndrome. Wiley, Chichester (Novartis Foundation Symposium 266) p 251–285

---

[1]The views expressed in this chapter are those of the author and do not necessarily reflect the views or opinions of the MHRA, other regulatory authorities or any of their advisory bodies.

Prolongation of the QT interval on the surface electrocardiogram (ECG) results principally from a decrease in the outward repolarizing current of the action potential, mediated by the rapid component of the delayed rectifier potassium channel (I_{Kr}). Mutations in the genes encoding this channel protein result in congenital long QT syndromes and lead to cardiac arrhythmias (Keating & Sanguinetti 2001). However, by far the most frequent cause of decrease in I_{Kr} function these days is clinical administration of drugs. Drug-induced QT interval prolongation is a mechanism-based effect, resulting usually from concentration-dependent inhibition of I_{Kr}. Native I_{Kr} is a co-assembly of hERG (human ether-a-go-go related gene) α-subunits encoded by the *KCNH2* gene and MiRP1 β-subunits encoded by the *KCNE2* gene. These subunits co-assemble to form a functional heteromultimeric channel complex. All the drugs that reduce I_{Kr} current have been shown to exert their effect by binding to the hERG α-subunits (Sanguinetti et al 1995, Taglialatela et al 1998a, Mitcheson et al 2000, Tseng 2001, Kongsamut et al 2002, Chachin & Kurachi 2002).

Although QT interval prolongation *per se* is not harmful, it can degenerate into a potentially fatal polymorphic ventricular tachyarrhythmia known as Torsade de Pointes (TdP). Drug-induced TdP is usually a transient tachyarrhythmia that results in palpitation. When TdP is sustained, there are symptoms of impaired cerebral circulation such as dizziness, syncope and/or seizures. Critically, however, TdP can subsequently degenerate into ventricular fibrillation in about 20% of cases (Salle et al 1985), often with a fatal outcome (Milon et al 1982). The overall mortality associated with TdP is of the order of 10–17% (Fung et al 2000, Salle et al 1985).

The apparently all-exclusive and persuasive association between QT interval prolongation and TdP is partly the consequence of the very definition of this unique polymorphic ventricular tachyarrhythmia. Even when meeting the morphological criteria of TdP, many physicians do not label these ventricular tachyarrhythmias as TdP unless preceded by QT interval prolongation. QT interval has therefore come to be recognized as a surrogate marker for the risk of TdP and indeed, numerous clinical and experimental reports have established that prolongation of QT interval is a major precursor of drug-induced TdP. However, TdP can occur in association with normal QT interval and TdP does not develop invariably in all individuals with equivalent prolongation of the QT interval. Neither do all drugs that prolong the QT interval to an equivalent duration carry the same risk of inducing TdP nor do all agents that have been linked with TdP cause significant lengthening of the QT interval. The torsadogenic risk following QT interval prolongation is modified by a number of factors (Shah 2002a, 2004, Roden 2004). Therefore, although QTc interval is widely used as a surrogate marker of the risk of TdP, it is also recognized to be an imperfect marker, albeit the best that is available at present.

Regulatory concerns over drug-induced QT interval prolongation

Drug-induced TdP is a modern 'epidemic' and a challenge in drug development and clinical therapeutics. Of the 2194 cases of TdP in the US Food and Drug Administration (FDA) database as of 1998, only 7.2% were reported between 1969 and 1988 in contrast to 92.8% between 1989 and 1998 (Fung et al 2000). Well over 10 antianginal and 90 non-cardiovascular drugs have now been reported to prolong the QT interval and/or induce TdP clinically (Shah 2002a) and their number continues to increase almost daily. Not surprisingly, the regulatory focus on drug-induced QT interval prolongation has therefore shifted from a potentially desirable antiarrhythmic mechanism to one of potentially fatal torsadogenic/proarrhythmic proclivity.

QT interval prolongation (with or without TdP) has been responsible for the withdrawal from the market, or denial of approval, of a number of drugs. Drugs withdrawn from the market because of their torsadogenic potential are shown in Table 1. Some of the new drugs have experienced significant delay in approval while the risk is being fully evaluated. For countless others, their 'QT liability' has resulted in severe prescribing restrictions being placed on their use. These restrictions impact on a number of sections in the labelling of a drug, notably the contraindications, dosing regimen and special warnings and precautions for use (Shah 2002b). Therefore, the significance of characterizing a new chemical entity (NCE) during its development for its potential to delay cardiac repolarization and prolong the QT interval cannot be overestimated.

Since the pharmacological responsiveness of native I_{Kr} is broadly recapitulated by the hERG channel, it is not surprising that hERG channel studies have been

TABLE 1 Drugs withdrawn from market because of their torsadogenic potential

Drug	Therapeutic class	Year withdrawn
Prenylamine	Antianginal	1988
Terodiline	Urinary incontinence	1991
Terfenadine	H_1-antihistamine	1998
Sertindole *	Atypical neuroleptic	1998
Astemizole	H_1-antihistamine	1999
Grepafloxacin	Fluoroquinolone antibacterial	1999
Cisapride	Gastric prokinetic	2000
Droperidol	Anti-emetic and neuroleptic	2001
Levacetylmethadol	Opiate analgesic	2001

*Now recommended for re-introduction to the market (see text)

used widely during early drug development for screening lead compounds for their QT-prolonging potential, often with a view to making 'go–no-go' decisions.

Regulatory guidance on investigation of 'QT liability'

European and international guidance

In December 1997, the Committee for Proprietary Medicinal Products (CPMP) of the European Union (EU) was the first regulatory authority to issue formal recommendations on the assessment of non-cardiovascular drugs for their potential to prolong the QT interval. It adopted a seminal document (CPMP/986/96), 'Points to consider: the assessment of the potential for QT interval prolongation by non-cardiovascular medicinal products', recommending pre-approval non-clinical and clinical strategies by which to characterize an NCE for its potential to prolong the QT interval. For regulatory purposes, the information required includes the potency of an NCE for its 'QT liability' in terms of magnitude and frequency of QTc interval changes, its dose–concentration–effect relationship, potential clinical risk of TdP and how this risk might be modulated by other genetic and non-genetic variables. Such is the regulatory concern regarding this life-threatening effect of drugs on ventricular cardiac repolarization and QT interval duration that currently, there are initiatives in progress at the level of the International Conference on Harmonization (ICH) aimed at harmonizing internationally the recommendations on non-clinical (ICH topic S7B) and clinical (ICH topic E14) strategies by which to evaluate an NCE for its effects on ventricular repolarization and QT interval during drug development. While the focus of S7B is ventricular repolarization and QT interval, E14 focuses on QT/QTc interval. At the time of writing this chapter (November 2004), both these ICH guidance notes are out to public consultation. The reader is advised to check the latest versions available at: *http://www.emea.eu.int/sitemap.htm*.

Non-clinical evaluation (ICH S7B)

Draft ICH S7B 'Guideline on Safety Pharmacology Studies for Assessing the Potential for Delayed Ventricular Repolarization (QT Interval Prolongation) by Human Pharmaceuticals' extends and complements the scope of the ICH S7A guideline (CPMP/ICH/539/00) 'Note for Guidance on Safety Pharmacology Studies for Human Pharmaceuticals'.

ICH S7B proposes an integrated approach to risk assessment. It outlines four non-clinical electrophysiological levels of integration that can uncover the evidence for delayed ventricular repolarization and potential sequelae. These levels include:

REGULATORY PERSPECTIVES ON hERG STUDIES AND QT INTERVALS   255

- Ionic currents measured in isolated animal or human cardiac myocytes, cultured cardiac cell lines, or heterologous expression systems for cloned human ion channels (*in vitro* I_{Kr} assay).
- Parameters of the action potential measured in isolated cardiac preparations or specific electrophysiological parameters indicative of action-potential duration (APD) in anaesthetized animals (repolarization assay).
- ECG parameters in conscious or anaesthetized animals *in vivo* QT assay.
- Proarrhythmic effects in isolated cardiac preparations or animals.

ICH S7B envisages the use of an *in vitro* I_{Kr} assay and an *in vivo* QT assay as the main practical components of integrated risk assessment. However, the guidance recognizes the complementary roles of *in vitro* and *in vivo* assays and recommends more than one type of assay for determining the risk of drug-induced delayed ventricular repolarization. Implicit in these recommendations is the conclusion that *in vitro* tests may indicate safety margins but may never be sufficient by themselves to declare safety.

Depending on the results of these assays and other non-clinical and clinical information, follow up non-clinical studies may be undertaken to provide greater understanding or additional knowledge regarding the observed effects and further information regarding the roles of metabolites, mechanisms of action, effects on other ion channels and pharmacological targets, potency, slope of the dose-response curve, or magnitude of the non-clinical response. As part of the integrated risk assessment and choice of reference compound(s), ICH S7B recommends that consideration also be given to the pharmacological or chemical class of the drug. Other important considerations include metabolic profile in man and animals and tissue distribution and accumulation.

Importantly, the guidance emphasizes that results from non-clinical S7B studies should be considered when determining the requirements for safety monitoring during clinical development.

Early *in vitro* I_{Kr} assay (including hERG channel study) provides valuable information relevant to planning early human studies. Slow repolarization triangulates the action potential and $APD_{90}-APD_{30}$ is a measure of triangulation. Proarrhythmia can be induced by triangulation. QT interval prolongation in the absence of triangulation is reported to be antiarrhythmic and triangulation even in the absence of QT interval prolongation can be highly proarrhythmic (Hondeghem & Hoffmann 2003). In order to investigate triangulation, effects on other ion channels and the presence of other powerful markers of the risk (e.g. early after depolarizations, EADs), there is an unquestionable need, regardless of the results from hERG studies, to study the effect of an NCE and its metabolites on action potential in isolated cardiac preparations. Studies with hERG channels are relatively cheap, rapid and simple and ideally they should be repeated as often as necessary. Early *in vitro* I_{Kr} assay (including hERG channel study) will almost certainly need to be repeated when additional relevant clinical information (for example, metabolic profile and the likely therapeutic dose) and/or non-clinical data become available. These *in vitro* I_{Kr} and *in vivo* QT assays should be completed well before commencing any clinical study that will generate core data on the effect of the NCE on ECG parameters (including the 'thorough clinical QT/QTc study') (see *Clinical evaluation* section below).

The concept of integrated risk assessment from non-clinical studies proposed in ICH S7B has intensified the debate on the extent to which non-clinical findings can serve as a signal of the potential clinical risk and whether negative non-clinical tests can exclude a clinical risk of QT/QTc interval prolongation. There are regional differences of opinion among regulatory authorities concerning the impact that these non-clinical studies should have on the need for a 'thorough clinical QT/QTc study'.

Clinical evaluation (ICH E14)

Draft ICH E14 ('The clinical evaluation of QT/QTc interval prolongation and proarrhythmic potential for non-antiarrhythmic drugs') anticipates that in general, drugs should receive an electrocardiographic evaluation, beginning early in clinical development, typically in a single trial in healthy volunteers dedicated to evaluating their effect on ECG parameters with a focus on cardiac repolarization (the 'thorough QT/QTc study'). The recommendations contained in this document are generally applicable to new non-antiarrhythmic drugs having systemic bioavailability. While this document is concerned primarily with the development of NCEs, the recommendations contained therein might also be applicable to drugs already approved when a new dose or route of administration is being developed that results in significantly higher peak plasma concentration (C_{max}) or systemic exposure (area under plasma concentration versus time curve, AUC), or when a new indication or patient population is being pursued. The evaluation of the effect of a drug on QT interval would be required if other members of its pharmacological or chemical class have been associated with QT/QTc interval prolongation, TdP, or sudden cardiac death during post-marketing surveillance.

All regulatory authorities believe that adequately conducted non-clinical studies provide important evidence of safety that requires to be corroborated by data from clinical trials. However, since there is no universal agreement at present on the extent to, or the findings by, which non-clinical testing can exclude a clinical risk, regulatory authorities in some regions of the world will require clinical evaluation of a new drug in a 'thorough QT/QTc study' (typically in healthy volunteers) for almost *all* new drugs. For other authorities, this single dedicated clinical trial is required only if the non-clinical studies are positive or equivocal and if other drugs in the same pharmacological or chemical class have been associated with QTc interval prolongation with or without TdP. In regions where non-clinical data are recognized to preclude a risk of QT/QTc prolongation in humans, the recommendations in ICH E14 for the clinical evaluation of QT/QTc could be modified. In the absence of a 'thorough QT/QTc study', provided the quality of ECG monitoring is acceptable and of high standard, these authorities would be content with evidence gathered from well-conducted clinical pharmacology studies in healthy volunteers and from an integrated summary of all ECG data across all trials as long as the information on the effect of the NCE on QTc interval includes data on its dose-concentration-response relationship. Other factors that could influence the need for, or the design of, such a study include durations of the pharmacodynamic action of a drug and treatment with it, the metabolic profile of the drug and previous experience with other members of the same pharmacological or chemical class.

Examples of drugs for which non-clinical data may be recognized to preclude a risk of QT/QTc prolongation in humans include those with short half-life and intended for short-term administration and for which there is adequate non-clinical evidence of an acceptable safety margin. For example, vardenafil (a phosphodiesterase 5 inhibitor indicated for use in male erectile dysfunction) is intended for intermittent use and has a short half-life (4–5 h). Its IC_{50} for the target enzyme is 0.89 nM and for hERG is 84 μM and it was shown in *in vivo* studies in dogs to have no effect on QT interval. These results suggested a low potential for QT interval prolongation in humans and it was given a positive opinion by the CPMP in November 2002 and approved by the European Commission in March 2003 without any dedicated QTc study in healthy volunteers. In contrast, the FDA approved it in August 2003 subject to the sponsor committing to a post-marketing study to evaluate the impact on QT interval prolongation of combining vardenafil with another drug with a similar QT effect size. Although a 'thorough QT/QTc study' when completed showed that vardenafil had only a very modest effect on Fridericia-corrected QTc interval (a mean placebo-corrected increase of 10 ms) even at eight times the recommended starting dose, the FDA contraindicated its use in patients who have pre-existing prolongation of the QT interval because of the possibility of producing an arrhythmia. In contrast, the CPMP were content to include only a descriptive narrative of the study and its results in the pharmacological properties section of the European labelling.

Even when required, the 'thorough QT/QTc study' is certainly not the study that is traditionally referred to as 'the first administration in human' because it is critical to have other basic pharmacological, especially pharmacokinetic, data in humans for its design and conduct. It is also vital to know the doses likely to be effective clinically in order that an appropriate range of exposures (C_{max} and AUC) is covered in the 'thorough QT/QTc study'. Although usually conducted in normal healthy volunteers, some drugs (e.g. neuroleptic and antineoplastic chemotherapeutic agents) will not be suitable for study in this population because of issues related to ethics and tolerability.

ICH E14 goes on to describe in detail the design and the statistical power requirements of the 'thorough QT/QTc study' (including the use of placebo and a positive control, blood sampling and ECG measurement times). An adequate drug development programme should ensure that the dose–response or concentration–response relationship of a drug's potential for QT/QTc prolongation has been characterized, with exploration of concentrations that are far higher than those achieved following the anticipated therapeutic doses. The duration of dosing should be sufficient to ensure that steady-state concentrations have been achieved.

If the 'thorough QT/QTc study' is negative, the collection of baseline and periodic on-therapy ECGs in subsequent clinical evaluation should be in accordance with the current standard investigational practices in each therapeutic field. If the 'thorough QT/QTc study' is positive, additional evaluation in subsequent clinical studies should be performed. Again, there are regional differences with regard to the implications of a negative 'thorough QT/QTc study' in healthy volunteers for subsequent drug development in patients.

Regulatory perspectives on hERG channel studies

Altered function of the ion channels involved in action potential alone is not sufficient to induce TdP. Therefore, when assessing the relevance of data from hERG channel studies for clinical risk of TdP, there are two components that should be evaluated separately. First, the relationship between hERG channel inhibition and QT interval prolongation and second, the relationship between QT interval prolongation and induction of TdP. For reasons that are discussed in detail below, hERG blockade is not synonymous with clinical risk of TdP. Given the implications for the approval and labelling of an NCE and for public safety, regulatory evaluation of the risk of TdP associated with clinical use of the NCE requires an integrated approach that takes into account:

- Pharmacological and chemical classes of the NCE
- Physicochemical characteristics of the NCE
- *In vitro* (ion channel and electrophysiology) and *in vivo* non-clinical studies
- High-dose pharmacology studies to steady state, with and without inhibitors of its elimination or inactivation
- Other ancillary pharmacology of the NCE

- Appropriate studies in 'at risk' groups
- Appropriate pharmacodynamic interaction studies
- Evaluations of mean increase in placebo-corrected QTc interval and outliers with categorical responses in clinical studies
- Genotype of the patients with respect to mutations of drug metabolizing enzymes, and sodium and potassium channels. Mutations of these ion channels are emerging to be an important pharmacogenetic risk factor in induction of TdP following administration of routine clinical doses of drugs with 'QT-liability' (Shah 2004, Roden 2004, Paulussen et al 2004).

As stated above, ICH S7B recommends ion channel studies (*in vitro* I_{Kr} assay) that include heterologous expression systems for cloned human ion channels. Although native I_{Kr} function can be studied in isolated cardiac myocyte preparations, a more typical test system is hERG channel heterologously expressed in mammalian cell lines because these are without contamination from other ion currents. The higher amplitude of current in hERG expression systems relative to native I_{Kr} current thus affords a better signal-to-noise ratio than can be expected when measuring native I_{Kr} in cardiac myocytes. In addition, there are the advantages of better voltage control and of investigating the actual human channel in these hERG systems. The hERG expression systems most often used are *Xenopus* oocytes, human embryonic kidney (HEK-293) cells, mouse fibroblasts (C cells), Chinese hamster ovary (CHO) cells or LTK cells, all stably transfected with hERG cDNA.

An indirect application of hERG channel studies providing data on relative hERG-blocking potency/affinity involves inhibition of radioligand (e.g. radio-labelled dofetilide) binding (Finlayson et al 2001). However, competition for radioligand-binding sites provides no information on agonistic or antagonistic effects of the drug on I_{Kr}. Neither does this assay identify drugs that bind to hERG at sites other than the radioligand-binding sites. For high-throughput screening and identifying lead compounds at a very early stage of the drug discovery process, other approaches to studying hERG binding in expression systems involve rubidium efflux studies and fluorescence assays using voltage-sensitive dyes. However, both these systems yield a number of false positives as well as false negatives. Rubidium efflux studies also show large discrepancies with electrophysiological studies (Netzer et al 2003). Despite the use of these novel methods, voltage-clamp measurements in mammalian hERG expression systems remain the 'gold standard' for testing hERG blockade by drugs (Witchel et al 2002a).

With regard to investigating the potential of a drug to prolong the QT interval in man by studying its effect on hERG channels, the present regulatory status of hERG channel studies is best discussed in terms of:

- Limitations of hERG channel studies
- Value of hERG channel studies
- Rates of false positives and
- Rates of false negatives

Limitations of hERG channel studies

A number of factors limit the value of hERG channel studies. Foremost among the limitations are the use of multiple expression systems that vary in their sensitivity, the use of non-standardized protocols (Witchel et al 2002a, Nakaya 2003) and inter-laboratory variability of results. The presence of a number of endogenous MiRPs suppresses hERG expression in *Xenopus* oocytes (Anantharam et al 2003a). Consequently, the *Xenopus* oocyte systems expressing hERG channels appear to be less sensitive to drug inhibition and the observed IC_{50} for drug effects may be a false overestimate, suggesting a lower than actual potential for hERG inhibition. Therefore, the use of a mammalian cell expression system is more desirable for studying drug effects. However, the increased sensitivity of these mammalian systems may falsely yield much lower estimates of IC_{50} values suggesting a higher than actual potential for hERG inhibition. Cavero & Crumb (2001) have compared data on loratadine using various expression systems (HEK293, SH-SY5Y, *Xenopus* oocytes) and experimental conditions (temperature, pulse duration and potassium concentration). The hERG IC_{50} values across these systems ranged from 0.17 μM in HEK293 at 36 °C to 101 μM in *Xenopus* oocytes at 22 °C. Similar discrepancy has been reported more recently for α-adrenoreceptor blocking drugs prazosin, doxazosin and terazosin, each tested prospectively for hERG blocking activities in *Xenopus* oocytes and HEK293 cell systems (Thomas et al 2004). Such disparity of data is bound to give rise to regulatory concerns on the safety margins claimed. Differences in hERG blocking pharmacology of a given drug may also arise from differences in stimulation protocol in the patch-clamp investigations. In addition, hERG block by different drugs displays different pH sensitivities (Dong et al 2004). There is therefore a need for a detailed assessment of various expression systems and pulse protocols (Witchel et al 2002a, Nakaya 2003), ideally using compounds with and without 'QT liability' and those known to be torsadogenic in human. This would greatly improve the regulatory utility of data from hERG channel inhibition studies.

As stated earlier, native I_{Kr} is a co-assembly of α-subunits (hERG) and β-subunits (MiRP1). Therefore, the other concern with hERG channel studies is that they disregard what might be potentially important modulatory interactions of auxiliary β-subunits with α-subunits (Petersen & Nerbonne 1999, Schroeder et al 2000, Abbott et al 2001, Teng et al 2003, Anantharam et al 2003a). The hERG channel represents only the α subunits of I_{Kr} and there is some controversy regarding the extent to which it fully recapitulates the native I_{Kr} function. A number of reports have emphasized important interactions between hERG and MiRP1 (Abbott et al 1999, Sesti et al 2000). In contrast, one comparison of pharmacological and selected biophysical properties of hERG with and without MiRP1 expression in CHO cells, with those of guinea-pig I_{Kr} under comparable conditions, suggested that the physiological role of MiRP1 might not be to act as an essential constituent of the hERG channel complex that carries the native I_{Kr} (Weerapura et al 2002). Another study evaluated the I_{Kr} blocking potency of a series of antiarrhythmic drugs associated with a range of clinical risks of TdP in two systems: mouse AT-1 cells (in which I_{Kr} is the major repolarizing current) and LTK cells transiently transfected with hERG. For each compound tested, the IC_{50} values for I_{Kr} block and hERG inhibition showed an excellent correlation in the two systems (Yang et al 2001). Additional comparative data on IC_{50} values for I_{Kr} block and hERG inhibition by non-antiarrhythmic drugs, gathered prospectively, are required before it can be concluded with confidence that MiRP1 does not influence either the inner pore binding or the toxicology of the hERG channel.

These interactions may also be system-specific or sensitive to mutation in the peptide subunits. For example, *KCNE2* has been shown to modulate *KCNQ1* current and gating properties in mammalian COS cells (Tinel et al 2000) but not in *Xenopus* oocytes expression system (Abbott et al 1999) and has very different effects from those of *KCNE1*. Certain mutations of *KCNE* have been shown to alter the activity of MinK or MiRP1 subunits with their corresponding α-subunit partners, KvLQT1 or hERG respectively (Abbott & Goldstein 2002). Amide local anaesthetics target hERG and hERG/MiRP1 channels with identical potency and these effects may significantly contribute to cardiac arrhythmia induced by these agents. However, MiRP1 (T8A) did not seem to confer an increased risk of severe cardiac side effects to carriers of this common polymorphism (Friederich et al 2004). This is in contrast to the observations on MiRP1(V65M) mutation. Expression studies in CHO with this novel missense mutation (V65M) in the *KCNE2* gene revealed that although mutant and wild-type MiRP1 co-localized with hERG subunits to form functional channels, the mutant hERG/MiRP1(V65M) channels mediated currents with an accelerated inactivation time course compared with wild-type channels (Isbrandt et al 2002).

Regardless of these complex, and often contradictory, *in vitro* data on the potential interactions of hERG with other auxiliary subunits, it is the case that MiRP1 co-assembles with hERG to constitute I_{Kr}, MiRP1 alters the binding kinetics between hERG and E-4031 and mutations of MiRP1 give rise to the clinical phenotype of LQT6 (Abbott et al 1999, Splawski et al 2000, Sesti et al 2000). It seems therefore intuitive that there most probably are some clinically relevant differences between hERG and native I_{Kr} in their responsiveness to pharmacological agents. This contention is further supported by the clinical observation that individuals who are silent carriers of *KCNE2* mutations are highly susceptible to acquired LQTS in response to therapeutic doses of some drugs (Sesti et al 2000).

Ehrlich et al (2004) have also reported a further interaction between hERG and the α-subunit of I_{Ks} (KvLQT1 encoded by *KCNQ1*), which modified the localization and current-carrying properties of hERG. Co-expression of KvLQT1 with hERG in a mammalian expression system resulted in almost doubling of hERG current density and significantly accelerated hERG current deactivation. Another recent study has shown that KCR1, a plasma membrane-associated protein that can be immunoprecipitated with hERG, reduces the sensitivity of the hERG channel to classical proarrhythmic hERG blockers in both cardiac and non-cardiac cell lines. It has been proposed that KCR1, when coupled to the hERG channel, may limit the sensitivity of hERG current to proarrhythmic drug blockade (Kupershmidt et al 2003). One common amino acid polymorphism K897T (A2690C) within the hERG channel has been shown, in one study in a German female population, to hasten hERG current and shorten the QTc interval (Bezzina et al 2003) while in another study in a Finnish female population, to prolong the QTc interval (Pietila et al (2002). In neither study was the genotype effect significant in the male population.

Even the most robust studies using hERG channels can only reveal the probability of whether or not a drug will block I_{Kr}. These studies provide no information on the effect of a drug on other ion currents and receptors relevant to induction of TdP. Drugs that prolong the QT interval do not necessarily induce TdP. Increase in transmural dispersion in repolarization, rather than prolongation of QTc interval, induced by a drug probably better predicts the risk of TdP associated with its use. Neither do hERG channel studies allow any observations on induction of EADs that are such powerful markers of TdP. Blockade of hERG channels is, therefore, not a wholly dependable surrogate of prolongation of the APD/QT interval *in vivo* and even less so for induction of TdP.

A further limitation in extrapolating hERG data to the *in vivo* performance of a drug is that hERG channel studies are indifferent to its other pharmacological properties that modulate the propensity of the drug to induce TdP. If a drug does not bind to myocardial cells or accumulate in the cardiac tissue, it is most unlikely that it will have any significant effect on ventricular myocytes. Although, for a series of antiarrhythmic drugs associated with a range of clinical risks of TdP, the IC_{50} values for I_{Kr} block in AT-1 cells and hERG inhibition in LTK cells transfected with hERG showed an excellent correlation. The relation between the liability of a drug to cause TdP appeared to be dissociated from I_{Kr} blocking potency. Not all drugs causing TdP were potent

I_{Kr} blockers, and I_{Kr} block was not necessarily associated with TdP (Yang et al 2001).

Thus, data from hERG channel studies are not adequate evidence because other factors modulate the susceptibility of patients with QT interval prolongation to develop TdP. Other ancillary properties of the drug (e.g. α-, β-adrenoceptor or calcium-channel blocking activities) greatly modify the torsadogenic risk of a drug for a given duration of QT interval (Ben-David & Zipes 1990, Lu et al 2000, Furushima et al 2001, Noda et al 2002).

For example, the incidence of TdP is lower with racemic sotalol in contrast to $(+)$-(S)-sotalol because of the β-adrenoceptor blocking activity of $(-)$-(R)-sotalol present in the former. Sertindole, a new atypical neuroleptic agent, further emphasizes the need for characterizing a drug for these ancillary pharmacological properties. Although sertindole is a powerful hERG inhibitor with an activity almost as powerful as that of dofetilide (Rampe et al 1998) and produces marked prolongation of the QT interval in many patients, there are hardly any well-documented reports of TdP associated with its use. Neither the parent drug nor its hERG-active metabolites induce EADs (Maginn et al 2000). The powerful α-blocking activity of sertindole seems to afford a relative protection against degeneration of prolonged QT interval into TdP (Arnt 1998, Eckardt et al 2002, Toumi et al 2002, Moore et al 2003). First approved and marketed in the EU in 1996, sertindole was removed from the market in 1998 because of concerns over its cardiac safety. Following an in-depth review of its electrophysiological and autonomic pharmacology, efficacy and safety in 2001, it is to be re-introduced, albeit under carefully monitored post-marketing surveillance. Modulation of α-adrenoreceptor activity seems to have a greater effect than that of β-adrenoreceptor activity (Urao et al 2004).

α-Adrenoreceptor blocking drugs such as prazosin, doxazosin, alfuzosin, terazosin and tamulosin are potent hERG blockers, and yet there are hardly any reports of TdP associated with their use. For example, there were a total of 110 million prescriptions for doxazosin, terazosin and tamulosin dispensed in the USA during the period from 1998–2003 and, yet, there were only six reports of TdP in total. Alfuzosin has been used in the EU, Australia and Canada since 1987 without any reports of TdP. As of 12 May 2004, the Medicines and Healthcare products Regulatory Agency (MHRA) in the UK had received a total of 2944 reports of adverse drug reactions in association with these five drugs. These had included only one report of QT interval prolongation, one of TdP, one of ventricular tachycardia and two of ventricular fibrillation.

Despite their hERG blocking activity (Thomas et al 2002), the arrhythmogenic potential of selective serotonin reuptake inhibitors (SSRIs), relative to neuroleptic drugs, is much lower in view of their other pharmacologic properties such as calcium channel blocking activity (Witchel et al 2002b). Tolterodine, a structural analogue of the highly torsadogenic drug terodiline, produces little QT prolongation clinically and there are no published reports of it inducing TdP. And yet, it blocks hERG (IC_{50} value 17 nM) with almost the same potency as dofetilide (IC_{50} value 11 nM), cisapride, terfenadine and pimozide. This discrepancy is most likely due to its low plasma therapeutic concentrations (total and free C_{max} 12–16 nM and < 1 nM) and calcium-channel blocking activity (Kang et al 2004). Despite prolonging the QT interval, some drugs (e.g. amiodarone) hardly ever induce TdP. In addition to blocking I_{Kr}, amiodarone has other ancillary properties such as its inhibitory effects on adrenoreceptors, sodium and calcium currents as well as I_{Ks}. It is therefore essential that an NCE be also characterized for its other ancillary pharmacological properties. Furthermore, beat-to-beat electrical alternans is a powerful predictor of the risk of TdP and it seems probable that proarrhythmic and non-arrhythmic hERG blockers may be differentiated by their magnitude of effect on rate-dependent alternans (Fossa et al 2004).

Value of hERG channel studies

Within the bounds of the above limitations, hERG channel studies are of considerable regulatory interest. They are easy to perform and because the cell is isolated, the drug has easy access to the channel. hERG channel studies, when complemented by APD assay in native cardiac cells, characterize better the effects of the drug on multiple ion channels and other receptors that modulate the clinical risk of TdP. When positive at low concentrations, the most direct benefit of hERG channel studies is the guidance they provide for ECG monitoring in early human studies. Furthermore, hERG studies are particularly useful in the evaluation of relative 'QT liabilities' of various lead compounds and determine the relative torsadogenic roles of the metabolites and enantiomers of a drug. These studies could also be useful in clarifying the mechanism of actions of drugs (parent drug and metabolites) and explaining the potential role of a co-medication or drug metabolizing genotype as a risk factor for proarrhythmicity.

The value of hERG studies in selecting lead compounds is illustrated by a hERG-based comparison of four different 5-HT_4 receptor agonists (cisapride, renzapride, prucalopride and mosapride) for their proarrhythmic potential. The IC_{50} values of cisapride, renzapride and prucalopride were 0.24, 1.8 and 5.7 μM, respectively. In contrast, mosapride had no significant effect in this assay. This suggested that mosapride might be a safer alternative to cisapride (Potet et al 2001). Mosapride lacks any electrophysiological effects in isolated guinea pig papillary muscles (Kii & Ito 1997) and anaesthetized guinea pigs (Carlsson et al 1997). Clinical studies with mosapride have also failed to reveal any significant effect on QT interval (Endo et al 2002). Despite a 1.6-fold increase in the C_{max} of mosapride when given concurrently with 1200 mg daily dose of erythromycin for 7 days, there were no significant differences in the QTc intervals between mosapride with (mean 376 ms) and without (mean 372 ms) erythromycin (Katoh et al 2003). Studies with hERG channels have also been put to similar use in searching opiate analgesics free from cardiotoxicity. Whereas levacetylmethadol and methadone were able to block the hERG channel at clinically relevant concentrations, fentanyl, meperidine, codeine, morphine, and buprenorphine required much higher concentrations. The ratio of hERG IC_{50} to plasma C_{max} at therapeutic doses was highest for codeine and morphine (>455 and >400 respectively) and lowest for levacetylmethadol and methadone (2.2 and 2.7 respectively) (Katchman et al 2002). The predictive value of these hERG studies is supported by clinical experience.

Terfenadine, astemizole, halofantrine and levacetylmethadol illustrate the appeal of hERG studies in investigating the role of metabolites and explaining drug interactions in torsadogenesis. The data on the relative hERG blocking potencies of these drugs and their main metabolites are summarized in Table 2. Since the H_1-antihistaminic activity of terfenadine resides in both the parent compound and its metabolite (fexofenadine), fexofenadine has now been developed and replaced terfenadine clinically. Since desmethylastemizole has a very long half-life relative to astemizole, plasma levels of desmethylastemizole are generally about 30-fold higher than that of astemizole and the clinically observed cardiotoxicity appears to be mainly due to desmethylastemizole. In one patient with astemizole-induced TdP, plasma desmethylastemizole and astemizole concentrations were 7.7–17.3 ng/ml and < 0.5 ng/ml, respectively (Volperian et al 1996). Since astemizole and both its metabolites are almost comparable in their potencies as H_1-antihistamines, norastemizole is being developed as a potential antihistamine with a much lower risk of TdP. Similarly, in view of the narrow difference between the IC_{50} values of halofantrine and its metabolite, N-desbutylhalofantrine (21.6 nM versus 71.7 nM), the development of the metabolite as an antimalarial agent will likely result in (only) a small gain in the safety margin since the metabolite and the parent drug are almost equipotent in their antimalarial activity. For levacetylmethadol, a long-acting narcotic analgesic, a better case can be made on the basis of hERG data for the development of its therapeutically active metabolite, noracetylmethadol.

Studies with hERG channels have often helped clarify otherwise inconsistent observations or the mechanism of action of drugs. The class III antiarrhythmic agent RP 58866 and its active enantiomer, terikalant, were thought to be selective blockers of only the inward rectifier potassium current (I_{K1}), although their effects were not consistent with a sole effect on this channel. Later studies that investigated their effects on I_{Kr} currents in systems expressing the hERG

TABLE 2 Role of hERG studies in evaluation of the I_{Kr}-blocking risk of some torsadogenic drugs and their main metabolite

Parent drug and main drug metabolizing enzyme(s)	Major metabolite(s)	Results of hERG study	Reference(s)
Terfenadine CYP3A4	Fexofenadine	hERG is insensitive to fexofenadine but highly sensitive to terfenadine	Roy et al 1996 Leishman et al 2001 Scherer et al 2002
Astemizole CYP2J2 and others	Desmethylastemizole Norastemizole	IC_{50} values for Astemizole: 0.9 nM Desmethylastemizole 1.0 nM Norastemizole 27.7 nM	Zhou et al 1999
Halofantrine CYP3A4	N-desbutyl-halofantrine	IC_{50} values for Halofantrine 21.6 nM N-desbutyl-halofantrine 71.7 nM	Mbai et al 2002
Levacetylmethadol CYP3A4	Noracetylmethadol	IC_{50} values for Levacetylmethadol 3 μM Noracetylmethadol 12 μM	Kang et al 2003

channel provided strong evidence that these drugs were potent blockers of I_{Kr} current. This evidence called for reinterpretation of previous studies that assumed these drugs to selectively block only the I_{K1} (Jurkiewicz et al 1996). Cetirizine produced a mild biphasic QT interval prolongation and was associated with EADs but not with TdP in isolated, electrically paced rabbit ventricles (Gilbert et al 2000). The appearance of potentially torsadogenic EADs is in contrast to the safety of cetirizine during its extensive clinical use (Delgado et al 1998, Yap & Camm 2002, Simons et al 2003). Among the 614 UK spontaneous reports for cetirizine received by MHRA as of 12 May 2004, there were just two reports of QT interval prolongation and none of TdP, but eight of serious ventricular tachyarrhythmias. The anomaly is explained by an earlier hERG channel study that found that cetirizine up to 30 μM did not inhibit this channel whereas terfenadine and astemizole effectively blocked this channel with IC_{50} values of 0.33 and 0.48 μM respectively (Taglialatela et al 1998b).

Sertindole and thioridazine are metabolized by a polymorphically expressed cytochrome P450 (CYP2D6) and are both associated clinically with marked prolongation of QTc interval. Poor metabolizers (PM) of CYP2D6 accumulate the parent drug and are at a greater risk of adverse reactions to a number of CYP2D6 substrates. Studies with hERG channels have shown that CYP2D6 generated metabolites of sertindole are equally active in inhibiting I_{Kr} (Maginn et al 2000), thus excluding PM genotype as a potential risk factor for sertindole-induced QT prolongation. Thioridazine has a hERG IC_{50} value of 1.25 μM but there are no data on the hERG blocking potency of its metabolites that are also known to prolong the QT interval in man (Hartigan-Go et al 1996a, Drolet et al 1999, Kongsamut et al 2002). Although CYP2D6 PMs attain higher levels of thioridazine and lower levels of its two major metabolites compared to extensive metabolizers (EM), there is highly contradictory evidence for the role of PM genotype as a risk factor for thioridazine-induced QT interval prolongation (Llerena et al 2002, Thanaccody et al 2003). hERG channel studies on the metabolites of thioridazine would greatly clarify the role of the CYP2D6 genotype as a potential risk factor.

Since interactions of a chirally active drug with its pharmacological target are often stereoselective, it is not surprising that the 'QT liability' of a chiral NCE may reside predominantly in only one of the enantiomers. For chiral drugs, hERG studies may be particularly suited to determining the relative I_{Kr} blocking potencies of the racemate and its enantiomers with a view to potentially improving the risk/benefit of the product to be developed.

Ropivacaine, a local anaesthetic, is believed to have a lower incidence of clinical cardiac side effects than bupivacaine. These cardiac side effects include atrioventricular (AV) conduction blocks and/or tachyarrhythmias including TdP. Studies with separate enantiomers of these two agents had confirmed stereoselectivity of the adverse pharmacological action at the AV node for the more fat-soluble bupivacaine. The overall assessment of the non-hERG data available at that time led to the development and marketing in 1999 of levobupivacaine as a potentially less cardiotoxic alternative to bupivacaine. Subsequent hERG channel studies explain why levobupivacaine has hardly any advantage over bupivacaine (ECG effects in 3.1% versus 3.8% of clinical trial subjects respectively).

Although (−)-(S)-enantiomers of both were safer than their corresponding (+)-(R)-antipodes, the (−)-(S)-enantiomer of the less fat-soluble ropivacaine at clinical concentrations had no advantage over its (+)-(R)-antipode in this respect (Graf et al 2002). With regard to their effects on I_{Kr}, hERG channel studies have shown that the block induced by bupivacaine is also stereoselective (Gonzalez et al 2002). Block induced by bupivacaine, ropivacaine or dextrobupivacaine [the (+)-(R)-enantiomer of bupivacaine] was similar while that induced by levobupivacaine [the (−)-(S)-enantiomer] was about twofold higher. Numerous non-clinical and clinical studies have compared levobupivacaine with bupivacaine and in most (but not all) studies, there is evidence that levobupivacaine is less toxic. Although levobupivacaine is found to cause smaller changes in QTc interval (Gristwood 2002), its clinical relevance is unclear. More recent data show that bupivacaine, levobupivacaine and ropivacaine inhibited hERG channels, with nearly comparable IC_{50} values of $20\,\mu M$, $10\,\mu M$ and $20\,\mu M$ respectively, at toxicologically relevant concentrations (Friederich et al 2004).

In vivo, the QT-prolonging effect of terodiline is mediated by (+)-(R)-terodiline (Hartigan-Go et al 1996b) and that of prenylamine probably by its (+)-(S)-enantiomer (Bayer et al 1988). Unfortunately, there are no hERG channel studies reported with prenylamine and enantiomers of either prenylamine or terodiline. Nevertheless, hERG studies with racemic terodiline *post hoc* provided compelling evidence that established the causal relationship between terodiline and proarrhythmias reported in association with its clinical use (Leishman et al 2001).

False positive hERG studies

Drugs with hERG IC_{50} values in low nanomolar or high micromolar concentrations rarely present problems when evaluating their potential clinical significance. From the industry as well as the regulatory perspectives, it must be a matter of concern that studies with hERG channels generate a high rate of equivocal or false positive results. One laboratory has investigated the effects of a large number of new drugs on HEK-293 cells stably transfected with hERG cDNA (data from a poster presentation by Lansdell K et al, Quintiles Ltd). Of the 391 test substances examined, 71% were shown to inhibit hERG current. In a separate series of tests, 27% of 265 drugs increased APD in Purkinje fibre preparations. Sixty drugs were tested in both these systems and 43–44% of these 60 drugs shortened APD regardless of their effect on the hERG channel. Of the hERG-positive drugs, 28% increased and 18% had no effect on APD in Purkinje fibre preparations. Of the hERG-negative drugs, 19% increased and 33% had no effect on APD in Purkinje fibre preparations. Martin et al (2004) have reported wide APD changes (158% prolongation–16% shortening) in association with nine of the 10 prominent hERG inhibitors (>50% inhibition). This poor correlation between hERG and repolarization assays suggests that the hERG assay oversimplifies drug effects. Thus, hERG blockade may be indicative of the mechanism of repolarization anomalies that may contribute to the risk, but hERG data by themselves may be insufficient for assessment of risk.

Provided the concentration of the drug is high enough, it seems that almost any drug will inhibit the hERG channel *in vitro*. These studies suffer from lack of ability to correlate *consistently* the hERG IC_{50} values to the maximum free therapeutic plasma concentrations. Although the available data suggest that a 30-fold margin between hERG IC_{50} and peak free therapeutic plasma concentrations may be adequate for most drugs, there is at present no consensus on the multiple that should be devoid of an effect in the hERG studies to exclude a clinical risk. It is not surprising therefore that data from three major pharmaceutical companies have shown a large proportion of drugs to be 'active' at hERG channels. Although this may simply reflect the nature of new drugs being developed, it is noteworthy that one company reported 86% of compounds to be active at $10\,\mu M$ while another reported that 76 (70%) of the 109 compounds tested were active at hERG. Of the 76 hERG inhibitors, 17% displayed an IC_{50} value $<1\,\mu M$, 51% had an IC_{50} value between 1–10 μM and 32% had an IC_{50} value $>10\,\mu M$. The third company reported that 25% of compounds were active at $1\,\mu M$ and expected about 75% to be active at $10\,\mu M$ (personal communication from various companies, 2003).

Perhexiline, a calcium antagonist, has been reported to block hERG current with an IC_{50} of 7.8 μM (Walker et al 1999). Interestingly, even carvedilol has been shown to block these channels (Karle et al 2001). Carvedilol at a concentration of only 10 μM blocked hERG potassium tail currents by 47% and yet, it is one of the few drugs that has been associated with a survival benefit in patients with congestive heart failure — one group of patients that is known to be at a particularly higher risk of TdP. The β-blocking activity of $(-)$-(S)-carvedilol and the α_1-adrenoceptor antagonist activity present in both the enantiomers of carvedilol (Bartsch et al 1990) almost certainly explain this paradox. Despite their use in patients with cardiac disease, neither perhexiline nor carvedilol has been clinically associated with TdP. Following more than 15 years of marketing, there is only one case of perhexiline-induced TdP reported in the literature (Kerr & Ingham 1990). The discrepancy between the hERG binding potency of tolterodine and its effect on QT interval or cardiac rhythm has already been referred to earlier. As of 12 May 2004, the UK adverse drug reactions database at MHRA included a total of 543 reports for perhexiline, 119 for carvedilol and 498 for tolterodine. Neither perhexiline nor carvedilol had QT interval prolongation or TdP reported in association with their use. Two reports of TdP with tolterodine were both confounded by other risk factors.

Phenytoin, an anticonvulsant drug thought to shorten the QT interval and often used to treat cardiac arrhythmias, has also been reported to block hERG channels (Danielsson et al 2003). Yet, ECGs recorded during toxicity in 52 (91%) of the 57 patients with severe oral phenytoin overdose revealed no clinically significant abnormalities attributable to phenytoin. Continuous single-lead electrocardiographic monitoring in 36 (63%) of these patients revealed no incidents of arrhythmias requiring treatment. ECGs during toxic and baseline states were available for detailed analysis in 15 (26%) cases. No change in heart rate, QRS duration, or corrected QT interval was observed (Wyte & Berk 1991).

Observations such as these support the contention that hERG screening of all drugs in the pharmacopoeia would unearth some potent hERG blockers that have never been associated clinically with QT interval prolongation or TdP (Redfern et al 2003). The exquisite sensitivity of hERG channels results from their rapid inactivation, the presence of a large cavity in its pore and the presence of multiple aromatic rings (particularly Tyr652 and Phe656) that line the pore (Mitcheson et al 2000, Chen et al 2002, Anantharam et al 2003b). These three properties facilitate stabilization of drug binding, accommodation of many large molecules and high affinity binding respectively. As stated earlier, KCR1 reduces the sensitivity of the hERG channel to classical hERG blockers (Kupershmidt et al 2003). Whilst KCR1 may be a rational *in vivo* target for modifying the proarrhythmic effects of otherwise clinically useful hERG blockers, the absence of KCR1 in heterologous expression systems may also contribute to high false positive results.

False negative hERG studies

The author is not aware of any drug that has been found to be torsadogenic in human and which, when investigated in adequate non-clinical studies, has been found not to declare its potential to alter cardiac repolarization.

Arsenic trioxide is an oncology drug with a very high QT-prolonging and torsadogenic potential in humans. Clinically, there is a delay of about 2 hours between peak serum arsenic concentration and maximum effect on QTc interval (Zhou et al 2003), suggesting a potential role for one of the metabolites in prolonging the QTc interval. It has often been cited as an example of a drug with a false negative result in hERG assays. Recent studies, however, have shown that the parent drug blocks both I_{Kr} and I_{Ks} at clinically relevant concentrations in CHO cells transfected with hERG or the KvLQT1/minK channel. In addition, it also activates I_{K-ATP}, which maintains normal repolarization (Drolet et al 2004). Clearly, these results need to be replicated and by similarly investigating the metabolites, reconciled with clinical suspicions on the role of one of the metabolites if hERG channel studies are to remain credible (see below).

Among the drug classes that have attracted considerable research interest in terms of clinical effects and possible false-negative non-clinical signals on cardiac repolarization are various fluoroquinolones, and in particular levofloxacin. At concentrations as high as 100 μM, sparfloxacin and moxifloxacin increased APD by 41% and 25%, respectively, while levofloxacin had little or no effect (Hagiwara et al 2001). In a hERG channel study, IC_{50} values for these three drugs were 18 μM, 129 μM and 915 μM respectively — the corresponding ratios of hERG IC_{50} to peak free plasma concentrations being 10, 22 and 70 (Kang et al

2001). Single doses of moxifloxacin and levofloxacin increase QTc interval in humans by mean values of 5–10 (some have reported up to 17.8) ms and 3.5–4.9 ms, respectively. This perceived cardiac safety of levofloxacin may appear to be rebutted by clinical experience. Clinically, both moxifloxacin and levofloxacin have been associated with a few reports of QTc interval prolongation and TdP, almost all of which were confounded by the presence of known risk factors (age >70 years, cardiac disease or concomitant drugs). hERG channel and clinical studies suggested that mosapride lacks any torsadogenic potential (Potet et al 2001, Katoh et al 2003) but Ohki et al (2001) have reported a case of TdP induced following prescription of mosapride to a patient receiving flecainide in the presence of hypokalaemia. Such reports further emphasize the problems of relying exclusively on hERG studies and the need to consider the role of other host factors when evaluating the potential clinical risk.

Drug-induced loss of function of hERG channels may be caused by several mechanisms, including altered current kinetics, altered ion selectivity, or defective intracellular protein trafficking. One emerging issue of concern is the effect of drugs on hERG trafficking. This has become a focus of particular interest recently, because some of the mutant hERG subunits display normal wild-type current properties when normal trafficking is restored and channels are inserted into the cell membrane *in vitro*. This has been observed with congenital LQT2 syndrome (Thomas et al 2003) and the question arises whether drugs can inhibit normal hERG trafficking. Cytosolic chaperones Hsp70 and Hsp90 are required for the biochemical maturation of hERG. Geldanamycin is an antitumour drug that specifically inhibits Hsp90 and it has been shown to prevent hERG maturation and to increase degradation of the hERG channel, while reducing hERG currents in heterologous expression systems (Ficker et al 2003).

A recent study has reported that arsenic trioxide interferes with hERG trafficking by inhibition of hERG–chaperone complexes and in addition, it also increases calcium currents (Ficker et al 2004). More recently, pentamidine (an antimicrobial drug) has also been shown to prolong the QT interval by inhibition of hERG trafficking (Kuryshev et al 2004). Just as mutations in the hERG channel can cause a trafficking deficiency (Furutani et al 1999, Ficker et al 2000), it seems now likely that drugs may cause trafficking deficiency as well. Classical hERG channel studies will not identify these drugs and this may seriously limit the values of even the negative hERG studies. In future, drugs may need to be screened for their potential to inhibit hERG trafficking.

Regulatory evaluation of risk/benefit and approval

Redfern et al (2003) have reviewed the relative value of various non-clinical cardiac electrophysiological data (*in vitro* and *in vivo*) for predicting risk of TdP in clinical use. In addition to cardiac APD and QT prolongation *in vivo* in dogs, they used

TABLE 3 Drug class and ratios between IC_{50} for hERG or I_{Kr} and maximum free therapeutic plasma concentrations (Redfern et al 2003)

Drug class	Ratios between IC_{50} for hERG or I_{Kr} and maximum free therapeutic plasma concentrations	Comments
Class Ia and III antiarrhythmic drugs ($n=11$)	0.1 to 31	All less than 30 except tedisamil and amiodarone
Drugs withdrawn from market due to TdP ($n=6$)	0.31 to 13	All less than 30
Drugs with measurable incidence/numerous reports of TdP ($n=6$)	0.03 to 35	All less than 30 except pimozide
Drugs with isolated reports of TdP ($n=14$)	0.13 to 35 700	All less than 30 except fluoxetine, amitriptyline, nifedipine, ciprofloxacin and diphenhydramine
Drugs with no reports of TdP ($n=16$)	23 to 3310	All more than 30 except cibenzoline, phenytoin, verapamil and ketoconazole

published data on hERG (or I_{Kr}) activity, as the basis for comparison against QT effects and reports of TdP in humans for a wide range of drugs. Their data on hERG or I_{Kr} IC_{50}, summarized in Table 3, suggest that a 30-fold margin between hERG IC_{50} and peak free therapeutic plasma concentrations may be adequate to confidently exclude a clinical effect on cardiac repolarization. This safety margin is also supported by data reviewed by Webster et al (2002). However, for the sake of greater certainty, Redfern et al (2003) recommend a higher safety margin in the future but correctly emphasize that the acceptable safety margin should depend on the lethality of the disease to be treated — ranging from 10-fold for a lethal disease to higher than 100-fold for (symptomatic treatment of) a benign condition. Based on their extensive review of hERG channel studies, Cavero & Crumb (2001) have suggested that the use of a potency parameter (potency for hERG channel and the potency at the desired site of action) of approximately 30.

This raises an additional regulatory concern with regard to the definition of safety margins. Termed the cardiac safety index (CSI), there are three immediately apparent variations to this parameter. It can be the ratio of *in vitro* hERG IC_{50} to (i) EC_{50} (drug concentration at the intended pharmacological target), (ii) EFPC (effective free plasma concentrations) or (iii) effective myocardial concentration that is so dependent on lipophilicity of the drug. The point is illustrated by terfenadine that is about 260 times more lipophilic than cetirizine. Depending on the method of calculation, the three CSI values are 0.5, >104 and 0.4 respectively for terfenadine, and >3, 28 and >56 respectively for cetirizine (Cavero & Crumb 2001).

When a number of drugs are investigated by a standard and uniform protocol, hERG studies have provided data consistent with clinical observations. For example, a dedicated clinical QT study comparing ziprasidone with five other neuroleptic agents identified thioridazine as the most potent in prolonging QT interval followed by ziprasidone ranking next with olanzapine being the least active. It would be interesting to investigate these six drugs for their hERG-blocking activities in a single study that uses a standard uniform protocol. Kongsamut et al (2002) found that the ratio of total plasma drug concentration to the IC_{50} value for hERG was indicative of the degree of QT prolongation observed with sertindole, pimozide, thioridazine ziprasidone, quetiapine, risperidone and olanzapine. These investigators too identified target receptor affinity and expected clinical plasma levels as important parameters by which to interpret hERG channel data.

In the regulatory framework, the non-clinical data (including the hERG channel studies) have to be seen in the context of all other available data. Any non-clinical signal, whether from hERG channels studies, tissue preparations or *in vivo* studies, has to be confirmed or repudiated in adequately powered and designed clinical trials. The clinical trials should be designed to investigate the QT prolonging potency of a drug in humans and to characterize the mechanisms by which the potential clinical risk may be modulated by other non-genetic and genetic risk factors. When ethically or practically impossible to do so clinically, risk factors should be investigated in non-clinical *in vitro* and *in vivo* models. Typical examples of risk factors that require non-clinical studies for their investigation are the effects of electrolyte imbalance, bradycardia and myocardial ischaemia, or coadministration of two QT-prolonging drugs.

With respect to drug-induced TdP, there is also a more philosophical question of the definition of 'clinical risk' and the level of risk that is unacceptable or tolerable. It seems inappropriate to categorize a drug with an incidence of TdP of 1 in 500 000 patients together with another that has an incidence of 1 in 3000. As with other potentially fatal adverse drug reactions such as myelotoxicity, gastrointestinal haemorrhage, hepatotoxicity or rhabdomyolysis, a level of risk may have to be tolerated. While an incidence of a potentially fatal event at the rate of 1 in 3000 may be unacceptable, an incidence of 1 in 500 000 may be considered acceptable with a whole range in between. This perceived risk has to be seen in the context of benefit and available alternatives. The risk of not treating a disease is also an important component in risk assessment.

Conclusions

The sponsors are now making great efforts to characterize NCEs for their proarrhythmic risk using a variety of non-clinical tests including hERG channel studies. Regulatory authorities carefully scrutinise the dossiers of these NCEs for any non-clinical and clinical evidence of proarrhythmic potential. There is now an urgent need for these hERG channel studies to be standardized in terms of expression systems and protocols. Studies are also required to investigate whether data on IC_{50} values for I_{Kr} block correlate well with data on IC_{50} values for hERG inhibition by non-antiarrhythmic drugs.

From the risk assessment perspective, it is reassuring to know that hERG channel studies have many false positives but hardly any false negatives. This should provide a degree of confidence in a negative hERG signal when planning clinical studies aimed at characterizing the potential of a drug to prolong the QT interval. Provided that the ancillary pharmacology of a drug, its relative affinity for the target receptor and its clinically expected free drug concentrations are taken into account, hERG channel studies provide additional complementary data in the evaluation of clinical risk. One disappointing feature of these studies in regulatory submissions is the investigation of chirally active drugs. For adequate evaluation of the proarrhythmic risk, data should be gathered on the racemic drug as well as each enantiomer and their metabolites. These data too should be presented in the context of therapeutically active free plasma concentrations.

In view of the exquisite sensitivity of the mammalian hERG channel assays, there is a real risk that valuable novel drugs may be lost if decisions are made solely on the basis of these studies. The sponsors of new drugs should repeat these studies frequently during drug development as more data are gathered. They should take into account other properties of the drug as well as data from all other non-clinical studies when evaluating the potential risk and making 'go–no-go' decisions. On the other hand, one troubling aspect of hERG channel studies is the probability that drugs such as arsenic trioxide and pentamidine may represent only the first two of many drugs that alter cardiac repolarization by inhibiting hERG trafficking. This is an area that will require further attention from scientists as well as the regulators.

Acknowledgements

I wish to extend my sincere appreciation to Professor Michael Sanguinetti (Eccles Institute of Human Genetics, University of Utah, USA) and Dr Harry Witchel (Cardiovascular Research Laboratories, University of Bristol, UK) for their very thoughtful and constructive comments on the previous draft of this paper. Any shortcomings and deficiencies, however, are entirely my own responsibility.

References

Abbott GW, Goldstein SA 2002 Disease-associated mutations in KCNE potassium channel subunits (MiRPs) reveal promiscuous disruption of multiple currents and conservation of mechanism. FASEB J 16:390–400

Abbott GW, Sesti F, Splawski I et al 1999 MiRP1 forms I_{Kr} potassium channels with HERG and is associated with cardiac arrhythmias. Cell 97:175–187

Abbott GW, Goldstein SAN, Sesti F 2001 Do all voltage-gated potassium channels use miRPs? Circ Res 88:981–983

Anantharam A, Lewis A, Panaghie G et al 2003a RNA interference reveals that endogenous Xenopus minK-related peptides govern mammalian K+ channel function in oocytes expression studies. J Biol Chem 278:11739–11745

Anantharam A, Markowitz SM, Abbott GW 2003b Pharmacogenetic considerations in diseases of cardiac ion channels. J Pharmacol Exp Ther 307:831–838

Arnt J 1998 Pharmacological differentiation of classical and novel antipsychotics. Int Clin Psychopharmacol 13(suppl 3):S7–14

Bartsch W, Sponer G, Strein K et al 1990 Pharmacological characteristics of the stereoisomers of carvedilol. Eur J Clin Pharmacol 38(suppl 2):S104–107

Bayer R, Schwarzmaier J, Pernice R 1988 Basic mechanism underlying prenylamine-induced torsade de pointes: differences between prenylamine and fendiline due to basic actions of the isomers. Curr Med Res Opin 11:254–272

Ben-David J, Zipes DP 1990 α-Adrenoceptor stimulation and blockade modulates cesium-induced early afterdepolarizations and ventricular tachyarrhythmias in dogs. Circulation 82:225–233

Bezzina CR, Verkerk AO, Busjahn A et al 2003 A common polymorphism in KCNH2 (HERG) hastens cardiac repolarization. Cardiovasc Res 59:27–36

Carlsson L, Amos GJ, Andersson B, Drews L, Duker G, Wadstedt G 1997 Electrophysiological characterization of the prokinetic agents cisapride and mosapride in vivo and in vitro: implications for proarrhythmic potential? J Pharmacol Exp Ther 282:220–227

Cavero I, Crumb W 2001 Native and cloned ion channels from human heart: laboratory models for evaluating the cardiac safety of new drugs. Eur Heart J Suppl 3(suppl K):K53–K63

Chachin M, Kurachi Y 2002 Evaluation of pro-arrhythmic risk of drugs due to QT interval prolongation by the HERG expression system [Japanese]. Nippon Yakurigaku Zasshi 119:345–351

Chen J, Seebohm G, Sanguinetti MC 2002 Position of aromatic residues in the S6 domain, not inactivation, dictates cisapride sensitivity of HERG and eag potassium channels. Proc Natl Acad Sci USA 99:12461–12466

Danielsson BR, Lansdell K, Patmore L, Tomson T 2003 Phenytoin and phenobarbital inhibit human HERG potassium channels. Epilepsy Res 55:147–157

Delgado LF, Pferferman A, Sole D, Naspitz CK 1998 Evaluation of the potential cardiotoxicity of the antihistamines terfenadine, astemizole, loratadine, and cetirizine in atopic children. Ann Allergy Asthma Immunol 80:333–337

Dong DL, Li Z, Wang HJ, Du ZM, Song WH, Yang BF 2004 Acidification alters antiarrhythmic drug blockade of the ether-a-go-go gene (HERG) channels. Pharmacol Toxicol 94:209–212

Drolet B, Vincent F, Rail J et al 1999 Thioridazine lengthens repolarization of cardiac ventricular myocytes by blocking the delayed rectifier potassium current. J Pharmacol Exp Ther 288:1261–1268

Drolet B, Simard C, Roden DM 2004 Unusual effects of a QT-prolonging drug, arsenic trioxide, on cardiac potassium currents. Circulation 109:26–29

Eckardt L, Breithardt G, Haverkamp W 2002 Electrophysiologic characterization of the antipsychotic drug sertindole in a rabbit heart model of torsade de pointes: low torsadogenic potential despite QT prolongation. J Pharmacol Exp Ther 300:64–71

Ehrlich JR, Pourrier M, Weerapura M et al 2004 KvLQT1 modulates the distribution and biophysical properties of HERG: a novel alpha-subunit interaction between delayed-rectifier currents. J Biol Chem 279:1233–1241

Endo J, Nomura M, Morishita S et al 2002 Influence of mosapride citrate on gastric motility and autonomic nervous function: evaluation by spectral analyses of heart rate and blood pressure variabilities, and by electrogastrography. J Gastroenterol 37:888–895

Ficker E, Thomas D, Viswanathan PC et al 2000 Novel characteristics of a misprocessed mutant HERG channel linked to hereditary long QT syndrome. Am J Physiol Heart Circ Physiol 279:H1748–H1756

Ficker E, Dennis AT, Wang L, Brown AM 2003 Role of the cytosolic chaperones Hsp70 and Hsp90 in maturation of the cardiac potassium channel HERG. Circ Res 92:87–100

Ficker E, Kuryshev YA, Dennis AT et al 2004 Mechanisms of arsenic-induced prolongation of cardiac repolarization. Mol Pharmacol 66:33–44

Finlayson K, Turnbull L, January CT, Sharkey J, Kelly JS 2001 [^{3}H]dofetilide binding to HERG transfected membranes: a potential high throughput preclinical screen. Eur J Pharmacol 430:147–148

Fossa AA, Wisialowski T, Wolfgang E et al 2004 Differential effect of HERG blocking agents on cardiac electrical alternans in the guinea pig. Eur J Pharmacol 486:209–221

Friederich P, Solth A, Schillemeit S, Isbrandt D 2004 Local anaesthetic sensitivities of cloned HERG channels from human heart: comparison with HERG/MiRP1 and HERG/MiRP1 T8A. Br J Anaesth 92:93–101

Furushima H, Chinushi M, Washizuka T, Aizawa Y 2001 Role of α1-blockade in congenital long QT syndrome — investigation by exercise stress test. Jpn Circ J 65:654–658

Furutani M, Trudeau MC, Hagiwara N et al 1999 Novel mechanism associated with an inherited cardiac arrhythmia: defective protein trafficking by the mutant HERG (G601S) potassium channel. Circulation 99:2290–2294

Fung MC, Hsiao-hui Wu H, Kwong K, Hornbuckle K, Muniz E 2000 Evaluation of the profile of patients with QTc prolongation in spontaneous adverse event reporting over the past three decades — 1969-1998. Pharmacoepidemiol Drug Safety 9(suppl 1):S24–25

Gilbert JD, Cahill SA, McCartney DG, Lukas A, Gross GJ 2000 Predictors of torsades de pointes in rabbit ventricles perfused with sedating and nonsedating histamine H1-receptor antagonists. Can J Physiol Pharmacol 78:407–414

Gonzalez T, Arias C, Caballero R et al 2002 Effects of levobupivacaine, ropivacaine and bupivacaine on HERG channels: stereoselective bupivacaine block. Br J Pharmacol 137:1269–1279

Graf BM, Abraham I, Eberbach N, Kunst G, Stowe DF, Martin E 2002 Differences in cardiotoxicity of bupivacaine and ropivacaine are the result of physicochemical and stereoselective properties. Anesthesiology 96:1427–1434

Gristwood RW 2002 Cardiac and CNS toxicity of levobupivacaine: strengths of evidence for advantage over bupivacaine. Drug Saf 25:153–163

Hagiwara T, Satoh S, Kasai Y, Takasuna K 2001 A comparative study of the fluoroquinolone antibacterial agents on the action potential duration in guinea pig ventricular myocardia. Jpn J Pharmacol 87:231–234

Hartigan-Go K, Bateman DN, Nyberg G, Martensson E, Thomas SH 1996a Concentration-related pharmacodynamic effects of thioridazine and its metabolites in humans. Clin Pharmacol Ther 60:543–553

Hartigan-Go K, Bateman ND, Daly AK, Thomas SHL 1996b Stereoselective cardiotoxic effects of terodiline. Clin Pharmacol Ther 60:89–98

Hondeghem LM, Hoffmann P 2003 Blinded test in isolated female rabbit heart reliably identifies action potential duration prolongation and proarrhythmic drugs: importance of triangulation, reverse use dependence, and instability. J Cardiovasc Pharmacol 41:14–24

Isbrandt D, Friederich P, Solth A et al 2002 Identification and functional characterization of a novel KCNE2 (MiRP1) mutation that alters HERG channel kinetics. J Mol Med 80:524–532

Jurkiewicz NK, Wang J, Fermini B, Sanguinetti MC, Salata JJ 1996 Mechanism of action potential prolongation by RP 58866 and its active enantiomer, terikalant. Block of the rapidly activating delayed rectifier K^+ current, I_{Kr}. Circulation 94:2938–2946

Kang J, Wang L, Chen XL, Triggle DJ, Rampe D 2001 Interactions of a series of fluoroquinolone antibacterial drugs with the human cardiac K^+ channel HERG. Mol Pharmacol 59:122–126

Kang J, Chen XL, Wang H, Rampe D 2003 Interactions of the narcotic l-alpha-acetylmethadol with human cardiac K^+ channels. Eur J Pharmacol 458:25–29

Kang J, Chen XL, Wang H et al 2004 Cardiac ion channel effects of tolterodine. J Pharmacol Exp Ther 308:935–940

Karle CA, Kreye VA, Thomas D et al 2001 Antiarrhythmic drug carvedilol inhibits HERG potassium channels. Cardiovasc Res 49:361–370

Katchman AN, McGroary KA, Kilborn MJ et al 2002 Influence of opioid agonists on cardiac human ether-a-go-go-related gene K^+ currents. J Pharmacol Exp Ther 303:688–694

Katoh T, Saitoh H, Ohno N et al 2003 Drug interaction between mosapride and erythromycin without electrocardiographic changes. Jpn Heart J 44:225–234

Keating MT, Sanguinetti MC 2001 Molecular and cellular mechanisms of cardiac arrhythmias. Cell 104:569–580

Kerr GD, Ingham G 1990 Torsade de pointes associated with perhexiline maleate therapy. Aust N Z J Med 20:818–820

Kii Y, Ito T 1997 Effects of 5-HT4-receptor agonists, cisapride, mosapride citrate, and zacopride, on cardiac action potentials in guinea pig isolated papillary muscles. J Cardiovasc Pharmacol 29:670–675

Kongsamut S, Kang J, Chen XL, Roehr J, Rampe D 2002 A comparison of the receptor binding and HERG channel affinities for a series of antipsychotic drugs. Eur J Pharmacol 450:37–41

Kupershmidt S, Yang IC, Hayashi K et al 2003 I_{Kr} drug response is modulated by KCR1 in transfected cardiac and noncardiac cell lines. FASEB J 17:2263–2265

Kuryshev YA, Ficker E, Wang L et al 2005 Pentamidine-induced long QT syndrome and block of hERG trafficking. J Pharmacol Exp Ther 312:316–323

Leishman DJ, Helliwell R, Wakerell J, Wallis RM 2001 Effect of E-4031, cisapride, terfenadine and terodiline on cardiac repolarization in canine Purkinje fibre and HERG channels expressed in HEK293 cells. Br J Pharmacol 133:130P

Llerena A, Berecz R, de la Rubia A, Dorado P 2002 QTc interval lengthening is related to CYP2D6 hydroxylation capacity and plasma concentration of thioridazine in patients. J Psychopharmacol 16:361–364

Lu HR, Remeysen P, De Clerck F 2000 Nonselective IKr-blockers do not induce torsades de pointes in the anesthetized rabbit during alpha1-adrenoceptor stimulation. J Cardiovasc Pharmacol 36:728–736

Maginn M, Frederiksen K, Adamantidis MM, Bischoff U, Matz J 2000 The effects of sertindole and its metabolites on cardiac ion channels and action potentials. J Physiol 525:79P

Martin RL, McDermott JS, Salmen HJ, Palmatier J, Cox BF, Gintant GA 2004 The utility of hERG and repolarization assays in evaluating delayed cardiac repolarization: influence of multi-channel block. J Cardiovasc Pharmacol 43:369–379

Mbai M, Rajamani S, January CT 2002 The anti-malarial drug halofantrine and its metabolite N-desbutylhalofantrine block HERG potassium channels. Cardiovasc Res 55:799–805

Milon D, Daubert JC, Saint-Marc C, Gouffault J 1982 Torsade de pointes. Apropos of 54 cases [French]. Ann Fr Anesth Reanim 1:513–520

Mitcheson JS, Chen J, Lin M, Culberson C, Sanguinetti MC 2000 A structural basis for drug-induced long QT syndrome. Proc Natl Acad Sci USA 97:12329–12333

Moore N, Hall G, Sturkenboom M, Mann R, Lagnaoui R, Begaud B 2003 Biases affecting the proportional reporting ratio (PRR) in spontaneous reports pharmacovigilance databases: the example of sertindole. Pharmacoepidemiol Drug Safety 12:271–281

Nakaya H 2003 Electropharmacological assessment of the risk of drug-induced long-QT syndrome using native cardiac cells and cultured cells expressing HERG channels [Japanese]. Nippon Yakurigaku Zasshi 121:384–392

Netzer R, Bischoff U, Ebneth A 2003 HTS techniques to investigate the potential effects of compounds on cardiac ion channels at early-stages of drug discovery. Curr Opin Drug Discovery Dev 6:462–469

Noda T, Takaki H, Kurita T et al 2002 Gene-specific response of dynamic ventricular repolarization to sympathetic stimulation in LQT1, LQT2 and LQT3 forms of congenital long QT syndrome. Eur Heart J 23:975–983

Ohki R, Takahashi M, Mizuno O et al 2001 Torsades de pointes ventricular tachycardia induced by mosapride and flecainide in the presence of hypokalemia. Pacing Clin Electrophysiol 24:119–121

Paulussen AD, Gilissen RA, Armstrong M et al 2004 Genetic variations of KCNQ1, KCNH2, SCN5A, KCNE1, and KCNE2 in drug-induced long QT syndrome patients. J Mol Med 82:182–188

Petersen KR, Nerbonne JM 1999 Expression environment determines K^+ current properties: Kv1 and Kv4 alpha-subunit-induced K^+ currents in mammalian cell lines and cardiac myocytes. Pflügers Arch 437:381–392

Pietila E, Fodstad H, Niskasaari E et al 2002 Association between HERG K897T polymorphism and QT interval prolongation in middle-aged Finnish women. J Am Coll Cardiol 40:511–514

Potet F, Bouyssou T, Escande D, Baro I 2001 Gastrointestinal prokinetic drugs have different affinity for the human cardiac human ether-a-gogo K^+ channel. J Pharmacol Exp Ther 299:1007–1012

Rampe D, Murawsky MK, Grau J, Lewis EW 1998 The antipsychotic agent sertindole is a high affinity antagonist of the human cardiac potassium channel HERG. J Pharmacol Exp Ther 286:788–793

Redfern WS, Carlsson L, Davis AS et al 2003 Relationships between preclinical cardiac electrophysiology, clinical QT interval prolongation and torsade de pointes for a broad range of drugs: evidence for a provisional safety margin in drug development. Cardiovasc Res 58:32–45

Roden DM 2004 Drug-induced prolongation of the QT interval. N Engl J Med 350:1013–1022

Roy M, Dumaine R, Brown AM 1996 HERG, a primary human ventricular target of the nonsedating antihistamine terfenadine. Circulation 94:817–823

Salle P, Rey JL, Bernasconi P, Quiret JC, Lombaert M 1985 Torsades de pointe. Apropos of 60 cases [French]. Ann Cardiol Angeiol (Paris) 34:381–388

Sanguinetti MC, Jiang C, Curran ME, Keating MT 1995 A mechanistic link between an inherited and an acquired cardiac arrhythmia: HERG encodes the I_{Kr} potassium channel. Cell 81:299–307

Scherer CR, Lerche C, Decher N et al 2002 The antihistamine fexofenadine does not affect IKr currents in a case report of drug-induced cardiac arrhythmia. Br J Pharmacol 137:892–900

Schroeder BC, Waldegger S, Fehr S et al 2000 A constitutively open potassium channel formed by KCNQ1 and KCNE3. Nature 403:196–199

Sesti F, Abbott GW, Wei J et al 2000 A common polymorphism associated with antibiotic-induced cardiac arrhythmia. Proc Natl Acad Sci USA 97:10613–10618

Shah RR 2002a The significance of QT interval in drug development. Br J Clin Pharmacol 54:188–202

Shah RR 2002b Drug-induced prolongation of the QT interval: regulatory dilemmas and implications for approval and labelling of a new chemical entity. Fundam Clin Pharmacol 16:147–156

Shah RR 2004 Pharmacogenetic aspects of drug-Induced torsade de pointes: potential tool for improving clinical drug development and prescribing. Drug Saf 27:145–172

Simons FE, Silas P, Portnoy JM, Catuogno J, Chapman D, Olufade AO 2003 Safety of cetirizine in infants 6 to 11 months of age: a randomized, double-blind, placebo-controlled study. J Allergy Clin Immunol 111:1244–1248

Splawski I, Shen J, Timothy KW et al 2000 Spectrum of mutations in long QT syndrome genes KVLQT1, HERG, SCN5A, KCNE1 and KCNE2. Circulation 102:1178–1185

Taglialatela M, Castaldo P, Pannaccione A, Giorgio G, Annunziato L 1998a Human ether-a-gogo related gene (HERG) K$^+$ channels as pharmacological targets: present and future implications. Biochem Pharmacol 55:1741–1746

Taglialatela M, Pannaccione A, Castaldo P et al 1998b Molecular basis for the lack of HERG K$^+$ channel block-related cardiotoxicity by the H1 receptor blocker cetirizine compared with other second-generation antihistamines. Mol Pharmacol 54:113–121

Teng S, Ma L, Zhen Y et al 2003 Novel gene hKCNE4 slows the activation of the KCNQ1 chain. Biochem Biophys Res Commun 303:808–813

Thanaccody R, Daly AK, Thomas SH 2003 Influence of CYP2D6 genotype on the QTc interval and plasma concentrations of thioridazine and its metabolites in psychiatric patients taking chronic therapy. Clin Pharmacol Ther 73:77

Thomas D, Gut B, Wendt-Nordahl G, Kiehn J 2002 The antidepressant drug fluoxetine is an inhibitor of human ether-a-go-go-related gene (HERG) potassium channels. J Pharmacol Exp Ther 300:543–548

Thomas D, Kiehn J, Katus HA, Karle CA 2003 Defective protein trafficking in hERG-associated hereditary long QT syndrome (LQT2): molecular mechanisms and restoration of intracellular protein processing. Cardiovasc Res 60:235–241

Thomas D, Wimmer AB, Wu K et al 2004 Inhibition of human ether-a-go-go-related gene potassium channels by alpha1-adrenoceptor antagonists prazosin, doxazosin and terazosin. Naunyn Schmiedebergs Arch Pharmacol 369:462–472

Tinel N, Diochot S, Borsotto M, Lazdunski M, Barhanin J 2000 KCNE2 confers background current characteristics to the cardiac KCNQ1 potassium channel. EMBO J 19: 6326–6330

Toumi M, Auquier P, Francois C 2002 The safety and tolerability of sertindole in a patient name use program. Pharmacoepidemiol Drug Safety 11(suppl 1):S115

Tseng GN 2001 IKr: the hERG channel. J Mol Cell Cardiol 33:835–849

Urao N, Shiraishi H, Ishibashi K et al 2004 Idiopathic long QT syndrome with early after depolarizations induced by epinephrine. Circ J 68:587–591

Volperian VR, Zhou Z, Mohammad S, Hoon TJ, Studenik C, January CT 1996 Torsade de pointes with an antihistamine metabolite: potassium channel blockade with desmethylastemizole. J Am Coll Cardiol 28:1556–1561

Walker BD, Valenzuela SM, Singleton CB et al 1999 Inhibition of HERG channels stably expressed in a mammalian cell line by the antianginal agent perhexiline maleate. Br J Pharmacol 127:243–251

Webster R, Leishman D, Walker D 2002 Towards a drug concentration effect relationship for QT prolongation and torsades de pointes. Current Opinion Drug Disc Dev 5:116–126

Weerapura M, Nattel S, Chartier D, Caballero R, Hebert TE 2002 A comparison of currents carried by HERG, with and without coexpression of MiRP1, and the native rapid delayed rectifier current. Is MiRP1 the missing link? J Physiol 540:15–27

Witchel HJ, Milnes JT, Mitcheson JS, Hancox JC 2002a Troubleshooting problems with in vitro screening of drugs for QT interval prolongation using HERG K^+ channels expressed in mammalian cell lines and Xenopus oocytes. J Pharmacol Toxicol Methods 48:65–80

Witchel HJ, Pabbathi VK, Hofmann G, Paul AA, Hancox JC 2002b Inhibitory actions of the selective serotonin re-uptake inhibitor citalopram on HERG and ventricular L-type calcium currents. FEBS Lett 512:59–66

Wyte CD, Berk WA 1991 Severe oral phenytoin overdose does not cause cardiovascular morbidity. Ann Emerg Med 20:508–512

Yang T, Snyders D, Roden DM 2001 Drug block of IKr: model systems and relevance to human arrhythmias. J Cardiovasc Pharmacol 38:737–744

Yap YG, Camm AJ 2002 Potential cardiac toxicity of H1-antihistamines. Clin Allergy Immunol 17:389–419

Zhou Z, Volperian VR, Gong Q, Zhang S, January CT 1999 Block of HERG potassium channels by the antihistamine astemizole and its metabolites desmethylastemizole and norastemizole. J Cardiovasc Electrophysiol 10:836–843

Zhou J, Meng R, Li X, Lu C, Fan S, Yang B 2003 The effect of arsenic trioxide on QT interval prolongation during APL therapy. Chin Med J (Engl) 116:1764–1766

DISCUSSION

Hancox: Early on you mentioned that you no longer thought of QT prolongation as at all therapeutic. So where does this leave novel class III drugs?

Shah: The pharmaceutical industry have themselves acknowledged this dilemma. In the last few years, class III drugs such as dofetilide, ibutilide and azimilide have all been developed for atrial and not ventricular tachyarrhythmias. Even for atrial arrhythmias, in the EU at least, dofetilide is the second-line drug. You are right: where does this leave class III drugs, particularly after the SWORD study (Waldo et al 1996)?

Schwartz: Towards the end you raised a critical issue in terms of numbers. I have always been puzzled by the attitude of giving a lot of weight to average changes on the basis of younger, healthy individuals, and very often disregarding gender. Personally, I think the average values mean very little. From a biological perspective what matters is the percentage of individuals with large changes, not the single outlier. What matters is the number of individuals who have a QTc prolongation greater than 30 ms. An average change of 10 ms in either direction may be totally irrelevant if we give these drugs to a population with a normal QT. It may matter when you give a drug to someone who is already outside the two standard deviations, because of a silent mutation or a prolonged QT.

Shah: When we wrote our CPMP document in 1997 on investigation of QT interval prolongation induced by drugs, we did not include the predictive utility of mean population-based changes in QTc interval for the very reason that you mention. The CPMP strategy on risk assessment is based on categorical responses of (1) an increase in QTc interval of more than 60 ms from baseline and/or (2) regardless of the magnitude of increase from baseline, a new absolute QTc interval of 500 ms or more. We looked at the data from Pratt et al (1996) from their study on terfenadine. These data show that the probability of a 50 ms increase of QTc interval in an individual being a spontaneous change is very low. It is almost always drug-induced, so we opted for a value of 60 ms increase in QTc interval from baseline as seriously indicative of a potential clinical risk. In the new ICH guideline that we are working on, we are going to be talking about mean population-based changes as well as categorical responses. Traditionally, in the USA, the FDA have relied more on mean changes. Even they now appear to me to recognize that categorical response may be a more powerful predictor of risk than mean changes. We are all going to evaluate drugs by both these parameters, though. When you put together your dossier and submit it to the regulators, it is possible that the FDA will look more closely at the mean changes and make one decision, and the Europeans and the Japanese might look more closely at categorical responses and come to the same or a different view.

Schwartz: If one wants to use the mean changes, they should be adjusted for the starting point. We always have regression towards the mean. If your normal population starts from 385 it is more likely that you will have a 25 ms increase than if your baseline is 415.

Shah: Regression to a mean is a problem with drug-induced QTc interval changes. Regardless of what is the change from baseline, an important parameter is how many subjects hit 500 ms duration of QTc interval. When you analyse the two categorical responses together, one would compensate for the other and circumvent any bias arising from regression to a mean. You can have a change from 400 to 470 ms, which is a big increase, but it hasn't hit the 500 ms proarrhythmic threshold. On the other hand, someone who starts at 470 ms and following an increase of only 35 ms, hits the magic number of 500 ms. The latter response would constitute a more powerful signal of clinical risk than the former. You are right about the problems of these studies being conducted in healthy volunteers and the extent to which you can extrapolate the results to the patient population. If you do a QT study with dofetilide in healthy volunteers and then extrapolate the data to cardiac failure patients without regard to the unique susceptibility of these patients, you could encounter a serious safety issue. This is another area that we intend to discuss in detail with our regulatory colleagues from US and Japan when we meet in Washington in June 2004 for the next ICH meeting.

Recanatini: You mentioned that at the committee of the ICH you take into consideration *in vitro* and *in vivo* studies. What about *in silico* studies? Are they excluded a priori? Or will they eventually be taken into consideration?

Shah: I am afraid I have some bad news for you. The ICH document, S7B, on non-clinical studies is fairly conservative; to the extent that even perfused heart preparations are pretty well excluded. *In silico* modelling therefore has little chance at the moment. The closest we have come to is specifying chemical or pharmacological classes that might carry the risk.

Gosling: Can you explain the rationale behind accepting data from a recombinant channel in a heterologous expression system and not taking data from an intact organ system? I find that difficult to understand.

Shah: I agree. I don't fully understand the basis for this decision either. I think we should use all the available methods. One problem is that when we see data from some of studies that use these preparations or whole-animal models, the investigators often use methoxamine to sensitize the preparations and then they test the drug in the expectation that there will be no QTc-related changes. If there are no changes, that is fine. However, if they do see changes, they often attribute these to methoxamine and not the test drug.

Netzer: I wish you luck with your difficult discussions in Washington. Concerning the conservative nature of S7B, I remember when Luc Hondeghem presented on this at the expert working group in Brussels last year, the only effect was that the wording of the last paragraph changed to say that interested parties are encouraged to do such experiments. We have heard that all kinds of chemicals can interact with the hERG channel. We had a drug that we wanted to develop as an antipsychotic, which then became an anti-Parkinson drug

Shah: This happens frequently. If we look at the history of terfenadine, it started its life as a potential neuroleptic agent that was found to be more potent at H1-histamine receptor. So, it was developed as an H1-antihistamine for allergic conditions. Not surprisingly, both H1-antihistamines and neuroleptics (as therapeutic drug classes) are notorious in terms of their effect on QT interval. I have a list of about 20 drugs where the therapeutic class has changed. This is why we keep referring to chemical class, which can't change.

Hoffmann: How often have you seen TdP in healthy volunteers in phase I study?

Shah: Never. I dread this. If there is one report of TdP in a healthy volunteer, I suspect the entire program would be halted or considerably slowed pending full investigation of the mechanism.

Brown: I saw that you had moxifloxacin listed with labelling restrictions. Is this true for the USA?

Shah: This is just for the EU. I am not familiar with the latest US labelling.

Brown: As I understand E14, the positive control is moxifloxacin. You will have a product liability issue here.

Shah: E14 does not mention moxifloxacin specifically. When moxifloxacin is given as a single dose to healthy volunteers, it is true that it produces mean changes in the range of 5–10 ms. We would be content to say that this is a safe positive control. The labelling for moxifloxacin in the EU is based on the clinical trials data in approximately 3000 patients, who were administered a range of doses, of whom 10–15 patients were found to have developed categorical responses described in the CPMP document. The mean change observed in this heterogeneous data set of patients was 17–18 ms.

Brown: If I was an IRB board member I would have difficulty passing this.

Shah: Our colleagues in Japan feel the same way as you do. They are asking why it has to be a pharmacological positive control. If the use of a positive control is intended simply as an assay of sensitivity and the validity of the trial, why can't the positive control be a validated non-pharmacological intervention? Their next question was, if the assay sensitivity is the core issue, why does the intervention have to increase QTc interval to be valid? Why can't it decrease QTc interval by 5 ms?

Brown: I spent a fair bit of time trying to show yesterday that if you do all of these things according to certain protocols, inter- and intra-lab variability can be greatly reduced. But one has to acknowledge that these are possibly not going to be adopted. I would encourage the regulators to at least have 95% confidence limits from any lab, and see whether there is a greater than threefold variation in the results in those confidence limits. Otherwise we will constantly be faced with the issue of absolute numbers.

Shah: We need people like you to comment on these guidelines when released for consultation. At the moment, we only get a few comments, which is a bit depressing because we go to great lengths to make sure that these guidelines are widely distributed and commented on. We don't want to impose anything. We are looking for input.

Brown: Are comments requested on E14?

Shah: They will be requested after June 2004 if, and when, we sign off Step 2 draft in Washington.

Brown: Isn't it a good idea to do it before June?

Shah: No, there are good reasons to wait. It is important to establish first that the regulators and the industry associations from the three regions are in broad agreement.

Brown: The other thing I have heard regulatory people say is that when we are dosing in a hERG assay we should dose up to the solubility limit. This is rubbish. There are plenty of lipophilic drugs, and if you go to the solubility limit with these the membrane will melt. You are going to see all kinds of non-specific effects, including a hERG one.

Traebert: It is written in the ICH S7B guideline that either you have to go up to the solubility limit, or go up to create a full concentration-response curve. If you can create a full concentration-response curve and get a 100% inhibition you don't have to test higher concentrations.

Brown: That is not what some of them ask you to do.

Shah: My view is that we should have a reasonable safety margin. If an occasional drug slips through a net of this safety margin, of say 30 times the free clinical concentrations in non-clinical studies, we will pick it up in the clinical QT study. Even if this clinical study is clean, it doesn't relieve the sponsor from monitoring ECGs in patients in phase 3 studies. There are various hurdles on the route to regulatory submission, and we don't need to put such a draconian hurdle of a very large safety margin at the beginning. There are these small areas of disagreement between the US on one side and the EU and Japan on the other, but I think they can be overcome. The big difference we have is over the utility of non-clinical data generally in excluding a clinical risk and guiding the clinical programme.

Sanguinetti: Can you say something about outlier analysis?

Shah: In a clinical trial dossier, we might see 2000–4000 patients. We want to know how many of these patients, on administration of a drug, respond with a QTc change from baseline of more than 30 ms, 30–60 ms and more than 60 ms. The other measurement is how many patients respond with a 500 ms value of QTc interval, regardless of the magnitude of change from baseline. Outlier analysis refers to the number and/or percentage of subjects with these categorical responses in each treatment arm.

Sanguinetti: How much additional clinical investigation is pursued with these outliers to determine the cause of abnormal QT intervals?

Shah: At present, very little. This is another issue. Do subjects who respond, for example, with a 60 ms increase have one of the variant alleles of drug metabolizing enzymes or ion channels? Should we follow this up by genetic profiling of these subjects? There is an increasing trend now for integrating genetics into clinical trials. This is one area where it could usefully be integrated.

Schwartz: What we are considering is re-exposure of outliers to see whether this effect is real, kinetic profiling and now we are also looking at mutations in the channels.

Traebert: You said that patch clamp is the gold standard for testing hERG activity. If someone makes a submission with fluorescence or binding data: would you request additional electrophysiological information?

Shah: Regulators would accept what you have provided, but if there are any doubts about the robustness of the data, it would work against you with a request for additional studies.

Netzer: Would data on hERG without patch-clamp data be accepted?

Shah: Yes.

Netzer: I think that the data generated on hERG that are submitted for safety reasons, should be obtained from patch-clamp measurements. It is the only clear investigation of the interaction of a compound with an ion channel. There are experiments that should be performed to protect the patients and these should be listed in the guidelines. Others are performed by companies on their own to save money.

References

Pratt CM, Ruberg S, Morganroth J et al 1996 Dose-response relation between terfenadine (Seldane) and the QTc interval on the scalar electrocardiogram: distinguishing a drug effect from spontaneous variability. Am Heart J 131:472–480

Waldo AL, Camm AJ, deRuyter H et al for the SWORD Investigators 1996 Effect of d-sotalol on mortality in patients with left ventricular dysfunction after recent and remote myocardial infarction. Lancet 348:7–12

Closing remarks

Michael C. Sanguinetti

Department of Physiology, Nora Eccles Harrison Cardiovascular Research & Training Institute, University of Utah, Salt Lake City, UT 84112, USA

One of the goals of this meeting was to reach a consensus on several issues concerning the structural basis of hERG function, the role of auxiliary subunits and the molecular mechanisms of drug-induced long QT syndrome. I think we have been fairly successful at achieving this goal, but some issues remain unresolved. Nonetheless our discussions and arguments over the details should provide the impetus for future studies.

There were several highlights worth recapping. Gail Robertson showed that hERG1b has re-emerged as an important component of I_{Kr}. The questions remain as to where this isoform is expressed and whether it is altered in disease states. Gea-Ny Tseng proposed a novel mechanism for hERG inactivation involving movement of the S5–P linker. The specific mechanism of hERG inactivation is still obscure, but there is agreement that it differs from Shaker and other channels that have been studied extensively and that it merits further study. The PAS domain still has a rather confusing role in gating; hopefully there will be protein partners discovered for this domain in the near future. We heard from Enzo Wanke that toxins can be used to differentiate hERG 1, 2 and 3 in neurons. This is important because the known organic blockers are not isoform specific. The potential use of chemical chaperones for the treatment of LQTS and the realization that some drugs decrease I_{Kr} by interfering with hERG trafficking was discussed by Eckhard Ficker. An unresolved issue is how to reconcile the differences in hERG channel properties characterized in heterologous cell expression systems versus native tissue and what might account for species and tissue variability, especially with regard to modulation by regulatory subunits. Annarosa Arcangeli has demonstrated that hERG is up-regulated in cancer; but the question of whether hERG blockers will be used as therapy was unanswered. Pharmacophore models for *in silico* prediction of hERG blocker potency are not yet ready for prime time, except perhaps within a chemical series. It is unclear if a modelling approach is being used routinely within the pharmaceutical industry because little has been published. Arthur Brown reported that when cellular assays are done very carefully there is not much variation in the IC_{50} values for hERG blockers. This is an important point, but different labs are not likely to perform the assays in exactly the same way. Another topic discussed was drug binding to sites on hERG other than the well-characterized central cavity site. There is no experimental evidence for other sites yet, with the exception of peptide toxins which clearly bind to an external site. Regardless of the specific binding mechanisms involved, in the near future it is likely that high-capacity voltage clamp assays will be available to screen for hERG block.

The mechanisms of repolarization inhomogeneities associated with sustained Torsade de Pointes are being investigated. Most agreed that the transgenic rodent models are inappropriate for the study of Torsade de Pointes, but it appears that transgenic rabbit models will be available soon. Luc Hondeghem talked about a new method to assess the proarrhythmic potential of drugs, the so-called 'TRIad'. This test was shown to be a better predictor of drug liability than measures of hERG block.

Finally, Rashmi Shah told us about how requirements for evaluation of drug-induced Torsade de Pointes by government regulators seem to be a moving target. New guidelines will be available soon. Hopefully, the guidelines will reflect the advances made in the past five years of intense study of the mechanisms of Torsade de Pointes and hERG channel biology.

Index of contributors

Non-participating co-authors are indicated by asterisks. Entries in bold indicate papers; other entries refer to discussion contributions.

A

Abbott, G. W. 39, 44, 96, **100**, 112, 113, 114, 115, 116, 155, 202, 223
*Anantharam, A. **100**
Arcangeli, A. 18, 73, 74, 155, 200, **225**, 232, 233, 234

B

Brown, A. M. 17, 38, **57**, 71, 96, 98, 116, **118**, 131, 132, 133, 134, 135, 150, 151, 153, 155, 156, 157, 158, 182, 183, 184, 185, 200, 219, 223, 246, 247, 249, 281, 282, 283

C

*Cavalli, A. **171**
*Chen, J. **159**

D

*Dennis, A. **57**

F

*Fernandez, D. **159**
Ficker, E. 17, 39, 41, 55, 56, **57**, 70, 71, 72, 73, 74, 152, 153, 169, 233

G

Gosling, M. 15, 38, 71, 73, 74, 98, 115, 132, 156, 157, 158, 182, 183, 184, 185, 200, 232, 281
*Guy, H. R. **19**

H

Hancox, J. C. 16, 71, 72, 97, 98, 114, 133, **136**, 153, 157, 169, 201, 279
Hebert, T. E. 18, 112, 113, 115, 156
Hoffmann, P. 72, 133, 156, 157, 183, 199, 201, 202, 221, 223, 281
Hondeghem, L. M. 73, 132, 134, 150, 169, 198, 200, 220, 221, **235**, 245, 246, 247, 248, 249, 250

J

January, C. T. 71, 72, 73, 93, 95, 96, 97, 133, 156, 157, 169, 170, 200, 201, 202, 223, 249, 250
*Jones, E. M. C. **4**

K

*Kagan, A. **75**
*Kamiya, K. **159**
*Kuryshev, Y. **57**

M

*Masetti, M. **171**
McDonald, T. V. 15, 16, 35, **75**, 89, 90, 92, 93, 114, 115, 117
Mitcheson, J. 36, 41, 55, 56, 97, 116, **136**, 150, 151, 152, 153, 154, **159**

N

Netzer, R. 151, 152, 156, 157, 169, 184, 199, 233, 281, 284
Nicklin, P. 182, 201, 202
Noble, D. 131, 135, 153, 167, 184, 219, 220

P

*Perry, M. **136**
*Piper, D. R. **46**
*Poelzing, S. **204**

R

Recanatini, M. 36, 150, 152, 167, 168, 169, **171**, 182, 183, 184, 281
Robertson, G. A. **4**, 15, 16, 17, 18, 35, 36, 37, 40, 44, 45, 52, 53, 54, 56, 89, 90, 91, 92, 93, 97, 115, 152, 155, 166, 200, 218, 233
Rosenbaum, D. S. 134, 199, 200, **204**, 217, 218, 219, 220, 221, 222, 223, 224, 244, 245, 248

S

Sanchez-Chapula, J. A. 36, **159**, 168
Sanguinetti, M. C. **1**, 16, 37, 38, 39, 41, 44, 45, **46**, 53, 55, 56, 71, 91, 96, 98, 116, 131, **136**, 151, 153, 154, 155, 156, **159**, 166, 167, 168, 169, 181, 182, 184, 200, 221, 222, 224, 233, 234, 283, **285**

Schwartz, P. J. 89, 97, **186**, 198, 199, 200, 201, 202, 220, 222, 223, 234, 247, 248, 279, 280, 283
Shah, R. R. 56, 70, 71, 73, 133, 134, 150, 183, 198, 199, 201, 202, 233, 234, 249, **251**, 279, 280, 281, 282, 283, 284
*Stansfeld, P. **136**

T

Terrar, D. A. 89, 90, 222
Thomas, D. 91, 92
Traebert, M. 15, 70, 112, 116, 117, 132, 135, 156, 157, 167, 184, 185, 283, 284
Tristani-Firouzi, M. 35, 36, 40, 44, 45, **46**, 52, 53, 54, 55, 113, 114, 218
Tseng, G.-N. **19**, 35, 36, 37, 38, 39, 40, 41, 42, 44, 45, 52, 54, 96, 97, 98, 217, 218, 221

W

*Wang, J. **4**
Wanke, E. 42, 44, 45, 97, 98, 233, 250
*Wible, B. A. **57**
Witchel, H. 98, **136**

Subject index

A

A-kinase anchoring proteins (AKAPs) 85
action-potential duration
   epicardial cells 207–208
   instability 238–239
   MICE 125
   midmyocardial cells 208, 221
   repolarization assay 123–124
acute lymphoblastic leukaemia 228
acute myeloid leukaemia 227
acute promyelocytic leukaemia 59, 66, 73–74, 126
adaptor protein 74
adrenaline, Ca^{2+} currents 250
adrenergic stimulation 75–76, 79
alanine scanning
   drug binding 142–145
   S4 domain 50–51
alfuzosin, TdP 263
alternans 245, 263
Alzheimer's disease, COX-2 234
amiodarone 125, 133
apoptosis 226
arsenic trioxide
   acute leukaemia 59, 66, 73–74, 126
   chronic exposure 70
   false negative hERG studies 269
   hERG block 126
   hERG channel maturation 66–67
   hERG trafficking 66–67, 126, 223, 270
   I_{Kr} 67–68, 70
   metabolites 71
astemizole 58, 59, 72, 73, 119, 264–265
astrocytomas 225, 228, 230
auditory stimuli 76, 86, 191, 198, 200
auxiliary subunits 11
azimilide 279

B

BeKm-1 24
β-adrenergic regulation, hERG/I_{Kr} 76–77
β-blocker therapy, LQTS 195
bupivacaine 266–267
buprenorphine 264

C

Ca^{2+} channel 71
   adrenaline 250
cAMP regulation, hERG/I_{Kr} 76–77, 85
cancer 225–232
   channel blockers 226
   channelopathy 225–226, 232
   geldanamycin therapy 65, 125
   haemopoietic cells 227–228
   hERG gene expression 227–228
   hERG proteins 227–228
   metastases 228
   oncogenes, ion channel function 225
   solid tumours 228
   tumour angiogenesis 230, 233
   tumour invasion 226, 229–230
cardiac safety index 272
carotid body, oxygen sensing 2
carvedilol 268
Cd^{2+} bridge 25, 42
cell–cell coupling 210–213
cell–cell interactions 226
cell cycle 226
cell proliferation 228–229
cell shrinkage 226
central nervous system, hERG 200
cetirizine
   cardiac safety index 272
   ventricular tachyarrhythmias 266
channel blockers 58, 61, 226
chaperones 58, 62–65, 115
charge–charge interactions 30

SUBJECT INDEX

chiral drugs 266
chloroquine 162, 169
CHOs
  metabolic capabilities 156–157
  run-down 157–158
cisapride
  hERG as target 120
  IC_{50} 264
  kinetics 134
  QT/TdP liability 118, 249
  TdP risk 119
  withdrawal 119
clarithromycin
  intravenous 223
  Q9E 108
clofilium 145–146, 150, 152–153
cocaine 116
codeine 264
collagenase type 2 95
colorectal cancer 228, 229
CoMFA model 177
Committee for Proprietary Medicinal Products (CPMP) 254
CoMSIA model 173, 175, 177
connexins 211, 222
  Cx43 211–213
  Cx45 211
costs
  automated patch clamp 184
  drug development 119
COX-2, Alzheimer's disease 234
cyclic nucleotide binding 58, 93
CYP2D6 266
cysteine scanning mutagenesis 23
cytoskeletal reshaping 226

D

delayed after depolarizations 205
desmethylastemizole 264
dextrobupivacaine 267
dispersion 239–240
  Cx43 213
  gap-junction expression patterns 211
  spatial 240
  TdP 205, 207–210
  temporal 240
  transmural 207–210, 221, 240
DMSO 58
dofetilide 279

doxazosin
  limitations of hERG channel studies 260
  TdP 263
Drk1 168
Drosophila, ether-a-go-go 2
drug binding 142–146
drug development 119, 137, 201–202
drug-induced long QT syndrome 2
  approval 271–273
  clinical evaluation 256–258
  false negative hERG studies 269–271
  false positive hERG studies 267–269
  hERG studies 258–271
  I_{Kr} 252
  non-clinical evaluation 254–256, 281
  non-torsadogenic 236
  regulatory guidance 251–279
  risk/benefit evaluation 271–273
drug safety 67–68, 119
drug screening
  MinK/MiRP1 co-expression with hERG 116
  virtual 184
drug trapping 138–139
drug withdrawals 67, 119, 253

E

E-4031 58, 107–108, 116
early after depolarization (EAD) 205, 240–241
ectopics 240–241
electrocardiogram (ECG), LQTS 187–188, 236
emotional stress 76, 87, 188, 190
endometrial adenocarcinoma 228
enzyme effects 95–96
ERG1 11–12
ERG1 5
ERG1-KA antibody 12
ERG1a, N-terminus 5
ERG1b 5, 12
ERG1b 5
ERG1b-specific antibody 12
ErgTx1 24
erythromycin
  intravenous 223
  temperature dependence 135
ether-a-go-go (EAG)
  origins 2
  S6 domain 160–162
Ewing sarcoma 228

exercise, LQTS 76, 86, 190, 191, 192
expression system sensitivity 260

F

fentanyl 264
fever 223
fexofenadine
  replacement for terfenadine 264
  trafficking in LQT2 58
flecainide 201
fluorescence assays 127, 259
fluoroquinolones 270
'foot in the door' effect 138
14-3-3 proteins
  affinities 93
  cAMP-dependent phosphorylation of hERG 85
  cardiac plasticity 87
  ERG channel protein in myocardial tissue 83
  expression 77
  14-3-3ε 10, 77, 79
  functional effects 77, 81
  hERG interaction 77, 79
  insufficient 90
  interaction groove 77
  isoforms in heart 93
  KCNK3 surface expression 85
  phosphorylation-dependent binding with hERG1 10
  PKA-dependent phosphorylation of hERG 81, 83
  RAF-1 kinase 77
  stress response 87

G

gap junctions 211, 218
gastric cancer 228
geldanamycin
  cancer therapy 65, 125
  hERG–Hsp90 complex formation 62
  hERG trafficking 125, 223, 270
gender
  K897T polymorphism 262
  LQTS 194, 201
gigaseal patch clamp 120
glioblastoma multiformis 228
gliomas 228
glycerol 58
glycine hinge 55–56

H

H-bonds 22, 25
haemopoietic cells 227–228
halofantrine 264, 265
heart failure, I_{Kr} 217–218
HEKs
  HEK–hERG1 cells 230
  metabolic capabilities 156–157
  run-down 157–158
hERG
  adaptor protein 74
  β-adrenergic regulation 76–77
  blocking assay 67, 119–123, 126–129, 131–132, 259
  cAMP regulation 76–77, 85
  CNS 200
  compared to Shaker 22, 25, 53
  D540K 139–140, 151
  drug binding sites 142–145
  drug trapping 139
  enzyme effects 95–96
  F656 143, 145, 163, 167
  fast inactivation 22–25, 39, 40, 47
  G601S 59, 72
  G648A 143
  gating current 47–50
  glycine hinge 55–56
  H-bonds 22, 25
  high-throughput assays 126–129
  Hsp70 61–62, 63, 65, 270
  Hsp90 61–62, 63, 65, 270
  I_{gOFF} 47–48
  I_{gON} 47–48
  K897T polymorphism 262
  M^{2+} binding 30, 32
  modelling 175–180
  origins 2
  outer mouth 19, 22–25
  oxygen sensing 2
  Phe656 145, 160, 161, 163–165
  PKA-mediated modulation 91–92
  PKC-mediated modulation 91–92
  pore loop 39
  regulatory perspective 258–271
  role 2
  S4 domain 26–29, 35, 40, 46, 50–51, 53
  S5-P helix 22, 23–24, 25, 36, 38, 39–40, 179
  S6 domain 160–162
  S631A mutation 48–50
  Ser624 145–146, 150–151

SUBJECT INDEX

slow activation 22, 26–32, 35–36, 37, 47
template selection 177–179
Thr623 146
trafficking 57–69, 125–126, 270–271
turnover rate 71
Tyr652 145, 160, 161, 162–165
V659A 143
voltage-sensing 26–32, 35–36, 37, 40, 46–52
Y652A 143, 151, 162–163, 168
hERG-Lite™ 128–129
hERG1
 cancers 228
 MinK 11
 MiRP1 11
 N-terminus 7–8
 PKA 10
hERG1 4–5
 tumour cells 227, 228
hERG1a
 defining features 5
 N-terminal phosphorylation 8, 10
 N-terminus 5, 7–8, 10–11
 hERG1a/1b heteromeric channels 11–12, 18
hERG1b
 α-subunit of I_{Kr} 5–8
 cancer 227
 cell depolarization 227
 cell proliferation 229
 N-terminus 8, 11
hERG1B, tumour cells 227
hERG2, retinoblastoma 227
hERG3, mammary adenocarcinoma 227
heterotrimeric G proteins, voltage-dependent protein 232–233
Hsp70 61–62, 63, 65, 270
Hsp90 61–62, 63, 65, 270
hyaluronidase 95

I

ibutilide 145, 150, 279
IC$_{50}$s
 accuracy 120
 interpretation 124–125
 step-pulse/step-ramp 120
 therapeutic concentration 133
 use dependence curves 134
$I_{g_{OFF}}$ 47–48
$I_{g_{ON}}$ 47–48
I_{K}-osome 115

I_{Kr} 1, 2
 adrenergic stimulation 79
 α-subunit 5–8
 arsenic trioxide 67–68, 70
 β-adrenergic regulation 76–77
 cAMP regulation 76–77
 cell to cell/lab to lab differences 96
 defining features 5
 drug-induced LQT 252
 drug trapping 139
 heart failure 217–218
 hERG1a/1b heteromeric channels 11–12, 18
 hERG1b 5–8
 MinK 101–104
 MiRP1 104–110
I_{Ks} 2
implantable defibrillators 196
in silico modelling 137, 171–181, 281
infant deaths 198
instability 238–239, 247
integrins
 β subunit 74, 230
 hERG currents 233
 tumour cell invasion 226, 230
International Conference on Harmonisation (ICH) 254
 five steps 282–283
 ICH E14 256–258
 ICH S7B 254–256, 281
ischaemic heart disease 199
isoproterenol 76
I_{x1} 2
I_{x2} 2

K

KCNE2 261
KCNK3 85
KCNQ1, LQTS 101
KCR1
 MiRP1 11
 proarrhythmic blockade 261–262, 269
Kv channels
 charge–charge interactions 30
 N-termini 10
 Pro-X-Pro motif 141–142

L

lead compounds, hERG assay 123, 264

left cardiac sympathetic denervation 196
levacetylmethadol 264, 265
levobupivacaine 267
levofloxacin 270
limiting slope method 29
long QT syndrome 186–198
  β-blocker therapy 195
  clinical severity 194
  drug development 201–202
  drug therapy 195, 201
  ECG 187–188, 236
  flecainide 201
  gender 194, 201
  genes 187
  genotype–phenotype correlation 187–193
  Hsp90 inhibition 65
  implantable defibrillators 196
  infant deaths 198
  KCNQ1 101
  left cardiac sympathetic denervation 196
  MinK 101, 102–103
  MiRP1 105, 107
  mutation site 194–195
  natural history 194–196
  normal ECG 236
  onset of cardiac events 193
  pore/non-pore mutations 194–195
  prevalence 200
  risk stratification 194–196
  spironolactone 201
  stress-induced 76
  sudden death as first symptom 236
  therapy response 195–196
  triggers 188, 190–193
  see also drug-induced long QT syndrome
loratadine 260
LQT1
  β-blocker therapy 195
  frequency 187
  gender 194, 201
  physical exercise 76, 190, 192
  PKA 10
  swimming 191, 202
  sympathetic activity 192
  symptomatic age 193
LQT2
  auditory stimuli 76, 191
  β-blocker therapy 195
  channel blockers 58
  channel trafficking 57–58
  chaperones 58, 62–65

drug block model 222
ECG 187–188
emotional stress 188, 190
fever 223
frequency 187
gender 194, 201
hERG1a N-terminus 5
implantable defibrillators 196
PAS domain 58
physical exercise 190, 191, 192
pore/non-pore mutations 194–195
startle 76
symptomatic age 193
T wave 187–188
transmural optical mapping in
  experimental model 206–207
LQT3
  β-blocker therapy 195
  flecainide 201
  frequency 187
  gender 194
  physical exercise 190, 192
  sleep/rest 89, 188, 190–191, 192
  symptomatic age 193
LQT6 261
LY97241 152

M

M-cells 208–210, 221, 222
M^{2+} binding 30, 32
mammary adenocarcinoma 227
mefloquine 162
meperidine 264
metastatic cancers 228
methadone 264
methanesulfonanilide block 5
mexiletine 220
MinK
  cardiac episodes when swimming 202
  chaperone role with hERG 115
  drug screening 116
  hERG1 association 11
  I_{Kr} 101–104
  in vitro hERG association 101
  LQTS 101, 102–103
  polymorphisms 114
  versus MiRP expression 155
MiRP1
  chaperone role with hERG 115
  co-expression with hERG 104–105, 112

co-immunoprecipitation with rERG
    108–110, 112
drug screening 116
hERG interactions 260–261
hERG pharmacology 107–108, 112
hERG1 11
I_{Kr} 104–110
KCR1 11
LQTS 105, 107
MinK versus 155
polymorphisms 108, 114
Q9E polymorphism 108
T8A polymorphism 108, 112–113
variable effects on hERG 105
MiRP2 114, 115
mixed ion channel effects (MICE) 125, 132, 133
MK-499 140–141, 162, 169
morphine 264
mosapride
    safer alternative to cisapride 264
    TdP 270
moxifloxacin 270, 281–282
mutant cycle analysis 30
myocardial infarction, QT prolongation 97

N

neuroblastoma 226, 230
new chemical entity 253, 254
noise *see* auditory stimuli
norastemizole 59, 265

O

olanzapine 272
oncogenes, ion channel function 225
opiate analgesics 264
orphan drug law 202
outlier analysis 283
oxygen sensing 2

P

p125^{FAK} 230
'paddle' voltage-sensing 27
PAS domain
    LQT2 58
    N-terminal function 8, 10
patch clamp
    automated 126, 184
    gigaseal 120

gold standard 67, 120, 259, 284
pentamidine 70, 126, 223
peptide toxins 24, 97–98
perhexiline 268
pharmacophores 172–175, 176–177, 181–182
Phe656 145, 160, 161, 163–165
phenytoin 269
physical exercise, LQTS 76, 86, 190, 191, 192
pimozide 272
pore helix residues 145–146
pore loop 39
potency parameter 272
prazosin
    limitations of hERG channel studies 260
    TdP 263
prenylamine 249
    QT prolongation 267
Pro-X-Pro motif 141–142
proarrhythmia 240–242
propafenone 146–148
prostate cancer 232
protease 14 95
protease 24 95
proteases 95–96
protein kinase A (PKA)
    A-kinase anchoring proteins (AKAPs) 85
    cAMP 76
    14-3-3 81, 83
    hERG modulation 91–92
    hERG1a N-terminal phosphorylation 8, 10
    LQT1 10
protein kinase C (PKC)
    hERG modulation 91–92
    voltage-dependent protein 232–233
proteinase K 95
prucalopride 264
psychosis, K$^+$ levels 201

Q

QT liability 253
    clinical evaluation 256–258
    European guidance 254
    international guidance 254
    non-clinical evaluation 254–256
    prediction, cost savings 119
    regulatory guidance 254–258
QT prolongation
    antiarrhythmic 236, 242
    definitional problems 247–248
    myocardial infarction 97

QT prolongation (cont.)
  pathological 235
  physiological 235
  TdP 204, 218–219, 236, 252
  see also long QT syndrome
quantitative structure–activity relationship (QSAR) models 173
quaternary amines 59–61, 137–138
quetiapine 272
quinidine 162, 168
quinolines 162–163

R

R18 79
Rac1 230
radioligand binding 127, 259
RAF-1 kinase 77
re-entrant activity 205, 213–215, 219–220
regulatory guidance 251–279
renzapride 264
rest, LQTS 188, 190, 192
retinoblastoma 227
retinoic acid 73
reverse frequency dependence 238
reverse use dependence 123, 135, 237–238, 246, 247
risk assessment 254–255
risk/benefit evaluation 271–273
risperidone 272
ropivacaine 266–267
RP 58866 265–266
rubidium flux 127–128, 259

S

S4 domain 26–29, 35, 40, 46, 50–51, 53
S5-P helix 22, 23–24, 25, 36, 38, 39–40, 179
S6 domain 160–162
S283 phosphorylation 10
safety margins 125, 272–273
salt-bridges 29–30, 37
sarcomas 228
scanning ion-conductance microscopy 16
scorpion peptide toxins, hERG specificity 24, 97–98
selective serotonin reuptake inhibitors (SSRIs) 263
sertindole
  approval refusal 119
  CYP2D6 266
  QT prolongation 262–263, 272
  reintroduction 263
Shp-1 10
signalling proteins, voltage-dependent 232–233
sleep, LQT3 89, 188, 190–191, 192
sotalol 262
sparfloxacin
  action-potential duration 270
  voltage dependence 162
  withdrawal 119
spatial dispersion 240
spironolactone 201
Src tyrosine kinase 10
startle 76, 86, 89, 198, 200
stress 76, 87, 188, 190
structure–function studies 19–35
sudden deaths 198, 236
sulfamethoxazole 108, 112–113, 114
surface charge effect 32
swimming, LQT1 191, 202

T

T tubules 16
T waves, LQTS 187–188
tamulosin 263
TEA 39
  N-alkyl derivatives 59–61
temporal dispersion 240
terazosin
  limitations of hERG channel studies 260
  TdP 263
terfenadine 118
  action-potential duration 123
  cardiac safety index 272
  hERG blocking 264
  kinetics 134
  TdP risk 119
  therapeutic class 281
  withdrawal 119
terikalant 265–266
terodiline 133, 249, 267
thapsigargin 58
thioridazine
  CYP2D6 266
  QT prolongation potency 272
'thorough QT/QTc study' 256–258
tolterodine
  mixed ion channel effects 125
  QT prolongation 133, 263

SUBJECT INDEX

TdP 268–269
Torsade de Pointes (TdP) 204–217
  acidosis 218
  delayed after-depolarizations 205
  dispersion of repolarization 205, 207–210
  early after-depolarizations 205
  gap junctions 211
  healthy volunteers 281
  mechanisms 205
  mortality 252
  QT prolongation 204, 218–219, 236, 252
  QTc as surrogate marker 252
  re-entrant activity 205, 213–215, 219–220
  self-terminating 241
  single rotor 215, 219–220
  triggered activity 205
  ventricular activation 205
  ventricular fibrillation 199, 241–242, 252
transmural dispersion 207–210, 221, 240
transmural optical mapping 206–207
TRIad 236–239, 242, 245, 247, 255

triangulation 237, 246–247
trypsin 95, 96
tumours *see* cancer
Tyr652 145, 160, 161, 162–165

V

vardenafil 257
VEGF 230, 233
ventricular fibrillation, TdP 199, 241–242, 252
verapamil
  mixed ion channel effects 125
  triangulation 246–247
vesnarinone 163
virtual screening 184

Z

ziprasidone 272